Pregnancy Proteins in Animals

Editor
Jann Hau, MD, PhD
Associate Professor
Institute of Veterinary Pathology
Laboratory Animal Unit
Royal Veterinary and Agricultural University
Bülowsvej 13
DK-1870 Copenhagen
Denmark

Library of Congress Cataloging in Publication Data

Pregnancy proteins in animals.
„Based on papers presented at the International Meeting on Pregnancy Proteins in Animals at the Royal Veterinary and Agricultural University in Copenhagen"--Pref.
Bibliography: p.
Includes indexes.
1. Pregnancy proteins--Congresses. I. Hau, Jann. II. International Meeting on Pregnancy Proteins in Animals (1985 : Royal Veterinary and Agricultural University).
QP552.P65P734 1986 599'.016 86-591
ISBN 0-89925-213-3 (U.S.)

CIP-Kurztitelaufnahme der Deutschen Bibliothek

Pregnancy proteins in animals : proceedings of the internat. meeting, Copenhagen, Denmark, April 22 – 24, 1985 / ed. Jann Hau. - Berlin ; New York : de Gruyter, 1986.
ISBN 3-11-010520-9
NE: Hau, Jann [Hrsg.]

 Printing: Gerike GmbH, Berlin. – Binding: D. Mikolai, Berlin. – Printed in Germany.

Preface

This book is based on papers presented at the International Meeting on Pregnancy Proteins in Animals at the Royal Veterinary and Agricultural University in Copenhagen, April 22-24 1985.

The intention of this book is, for the first time, to attempt a survey of the present status and future trends in the research on the so-called "new pregnancy proteins" in animals and humans.

The past fifteen years have seen intense research activities on pregnancy proteins in humans as well as in animals, but meetings have been devoted to the human proteins.

The need for laboratory animal models for the study of the potential biological function of these proteins has been emphasized by many scientists. This book focuses on pregnancy proteins in animals, and it demonstrates a world-wide interest in these molecules. An interest which is often directed by the desire to elucidate the physiological significance of the proteins, and sometimes directed by the wish to develop reliable biochemical pregnancy diagnoses for domestic animals.

I want to thank all participants and contributors for creating a most stimulating meeting and this book, which I hope may serve as a relevant introduction for newcomers to the field as well as a practical collection of information about the animal pregnancy proteins for all of you who are already involved in this business.

I am indebted to Mrs. Ingelise Martens, Mrs. Regina Lund, Mrs. Pia Jensen, Miss Helle Milbo, Mrs. Ulla Salado-Jimena, Mrs. Bente R. Hau, Mr. David E. Hau, and Dr. O.M. Poulsen for their excellent help in the arranging of the meeting in Copenhagen.

Copenhagen, March 1986 Jann Hau

CONTENTS

COMPARATIVE STRUCTURAL AND FUNCTIONAL FEATURES OF THE MAMMALIAN PLACENTA

N. Björkman
Department of Anatomy
Royal Veterinary and Agricultural University, Copenhagen, Denmark

Definition

Pregnancy proteins are partly circulating in the blood. In the placenta the vascular systems of mother and offspring come in close proximity making exchange of elements and compounds possible. Accordingly Mossman (14) has given the universal definition of the placenta as "an apposition of fetal and parental tissues for purpose of physiological exchange". For mammals the definition is restricted to an apposition of fetal membranes to uterine mucosa. The latter situation will be considered in the following.

Functions

Although physiological exchange is the most important biologic activity, the placenta has also many other functions. It forms an anchoring device for the fetus, it provides bouyancy and freedom of movement and growth for the fetus in an aquatic environment. The surrounding (amniotic) fluid offers constant thermal, osmotic and chemical conditions (3, 4). Synthesis, storage and break down of compounds are vital functions in placental metabolism.

Albeit having similar functions different placentae vary considerably in structure. Indeed, there is no organ of comparable importance with such wide variation in morphology as the placenta. This may be an expression of the outcome in different species. The litter size varies from one to 35 (Dasyurus viversinus) young, the length of gestation ranges between 8 days (some marsupials) and almost 2 years (elephant) and the developmental stage of neonatal young varies from almost embryonic conditions (marsupials) to well developed ability of locomotion (many ungulates).

Pregnancy Proteins in Animals

Implantation and fetal membranes

The development of the placenta is initiated by the trophoblast. The free living blastocyst, consisting of trophoblast and an inner cell mass, is nourished by secretion from uterine glands (histiotrophe). With the development of the vascular system of the embryo a more efficient, hemotrophic nutrition is eventually established. The simple epithelial membrane formed by trophoblast, the chorion, grows attached to endometrium by the process of implantation. To be vascularized the chorion combines with other fetal membranes. The yolk sac and allantois are entodermal cavities in communication with the mid gut and hind gut, respectively. The entoderm rests on mesenchyme vascularized by paired vitelline (yolk sac) and umbilical (allantoic) arteries and veins. The fusion of the mentioned fetal membranes with the chorion gives raise to two fundamental types: choriovitelline (yolk sac) placentation and chorioallantoic placentation (2). The yolk sac placenta regresses early in most animals and is replaced by the chorioallantoic placenta, but they can also function simultaneously as in rodents. Most marsupials only have a yolk sac placenta. The yolk sac produces nucleated erythrocytes and primitive germ cells. In some species, e.g. rabbit they absorb immunoglobulins (6).
The allantois, in communication with the urinary bladder, forms a second fluid filled sac (partly fetal urin) in addition to the amniotic sac. In hemochorial placentae (se below) however, the allantoic entoderm is insignificant or absent and only the vascular mesenchyme contributes to the allantochorion.

Increase of absorptive area

The surface area of the chorion is increased in diffuse distribution over the endometrium or at restricted sites (chorion frondosum) (Review, 5). The chorion forms complex folds or branched villi, which fit into corresponding furrows (swine, 9) or crypts (horse, ruminants) of the endometrium. Villi can also be freely exposed to uterine blood (man). Another mode to increase the surface or contact area is the formation of a labyrinth of fetomaternal tissue complex (carnivores) or of chorion alone (rodents). A further increase in surface or contact area is brought about by free microvilli of trophoblast or interdigitating microvilli of trophoblast and endometrial surface epithelium (ungulates, 5).

Interhemal membrane

The feto-maternal physiological exchange takes place between fetal and maternal blood. The process goes on at sites of intimate adhesion or attachement of the allantochorion to endometrial tissue. The respective blood circulations are separated by an interhemal membrane, which constitutes a selective barrier and transport avenue with two way traffic under biological control.

The membrane contains 2-4 complete layers of cells, which participate in the formation of the placental barrier. In addition connective tissue, basal laminae and other noncellular substances may intervene between the circulations (20). Two cell layers are invariably present, trophoblast (chorionic epithelium) and fetal capillary endothelium. On the uterine side there can be one or two cell layers, or they may be absent. Considering these features Grosser (12) proposed a classification, which has been widely used and by and by modified as new morphological evidence has been gained.

In the original classification 3 layers of tissue in the maternal as well as the fetal parts of the placental membrane were considered: capillary endothelium, connective tissue and epithelium on the uterine side, and trophoblast, mesenchyme and capillary endothelium on the fetal side. A reduction of layers in the uterine placenta can limit the barrier to 5, 4 or 3 tissue layers. However, the connective tissue is not a cellular barrier. Furthermore, capillaries on both sides tend to bypass the connective tissue and mesenchyme leaving only basal laminae between the endothelium and epithelium when present. Therefore the connective tissue and mesenchyme should not be included in the barrier. Thus the epitheliochorial placenta has 4 layers: uterine endothelium and epithelium and fetal epithelium (trophoblast) and endothelium. The endotheliochorial placenta has 3 layers: uterine endothelium, trophoblast and fetal endothelium. The hemochorial placenta has two layers: trophoblast and fetal endothelium. In the case of the hemochorial placenta, however, there can be up to 3 layers of trophoblast (11). The situation is summarized in Table I.

The trophoblast can appear as individual cells, cytotrophoblast, or as a syncytium. The syncytiotrophoblast is formed by coalescense of cytotrophoblast, e.g. in man and carnivores. The trophoblast can also form giant cells with two or more nuclei, e.g. in ruminants (18).

The principal difference in this system is apparently found between the hemochorial placentation on one hand, where the interhemal membrane is entirely fetal and where the chorion is freely exposed to uterine blood,

TABLE I: Tissues separating maternal and fetal blood.

	Maternal tissue		Fetal tissue		
Designation	Endo-thelium	Epi-thelium	Tropho-blast	Endo-thelium	Typical examples
Epithelio-chorial	+	+	+	+	Ungulates
Endothelio-chorial	+	–	+	+	Carnivores
Hemo – chorial					
" mono "	–	–	+	+	Guinea pig, homo
" di "	–	–	+ +	+	Rabbit
" tri "	–	–	+ + +	+	Rat, mouse

Modified from Grosser (12) and Enders (11).

and the non-hemochorial placentation on the other hand, where the trophoblast abuts on uterine endothelium (endotheliochorial placenta) or epithelium (epitheliochorial placenta). It is a remarkable feature in the hemochorial placenta that functional blood channels or blood spaces are not lined with endothelium but with trophoblast. Most interesting in this context (pregnancy proteins) are obviously the human and laboratory animals. The human placenta is villous and hemochorial with discrete cytotrophoblast cells and continuous syncytial trophoblast. Among laboratory animals rodents and lagomorphs have labyrinthine hemochorial placentae with varying numbers of trophoblast layers (Table I). Domestic mammals have a labyrinthine, endotheliochorial placenta (dog, cat) or folded (swine) or villous (horse, ruminants) epitheliochorial placenta.

Transport

Several different transport mechanisms over the barrier are recognized (15).

1. Simple diffusion and ultrafiltration
2. Facilitated transport with the aid of receptors and carriers

3. Active transport against the concentration gradient.
   This can also involve coupled transport.
4. Endocytosis, especially of large molecules, e.g. proteins.

In these modes of transport the plasma membrane of the cells plays a decisive role. Receptors, carriers and enzymes are bound to the plasma membrane. The capacity is vastly increased by the formation of microvilli on trophoblast cells and uterine epithelial cells, when present. The barrier function is also dependent on plasma membranes, which number twice as many as the cell layers.

A possibility of transfer that does not necessarily involve plasma membranes is leakage through defects in the placental barrier. This is a means by which cells might pass with resulting risk of immune sentizitation. Barrier functions of the interhemal membrane are also maintained by basal laminae, which serve as filter for particles and high molecular weight compounds.

The transport potential of molecules is influenced by 1) molecular weight; 2) lipid solubility; 3) degree of ionization; 4) ability to bind tissue or plasma compounds; 5) susceptibility to chemical transformation by the placenta; 6) rate of clearance. There are also wide species differences (15).

Transported elements and compounds

The most acute demand is for oxygen. The exchange of O_2 and CO_2, or "lung" function, is therefore of utmost importance. The transfer is mainly a diffusion process. In addition to concentration gradients flow velocity of the blood and buffer capacity of the plasma influence on the gaseous exchange. In this respect blood plasma can be considered as part of the placental transport system. It is also noteworthy that embryonic and fetal hemoglobins possess higher affinity to oxygen than adult hemoglobin. This is suited to compensate for the low oxygen tension in the fetal circulation (8).

The major metabolic fuel for the fetus is sugar. Its transport is active. Amino acids, especially important for protein synthesis in the fetus, are actively transported against a substantial gradient (15). In addition diffusion is also a transport component. The transfer is stereospecific and different for acid and basic amino acid. In spite of their large molecular size maternal plasma proteins are transferred to the fetus, although not necessarily across the chorioallantoic placenta. Alternative routes are over the yolk sac, exocelome and amnion (rabbit and some rodents) (6). Purin bases and nucleotides are transported to the fetus for synthesis of nucleic acids. Water soluble vitamins are actively transported and lipid soluble vitamins may diffuse over the interhemal membrane. Fetal fat is produced from free fatty acids transported across the placenta by diffusion (13), and maternal phospholipids are hydrolyzed and resynthesized by the fetal liver (7).

Absorption of iron by the placenta is effected in different ways. In the hemochorial placenta iron bound to transferrin is picked up by receptors on the surface of the trophoblast and transported further (17). In the endotheliochorial placenta of the dog maternal erythrocytes are extravasated into marginal hematoma of the placental girdle, and broken down by lysosomal activity. The epitheliochorial placenta of sheep contains a hemotophagous structure in the center of the placentomes, where the erythrocytes are dissolved (16). In the epitheliochorial porcine placenta iron coupled to a glycoprotein forming uteroferrin is secreted by uterine glands and delivered to areolae, where it is absorbed by trophoblast (10).

Synthesis

Placental tissue (trophoblast, decidual cells, uterine epithelium) is involved in synthesis, and the cells concerned are provided with well developed organelles. Among the products are proteinaceous tropic hormons and steroid sex hormones. At the junction of the developing allantochorion and the regressing yolk sac of the equine placenta an annulate chorionic girdle is formed. Between days 36 and 38 the trophoblastic girdle cells invade the uterine mucosa and form endometrial cups, which elaborate equine chorionic gonadotropin (eCG) (1). Binucleate trophoblastic giant cells in ruminants partly migrate to the maternal side and fuse with endometrial epithelial cells and so deliver synthesized substances, e.g. gonodotropic hormons to the mother (19). Estrogen has been found in porcine trophoblast, and its concentration has been correlated with the occurrence of smooth endoplasmic reticulum (Dantzer and Svenstrup, personal communication).

The placenta as an organ

To sum up the placenta can be regarded as an organ with functions in common with several more specialized organs such as lung, gut, kidney, liver, hypophysis, genital gland, bone marrow.

References

1. Allen, W.R. 1982: Immunological aspects of the endometrial cup reaction and the effect of xenogeneic pregnancy in horses and donkeys. J.Reprod.Fert., Suppl. 31, 57.

2. Amoroso, E.C. 1952: Placentation. In: Marshall's Physiology of Reproduction (Ed. A.S. Parkes). Longmans,Green. London. 3rd Ed., Vol. 2, p. 127.

3. Björkman, N. 1968: Contribution of electron microscopy in elucidating placental structure and function. Int.Rev.gen.exp.Zool. 3, 309.

4. Björkman, N. 1970: An Atlas fo Placental Fine Structure. Baillère, Tindall & Cassell, London.

5. Björkman, N. 1973: Fine structure of the fetal-maternal area of exchange in the epitheliochorial and endotheliochorial types of placentation. Acta anat. 86, Suppl. 61.

6. Brambell, R.W.R. 1970: Transmission of Passive Immunity from Mother to Young. North-Holland Publ. Co., Amsterdam.

7. Brezenski, J.J., J. Canazza, J. Li. 1971: Role of placenta in fetal lipid metabolism. III. Formation of rabbit plasma phospholipids. Biochem.Biophys.Acta. 239, 92.

8. Carlson, R.M. 1981: Patten's Foundations of Embryology. Mc-Graw-Hill Book Company, New York. 4th Ed.

9. Dantzer, V. 1984: Scanning electron microscopy of exposed surface of the porcine placenta. Acta anat. 118, 96.

10. Dantzer, V., M.H. Nielsen. 1984: Intracellular pathways of native iron in the maternal part of the porcine placenta. European J. Cell Biol. 34, 103.

11. Enders, A.C. 1965: A comparative study of the fine structure of the trophoblast in several hemochorial placentas. Amer.J.Anat. 116, 29.

12. Grosser, O. 1909: Vergleichnede Anatomie und Entwicklungsgeschichte der Eihäute und der Placenta. Braumüller, Wien.

13. Longo, I.D. 1972: Disorders of placental transfer. In: Pathophysiology of Gestation (Ed. N. Assali). Academic Press, New York. Vol. 2.

14. Mossman, W. 1937: Comparative morphogenesis of the fetal membranes and accessory uterine structures. Contrib.Embryol.Carneg.Inst. 26, 129.

15. Miller, R.K., T.R. Koszalka, R.I. Brent. 1976: The transport of molecules across placental membranes. In: The Cell Surface in Animal Embryogenesis and Development. (Ed. G. Poste, G.L. Nicolson). Elsevier/North-Holland Biomedical Press.

16. Myagkaya, G., J.P.M. Schellens. 1981: Final stages of erythrophagocytosis in the sheep placenta. Cell Tissue Res. 214, 501.

17. Wild, A.E. 1983: Trophoblast cell surface receptors. In: Biology of Trophoblast. (Ed. Loke & Whyte). Elsevier Science Publishers B.V. p. 472.

18. Wimsatt, W.A. 1951: Observations on the morphogenesis, cytochemistry and significance of the binucleate giant cells of the placenta of rumunants. Amer.J.Anat. 89, 233.

19. Wooding, F.B.P. 1984: Role of binucleate cells in fetomaternal cell fusion at implantation in the sheep. Am.J.Anat. 170, 233.

20. Wynn, R.M. 1969: Noncellular components of the placenta. Amer.J. Obstet.Gynec. 103, 723.

Recommended literature

21. Beier, H.M., P. Karlson. 1982: Proteins and Steroids in Early Pregnancy. Springer-Verlag, Berlin Heidelberg New York.

22. Loke, Y.W., A. Whyte. 1983: Biology of Trophoblast. Elsevier, Amsterdam New York Oxford.

23. Ramsey, E.M. 1982: The Placenta. Human and Animal. Praeger Publishers. New York.

24. Steven. D.H. 1975: Comparative Placentation. Academic Press, New York.

PREGNANCY PROTEINS IN THE HUMAN - BIOLOGICAL AND CLINICAL SIGNIFICANCE

T. Chard

Department of Reproductive Physiology, and Obstetrics and Gynaecology, St. Bartholomew's Hospital Medical College and the London Hospital Medical College, London, UK

Introduction

During normal human pregnancy a number of proteins either appear for the first time, or are greatly increased in the maternal circulation. Many of these proteins are produced by the placenta and this includes the two most familiar products in terms of clinical application: human placental lactogen (hPL) and human chorionic gonadotrophin (hCG). This group of materials was at one time often referred to as the 'placental proteins'. However, it is now apparent that not all of these materials are of placental origin in the strict sense of the placenta as a fetal rather than a maternal organ. Some are produced by the maternal endometrium/decidua and some (notably the various plasma binding proteins) by the maternal liver. For this reason the term 'pregnancy-associated proteins' has been widely used as an alternative designation for the group. But this again has been overtaken by events. The demonstration of a range of 'pregnancy' proteins at relatively high concentrations in seminal plasma (20,24) and follicular fluid (see article by Hoyer and colleagues) clearly shows that these materials are unique to neither the placenta nor a pregnancy. Indeed, until further information accumulates, it might be more appropriate to refer to this entire range of products as 'reproductive proteins' since their only reasonably definitive common feature is a distribution in all sites associated with the process of reproduction.

Pregnancy Proteins in Animals

Another conceptual problem is the increasing number of 'new' compounds which in turn has generated a large and confusing literature on more than a dozen potential clinical tests. It seems intuitively unlikely that each and every one of these will be unique in terms of its biology and clinical application. Instead, it is probable that they can be categorised into a small number of groups, within each of which all tests would have some physiological and diagnostic features in common. This review will present a preliminary classification of this type and use it to describe the biological and clinical significance of the 'reproductive' proteins.

Classification of 'Reproductive' Proteins

This classification has been presented before though in rather less complete form (4) and proposes that there are three groups of materials according to a variety of physiological and clinical criteria (Tables 1-3). It should be emphasised that the classification is preliminary and subject to modification as a result of the accumulation of further information. The aim is to describe a simple conceptual framework which may serve to identify in advance the likely clinical applications of a newly discovered compound, and thus greatly simplify the process of experimental testing. It should also be emphasised that the classification is not based on the chemistry of these materials, but rather on their functional properties and clinical applications.

Table 1. Functional Classification of 'Reproductive' Proteins

It should be noted that attribution to a given group is sometimes tentative rather than final, and may be the subject of revision when further evidence has been accumulated. The placental steroid hormones (oestriol and progesterone) are not discussed here but would appear in Group 1. Recent evidence on the endometrial/decidual origin of PP12 and PP14 suggests that these proteins should probably be renamed (see Review by Bell in this volume)

Group 1	Group 2	Group 3
Enzymes	PP5	Binding proteins
hCG, hPL	PAPP-A	PZP
SP1		PP12, PP14

Table 2. Biological Features of the Three Groups of 'Reproductive' Proteins

The source specified here is that during pregnancy. The precise cellular origin of these materials in seminal plasma and follicular fluid is unknown.

Group 1

Source: Trophoblast

Control mechanism: Trophoblast mass and uteroplacental blood-flow

Postulated functions: Hormone/enzymatic

Group 2

Source: Trophoblast

Control mechanism: Trophoblast mass and uteroplacental blood-flow plus ? other factors

Postulated functions: Local (immune and coagulation systems)

Group 3

Source: Maternal liver or endometrium/decidua

Control mechanism: Placental oestrogens/progesterone

Postulated functions: Binding of small molecules (e.g. thyroxine, steroids); maternal response to pregnancy.

Table 3. Clinical Features of the Three Groups of 'Reproductive' Proteins.

Group 1: Low levels reflect placental pathology and thus fetal well-being at all stages of pregnancy

Group 2: Low levels of PAPP-A may reflect specific events at an early stage of spontaneous abortion.

Elevated levels may be associated with specific pathology of the placenta (severe pre-eclampsia, abruption).

Group 3: Reflect the maternal response to pregnancy

Group 1 Products

Group 1 products include all of those materials which were traditionally regarded as 'specific' to the placenta (Table 1). They have in common that many of their activities (hormonal and enzymatic) would be expected to be exerted at a site distant from the placenta. In the human it has proved remarkable difficult to demonstrate an essential function for any of the Group 1 products in vivo (demonstration of biological effects in experimental systems in vitro cannot necessarily be extrapolated to the intact organism). Indeed, it has been suggested that many or even all of these materials have no function whatsoever but rather are the biological 'waste-products' of some more fundamental and ill-understood systems (8,11). Strong evidence for this rather nihilistic hypothesis comes from the fact that pregnancies may be entirely normal in the complete absence of one or other of the Group 1 proteins. Cases of specific deficiency of hPL (7) and SP1 (9) have been described, but not, at the present time, of hCG.

The apparent absence of function of many of the placental products in the human does not necessarily imply the same situation in animals. Of all aspects of physiology, that concerning reproduction shows exceptionally wide variation among different species. A system may be of key functional importance in one species but not in another. For example, the placental synthesis of oestrogen and progesterone plays an essential part in the initiation of parturition in the sheep while the same products, at high levels, have no relation to the onset of labour in the human.

Control mechanisms of Group 1 placental products

The lack of obvious functions of Group 1 products is further reflected by an apparent absence of feedback control mechanisms. In cases where such mechanisms have been proposed - for example, the control of hPL release by maternal carbohydrates and lipids - the evidence is disputed and subject to other explanations (19). As a hypothesis which is compatible with most of the accepted facts about placental protein synthesis, it has been proposed that the potential for placental protein production is a direct function of the total mass of the syncytiotrophoblast; that the rate of release (and secondarily, of synthesis) is a function of the concentration of the protein in the maternal blood of the intervillous space; and that this, in turn, relates to the rate at which blood flows in the intervillous space (3). An essential feature of this hypothesis is the haemochorial placentation of the human, allowing for direct contact of the trophoblast and the maternal bloodstream.

If this hypothesis is true then the rate of placental production of placental proteins should relate closely to uteroplacental bloodflow. Compatible with this is the fact that most studies have shown no systematic time-to-time variation of Group 1 products in maternal blood (10). Furthermore, the decrease of placental products in the maternal circulation observed in complicated pregnancies could as well be due to changes in bloodflow as to specific damage to the trophoblast. Thus, it is not surprising that products which depend on fetal precursors, such as oestriol, and products of the placenta, such as hPL, usually show a similar pattern in abnormal states (6).

The requirement for direct contact with blood may also explain why non-placental areas of the trophoblast (i.e. chorion) only produce minute quantities of placental proteins. These areas, unlike the placental trophoblast, are not in direct contact with the intervillous space and thus there is no on-going stimulus to synthesis and release.

The absence of simple 'cause-and-effect' or purposive mechanisms is an unfamiliar concept in physiology, and therefore superficially unappealing. However, it is perfectly interpretable in the context of events which occur only once in the lifetime of an organism: for example, pregnancy and puberty. It accommodates the immense variability of mechanisms between different species and removes the need to postulate the existence of the specific triggers and control systems which govern other physiological situations.

Clinical applications of Group 1 products

Given the simple control mechanism described above it would be expected that the main 'functions' reflected by maternal levels of Group 1 products would be the mass of the trophoblast and uteroplacental bloodflow. In turn, these should be related to the adequacy of placental transfer and thence to the growth of the fetus. Numerous studies on the clinical application of the measurement of Group 1 products support this expectation; there is almost invariably

1. This association is far from absolute and there is extensive overlap between normal and abnormal. The implications of this for clinical efficiency of placental function tests have been discussed in detail elsewhere (13).

an association between low levels in the mother and retarded fetal growth (5). In addition this association is not simply a direct relation to fetal weight; instead the extent of the reduction is greater in cases with specific evidence of dysmaturity than in those in which the child is small but normal (31). There is also a weaker association between low levels of Group 1 products and specific evidence of reduced placental transfer in the form of acute fetal distress during labour (16). In the past there has been much dispute on the relative value of different Group 1 products in the detection of fetal risk but most recent studies suggest that there is little to choose amongst them in this respect (6,17).

As a general rule, it would appear that Group 1 products reflect the condition of the fetus rather than the pathology which is responsible for that condition. An exception to this statement is the observation of elevated levels of Group 1 substances in conditions associated with excessive trophoblast growth, such as diabetes, rhesus iso-immunisation, and pre-eclampsia in primigravidae (18).

The lack of specific 'function' reflected by placental function tests can be regarded as an important advantage of this type of investigation. Thus, it can be argued that the assessment of uteroplacental bloodflow without significant interference by other factors is a highly desirable objective. This is the parameter which would be most affected by different forms of placental pathology.

Group 2 Products

The only products which can be identified as falling within Group 2 are pregnancy-associated plasma protein A (PAPP-A) and placental protein 5 (PP5) (Table 1). At the molecular level the most striking feature held in common by these materials is their strong binding affinity for heparin (25,27). At the physiological level, their proposed activities (anticoagulant, anti-complementary, and anti-proteolytic) (1,25,28) could all be exerted locally at the placental site. None of these activities are, as yet, firmly proven, and whether or not they would be 'essential' is as uncertain as is the case with Group 1 products. The only specific deficiencies so far reported are of PAPP-A, and these are associated with the Cornelia de Lange syndrome in the child (29). But the pregnancies themselves are apparently uneventful - further evidence for the lack of immediately obvious function of these materials during gestation.

Control mechanisms of Group 2 placental products

As with Group 1 products it is probable that trophoblast mass and uteroplacental bloodflow again play an important role in time-to-time regulation under physiological conditions; no significant circadian or other rhythm has been identified (10). One unique and very striking finding is that injection of sub-therapeutic doses of heparin is followed by a dramatic systemic rise in PP5 levels, an effect not found with PAPP-A or any other placental product (Fig. 1) (15). This finding fits in well with the overall hypothesis that PP5 is analogous to antithrombin III (25). Furthermore, it suggests that there might be control by endogenous heparin-like substances, though this is unproven at the present time.

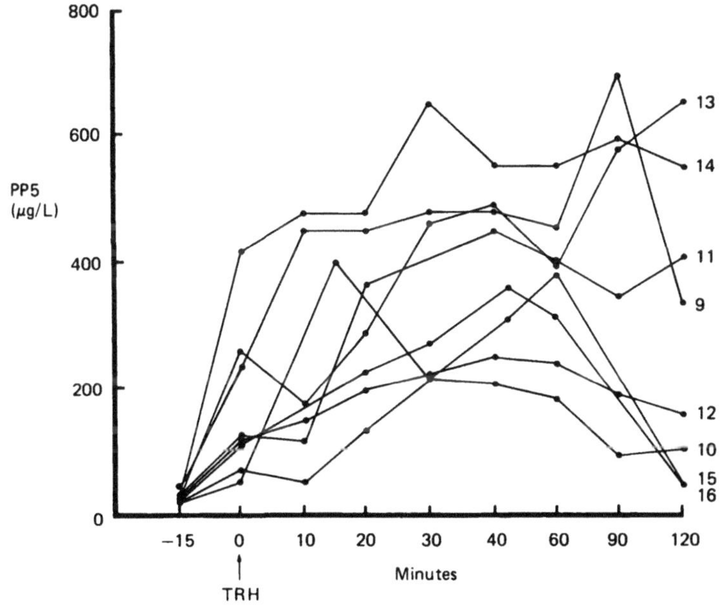

Figure 1. PP5 concentrations in serial blood samples taken before and after intravenous injection of TRH in 8 subjects. It should be noted that the increase occurred prior to the injection and was the result of the use of small quantities of heparin to maintain the potency of the cannula (modified from Reference 15).

<u>Clinical applications of Group 2 products</u>

In terms of clinical application Group 2 products appear to behave as rather inefficient versions of Group 1 products with an overall relationship to trophoblast function. In mid- and late pregnancy a number of observations suggest that elevated levels of Group 2 products may be associated with specific pathology such as pre-eclampsia and eclampsia (14) (Fig. 2). The hypothetical mechanism could be placental damage leading to the release of PP5 as a defence against further coagulation in the intervillous space. An earlier suggestion that PP5 levels in mid-trimester might be predictive of premature labour has not been confirmed by subsequent studies (21). PAPP-A may have a unique application in that low levels have been reported in cases of threatened abortion in which both other placental products and ultrasound examination suggest that the pregnancy is viable (30).

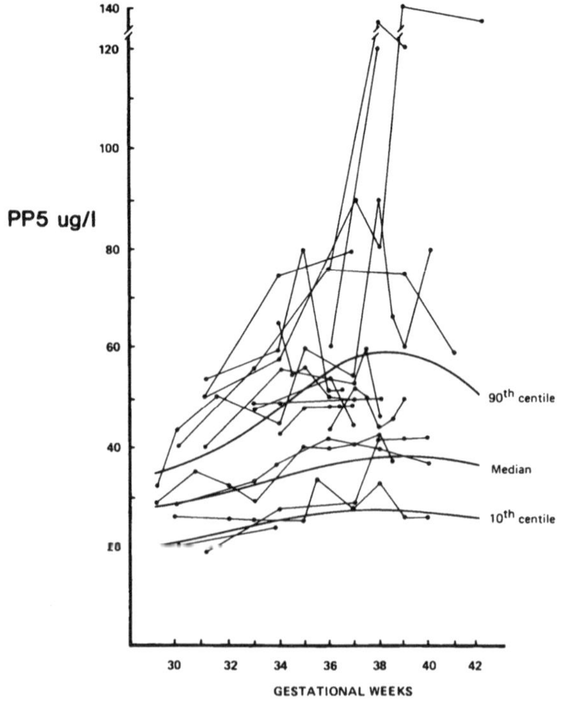

Figure 2. Circulating PP5 levels in 18 patients with pre-eclampsia.

Group 3 Products

This group includes materials which are not specific to pregnancy but which show a very striking rise in the maternal circulation during gestation. They are synthesised by maternal tissues such as the liver (26) or by the endometrium/decidua (PP12 and PP14) (2,22,23).

This latter group of materials is of particular interest because they may have a specific role in the process of implantation and the earliest stages of formation of the placenta. As with other 'reproductive' proteins they appear at very high levels in seminal plasma (see article by Bolton and colleagues).

Control mechanisms of Group 3 products

The increased production of Group 3 products during pregnancy is likely to reflect two different processes: in the case of the endometrial/decidual products the increased mass of the tissue of origin; and in all cases a stimulus from the rising levels of steroid hormones (oestrogen and progesterone).

An increase in Group 3 products is characteristic of oestrogen administration to non-pregnant subjects (e.g. the contraceptive pill). It is also notable that the pattern of increase of Group 3 products - a rapid rise initially, followed by a plateau at 8-12 weeks - is unlike that of the more specific materials, but similar to other parameters which show an oestrogen-dependent rise during gestation (e.g. cardiac output).

Clinical applications of Group 3 products

At the present time there is very little information on the clinical application of Group 3 products. However, it is likely their levels reflect not just the presence and activity of the fetus and placenta, but also the maternal response to the pregnancy. It is therefore attractive to speculate that variations in the local or general maternal response, as revealed by levels of Group 3 products, might yield important diagnostic information on abnormalities of gestation which are not revealed by more 'specific' tests. For example, the recent demonstration that maternal levels of PP12 are inversely related to fetal weight, and that the clinical predictive value of this observation in growth retardation is equal or superior to that of hPL (12) clearly opens a new range of diagnostic possibilities.

Conclusions

All of the 'pregnancy proteins' in the human can be found in the reproductive tract in the non-pregnant state (seminal plasma and follicular fluid). Consequently, it is suggested that they might be better termed the 'reproductive proteins'. In the human these proteins can be classified into three groups according to a variety of biological and clinical criteria. Comparable proteins are known to exist in a number of animals, but the diversity of reproductive phenomena between different species suggests that materials which are chemically analogous may have quite different biological implications.

References

1. Bischof, P. 1979. Arch.Gynaec. 227, 315.
2. Bolton, A.E., Dawkins, K.P., Rubenstein E., Obiekwe, B.C., Chard, T. 1985. Placenta 6, 1-4.
3. Chard, T. 1981. In: Hormones in Normal and Abnormal Tissues (K. Fotherby and B.S. Pal, eds). Walter de Gruyter, Berlin, p.490.
4. Chard, T. 1982. In: Pregnancy Proteins - Biology, Chemistry and Clinical Application (J.G. Grudzinskas, B. Teisner and M. Seppala, eds). Academic Press, Sydney, p.3.
5. Chard, T., Klopper, A. 1982. Placental Function Tests. Springer-Verlag, Heidelberg.
6. Chard, T., Obiekwe, B.C., Cockrill, B., Sturdee, D.W. 1985. Europ. J. Obstet. Gynec. Reprod. Biol. 19, 13.
7. Gaede, P., Trolle, D., Pedersen, H. 1978. Acta Obstet. Gynaec. Scand. 57, 203.
8. Gordon, Y.B., Chard, T. 1979. In: Placental Proteins (A. Klopper and T. Chard, eds). Springer-Verlag, Heidelberg, p.1.
9. Grudzinskas, J.G., Gordon, Y.B., Humphreys, J.D., Brudenell, M., Chard, T. 1979. Br J Obstet Gynaec 86, 978.
10. Houghton, D.J., Newnham, J.P., Lo, K., Rice A. 1982 . Brit. J. Obstet. Gynaec. 89, 831.

11. Houghton, D.J., Shackleton, P., Obiekwe, B.C., Chard, T. 1985. Placenta (in press).

12. Howell, R.J.S., Choglay, N.S., Perry, L.A., Bohn, H., Chard, T. 1985. Brit. J. Obstet. Gynaec. (in press).

13. Lilford, R.J., Obiekwe, B.C., Chard, T. 1983. Brit. J. Obstet. Gynaec. 90, 511.

14. Lee, J.N., Salem, H.T., Huang, S.C., Ouyang, P.C., Seppala, M., Chard, T. 1981. Int J Gynaecol Obstet. 19, 65.

15. Menabawey, M., Silman, R., Rice, A., Chard, T. 1985. Brit. J. Obstet. Gynaec. 92, 207.

16. Obiekwe, B.C., Chard, T. 1982. Europ. J. Obstet. Reprod. Biol. 14, 69.

17. Perry, L., Fattah, D., Cochrane, G., Obiekwe, B.C. 1984. J. Obstet. Gynaec. 4, 154.

18. Obiekwe, B.C., Sturdee, B.L., Cockrill, B.L., Chard, T. 1984. Brit. J. Obstet. Gynaec. 91, 1077.

19. Pavlou, T., Chard, T., Landon, J., Letchworth, A.T. 1973. Europ. J. Obstet. Gynec. Reprod. Biol. 3/2, 45.

20. Ranta, T., Siiteri, J.e., Koistinen, R., Salem, H.T., Bohn, H., Koiskimies, A.I., Seppala, M. 1981. J. Clin. Endocr. Metab. 53, 1087.

21. Rice, A., Chard, T., Hird, V., Grudzinskas, J.G., Nysenbaum, A.M. 1985. J. Obstet. Gynaec. (in Press).

22. Rutanen, E.-M., Bohn, H., Seppala, M. (1982). A.. J. Obstet. Gynec. 144, 460.

23. Rutanen, E.-M., Koistinen, R., Wahlstrom, T., Bohn, H., Seppala, M. 1984. Brit. J. Obstet. Gynaec. 91, 1240.

24. Salem., H.T., Menabawey, M., Seppala, M., Shaaban, M.M., Chard, T. 1984. Placenta, 5, 413.

25. Salem, H.T., Seppala, M., Chard, T. 1984. Placenta, 2, 205.

26. von Schoultz, B., Stigbrand, T. 1982. In: Pregnancy Proteins: Biology, Chemistry and Clinical Application. (J.G. Grudzinskas, B. Teisner, M. Seppala, eds). Academic Press, Sydney, p. 167.

27. Sinosich, M.J., Teisner, B., Davey, M., Grudzinskas, J.G. 1981. Aust. N.Z. J. Med. 11, 429.

28. Sinosich, M.J., Davey, M.W., Ghosh, P., Grudzinskas, J.G. 1982. Biochem. Int. 5, 77.

29. Westergaard, J.G., Chemnitz, T., Teisner, B., Poulson, H.K., Ipsen, L., Beck, B., Grudzinskas, J.G. 1983. Prenat. Diag. 3, 225.

30. Westergaard, J.G., Teisner, B., Sinosich, M.J., Madsen, L.T., Grudzinskas, J.G. 1985. Brit. J. Obstet. Gynaec. 92, 77.

31. Westergaard, J.G., Teisner, B., Hau, J., Grudzinskas, J.G., Chard, T. 1985. Obstet. Gynecol. 65, 316.

CLINICAL ASPECTS OF PLACENTAL PROTEIN MEASUREMENTS IN EARLY PREGNANCY AND ITS COMPLICATIONS.

J.G.Grudzinskas[1], J.G. Westergaard[2], B. Teisner[3]

1. Academic Unit of Obstetrics and Gynaecology, The London Hospital Medical College, UK; 2. Department of Obstetrics and Gynaecology and 3. Institute of Medical Microbiology, Odense University, Odense, Denmark

Introduction

The detection of human chorionic gonadotrophin (hCG) in the peripheral blood or urine is the cornerstone of the diagnosis of pregnancy by biochemical techniques, hCG being observed at the time of implantation. In addition, the systematic examination of human trophoblast and late pregnancy blood has led to the identification of a new generation of proteins which are likely to be used in the diagnosis of early pregnancy and its failure (Table 1) (1).

Table 1 Pregnancy-associated proteins which may be diagnostic markers (1)

	Detection after conception (days)
1. Human chorionic gonadotrophin (hCG)	6-10
2. Schwangerschaftsprotein 1 (SP1)	10-23
3. Pregnancy-associated plasma protein A (PAPP-A)	28
4. Placental protein 5 (PP5)	42

Schwangerschafts-protein 1 (SP1), also referred to as pregnancy-specific beta-1 glycoprotein ($PS\beta_1G$) measurements can be used in the biochemical diagnosis of pregnancy (2) and recent reports have described consistently depressed levels of PAPP-A in early pregnancy

Pregnancy Proteins in Animals

failure (3). This review deals with the clinical implications of measurements of hCG and other placental proteins in normal pregnancy and in complications such as imminent abortion , ectopic gestation and fetal congenital abnormality.

Diagnosis of pregnancy

The mainstay of the simple biochemical diagnosis of pregnancy is the detection of hCG in maternal urine, or a day or two earlier in the peripheral blood. However, the definitive diagnosis of pregnancy is dependent on the observation of the embryonic heart action, and ultrasonic findings prior to this only imply the presence of pregnancy. It is noteworthy that the latest developments in hCG assay technology have contributed little to the resolution of problems faced by the clinician using these tests. Firstly, given the possibility of non-specific effects, the interpretation of low levels of of hCG must be made cautiously. Secondly, assays of greater sensitivity have been developed using monoclonal antibodies without compromising the specificity of the assay. However, these assays have provided further evidence of the ubiquitous nature of hCG, albeit in low concentrations (≤5 IU/l). Consequently, in clinical practice, a single estimation of hCG should only be considered indicative of pregnancy if it is greater than 25 IU/l, a pregnancy can be inferred if a lower level of hCG is seen to increase two-fold at an interval of three days. Thirdly, hCG levels at or above which a diagnosis of pregnancy can be confidently made, should be determined in each unit so that assay background or nonspecific 'noise' levels can be accounted for. Fourthly, if hCG has been administered therapeutically, estimations should be delayed until clearance of exogenous hCG has occurred, possibly postponing the diagnosis by several days. In this situation, assays for other proteins of placental origin, namely SP1, may be appropriate. However, in common with hCG, the apparent detection of very low levels of SP1 may not necessarily be indicative of pregnancy; yet, assays for this molecule are likely to be as specific for pregnancy as is hCG in the post-implantation period. Other pregnancy-associated proteins, such as PAPP-A and PP5, are not serious contenders for diagnostic tests given their relatively late appearance (after six weeks amenorrhoea) in the maternal peripheral blood. Finally, the interpretation of hCG assay results, in particular the immediate post-implantation period should be conducted with care, emphasising that the earlier the diagnosis is made, the less likely will the outcome be normal (Table 2) given the

Table 2. Likely outcome of pregnancy in relation to time of diagnosis[1].

Time of diagnosis	Chance of normal outcome (%)
Pre-implantation	25-30
Post-implantation	46-60
Six-weeks pregnancy	85-90
Second trimester	95
Third trimester	98

very high spontaneous rate of pregnancy failure in the peri-implantation period (1). It seems logical, therefore, to restrict the usage of the most sensitive hCG tests to specific clinical situations, e.g. subfertility, suspected ectopic gestation, rather than increasing the availability for diagnosis at this early state in the normal population.

Determination of gestational age

We and others have recently confirmed the validity of a new method of assessing gestational age at 4 to 6 weeks of pregnancy (4). A single determination can be made of human placental lactogen (hPL), hCG and SP1 on a maternal blood sample and the result applied in a calculation that relates the logarithm of the hormone or protein level to the day of gestation (Fig. 1). The resultant estimation of

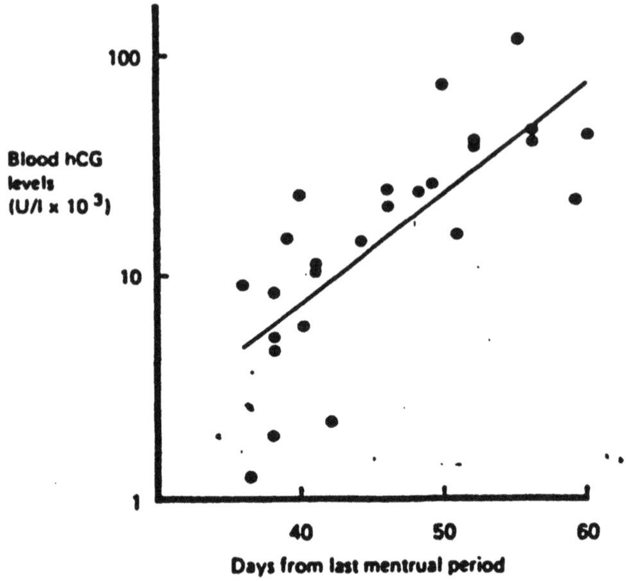

Fig. 1. Relationship between normal serum concentration of hCG and day of gestation as judged by LMP (reproduced by permission from J. Reprod. Med.).

gestational age has an error range similar to that which can be made from an accurate menstrual history. We have shown that there is a highly significant linear correlation between the levels of hCG and SP1 and (a) the stage of gestation and (b) the day of spontaneous delivery at term. Since virtually all diagnostic and therapeutic measures in antenatal care depend on knowledge of the gestational age for efficient implementation, a good case can be made for the routine application of these tests in this manner to all pregnancies that present at the appropriate time.

Threatened, inevitable and missed abortion

Recent studies in which ultrasonic examination was performed at the time of blood sampling in these abnormalities, have critically evaluated the usefulness of estimation of protein and hormones of fetal, placental and ovarian origin in comparison with biophysical measurements and observations (5,6). Their design addressed many of the previously unanswered questions concerning the value of biochemical tests in the clinical context with particular reference to their relative merits compared to ultrasound. Firstly, only abnormal PAPP-A levels distinguished those pregnancies which miscarried from those which did not, even when there was ultrasonic evidence of fetal heart action at the time of blood sampling (Fig. 2). Secondly,

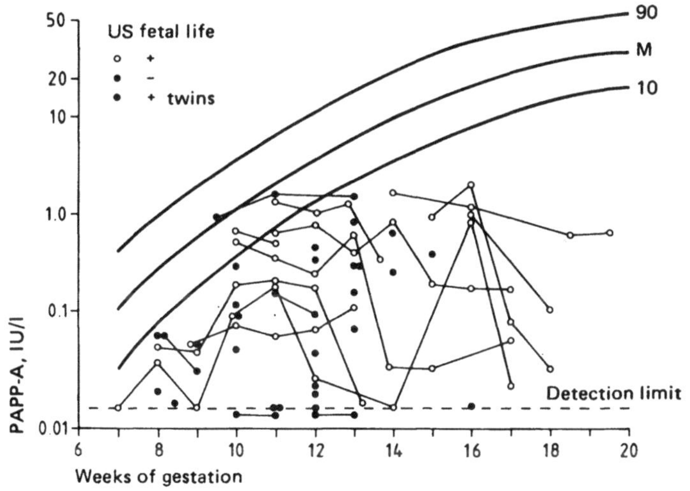

Fig. 2. Serum levels of PAPP-A in 42 women with spontaneous abortion. • = no ultrasonic evidence of fetal hear action; o = evidence of fetal heart action; • = twin pregnancy with life fetuses; 10, M, 90 = 10th, 50th, 90th centiles of the normal range (reproduced by permission from Brit. J. Obstet. Gynaec.).

depressed PAPP-levels were seen weeks before spontaneous miscarriage occurred in some patients. Thirdly, if the heart action was not evident ultrasonically on repeated examination, depressed levels of placental (hCG, SP1, hPL, PAPP-A), ovarian (oestradiol, progesterone) or fetal (AFP) were consistently seen. Fourthly, it abortion occurred after the detection of the fetal heart action, the levels of fetal, ovarian and placental molecules were generally in the normal range, the sole exception being PAPP-A. Finally, if the heart action was detected and the serum PAPP-A levels were normal, the chance of a normal outcome for that pregnancy was in excess of 98%. These observations, which are discussed in detail elsewhere (6), although apparently discordant with recent reports on the usefulness of biochemical tests in threatened abortion (7) substantiate that ultrasound has a useful role in the management of these disorders. On closer review, these data are in accord with many reports on this subject, substantially expanding the information on women with recurrent bleeding. Our recent data further support the contention that PAPP-A measurements in women with vaginal bleeding at 8-16 weeks of gestation are capable of providing the clinician with information hitherto beyond the reach of any diagnostic tool. diagnostic tool.

In the earliest weeks of pregnancy, when ultrasonic examination is less reliable, it is likely that measurements of hCG and SP1 will largely reflect the normality of a pregnancy. However, the extensive information on this subject recently reviewed by Sinosich and his colleagues (8) suggests that such observations are considerably less precise at this time given the multifactorial nature of early intrauterine pregnancy.

Ectopic gestation

We and others have extensively investigated the clinical usefulness of biochemical measurements, in addition to hCG, in this condition (9,10). Circulating hCG can be detected in more than 90% of women with this condition, the levels being consistently depressed when

PAPP-A is actually detected (3,10) (Fig. 3). These findings suggest

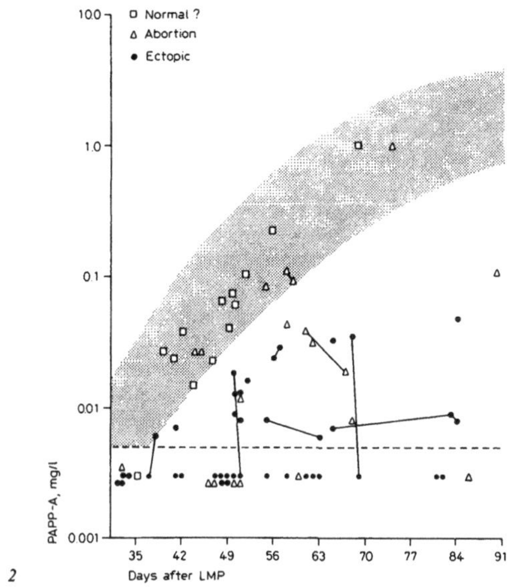

Fig. 3. Circulating PAPP-A in 50 women with lower abdominal pain. ● = ectopic gestation; Δ = inevitable abortion; ◘ = threatened abortion, normal outcome. (Reproduced by permission from Karger, Basel).

that PAPP-A synthesis is severely compromised in this condition evidenced by the frequent absence of PAPP-A in trophoblastic tissue obtained from the fallopian tube (11) and that depressed levels of PAPP-A in conjunction with positive hCG measurements may be indicative of pregnancy failure, notably ectopic gestation.

Fetal congenital abnormality

PAPP-A could not be detected either in the maternal circulation or the placental tissue in two patients with a rare but serious condition called 'Cornelia de Lange syndrome (12). This condition is characterised by severe mental retardation and facial and limb dysmorphic features. Since circulating and tissue hPL, SP1 and oestriol were detected and distributed normally, it is likely that the absence of PAPP-A may be of value in the pre- and post-natal diagnosis of this condition, as the diagnosis is currently made by the recognition of phenotypic features. A third case in which we failed to detect PAPP-A in trophoblastic tissues has been obtained, confirming our hypothesis that if PAPP-A is biologically active that its function is expressed during embryogenesis or earlier.

References

1. Grudzinskas, J.G., Nysenbaum, A.M. 1985. Ann. N.Y. Acad. Sci. 442, 38.

2. Grudzinskas, J.G., Gordon, Y.B., Jeffrey, D. Chard, T. 1977. Lancet, i, 333.

3. Grudzinskas, J.G., Westergaard, J.G., Teisner, B. 1985. In: Proteins of the Placenta (P. Bischof, A. Klopper, eds). Karger, Basel, 184.

4. Westergaard, J.G., Teisner, B., Grudzinskas, J.G., Chard, T. 1985. J.Reprod.Med. 30, 57.

5. Westergaard, J.G., Sinosich, M.J., Bugge, M., Masden, L.T., Teisner, B., Grudzinskas, J.G. (1985) Amer. J. Obstet. Gynaec. 145, 67.

6. Westergaard, J.G., Teisner, B., Masden, L.T., Grudzinskas, J.G. 1985. Brit. J. Obstet. Gynaec. 92, 77.

7. Huisjes, H.J. 1984. In: Current Reviews in Obstetrics and Gynaecology (T. Lind, ed). Churchill Livingstone, London, 132.

8. Sinosich, M.J., Grudzinskas, J.G., Saunders, D.M. 1985a. Obstet. Gynecol. Surv. 40, 273.

9. Lindstedt, G., Lundberg, P.A., Janson, P.O. et al. 1982. Scand. J.Clin.Lab.Invest. 42, 201.

10. Sinosich, M.J., Ferrier, A., Teisner, B. et al. 1985b. J. Clin. Reprod. Fertil. In press.

11. Chemnitz, J., Tornehave, D., Teisner, B. et al. 1984. Placenta, 5, 489.

12. Westergaard, J.G., Chemnitz, J., Teisner, B., Poulsen, H.K., Ipsen, L., Beck, B., Grudzinskas, J.G. 1983. Perinat. Diag. 3, 225.

"NORMAL" PLASMA PROTEINS DURING PREGNANCY

B. Teisner
Institute of Medical Microbiology, University of Odense, Denmark

In a cross-sectional study the levels of circulating enzyme inhibitors (antithrombin III (AT III), α_2-macroglobin (α_2M), α_1 antitrypsin), complement factors (C3, C4, factor B), and split products of C3 (C3c and C3d) were quantified during normal pregnancy. The concentrations of AT III were significantly lower, α_1 antitrypsin significantly higher and α_2M no different when compared with those of the non pregnant and puerpiral states. Increased levels of C3 and the plit product C3d were observed during pregnancy, whereas the concentrations of C4, factor B and C3c remained unchanged.

The levels of C3d and substances which have been suggested to be involved in the regulation of complement activation such as protease inhibitors (α_1 antitrypsin, α_2M and AT III) and pregnancy associated proteins (PZP, PAPP-A AND SP-1) were compared in 14 women with severe pregnancy induced hypertension (PIH) and matched controls. The only significant finding was depressed concentrations of AT III ($p = 0.013$) in the PIH group.

Increased concentrations of corticosteroid binding globulin (CBG) have been reported in PIH. The concentration was found to increase 3 times during normal pregnancy. However, the CBG concentrations found in a group with PIH ($n = 32$) did not differ significantly from those of matched controls.

No evidence was found to support the hypothesis that circulating immunecomplexes are involved in the pathogenesis of PIH.

Matched maternal and fetal plasma, as well as amniotic fluid (AF) were obtained during weeks 14 to 26 of retrospectively defined normal pregnancies ($n = 70$) and complement components (C3, C4, C5,

Pregnancy Proteins in Animals

factors B, I and H) measured by electroimmunoassay. During the observation period (weeks 14 to 26) no significant increase in the concentrations of maternal C-components was seen. By contrast, concentrations in the fetal circulations of C3, C4, and factor H showed a significant increase, whereas only C3 and factor B were found to increase in AF.

PLACENTAL PROTEINS IN THREATENED ABORTION

J.B. Hertz, P. Schultz-Larsen

Departments of Obstetrics and Gynecology, Gentofte and Herlev Hospitals, and Department of Clinical Chemistry, Herlev Hospital, University of Copenhagen, Denmark

Introduction

Threatened abortion is the most frequently encountered complication in early human pregnancy. The exact incidence, however, is unknown. South & Naldrett (1) found an incidence of vaginal bleeding before the 28th gestational week on approximately 16% among 7824 pregnancies. The study, however, only included singleton pregnancies carried beyond the 28th week and not pregnancies complicated by vaginal bleeding ending in spontaneous abortion. By means of different experimental models incidences of spontaneous abortion on 27% (2), 28% (3) and 43% (4) have been reported. The pregnancies were recognized clinically and/or biochemically. Since all patients who ultimately abort must threaten to abort for some time before the abortion is inevitable or completed, the incidence of threatened abortion must be at least as great as the incidence of spontaneous abortion. Followingly an incidence of threatened abortion on at least 20 to 25% seems reasonable.

In case of admission to hospital approximately 50% of the threatened pregnancies terminate in abortion (5,6). It is difficult on clinical grounds alone to determine whether the pregnancy is doomed or capable of continuing. In order to identify the failing gestations and thus avoid prolonged anxiety and an unnecessary and costly stay in hospital, reliable indicators of the viability of the conceptus are needed. A number of diagnostic tests have been applied to the evaluation of threatened abortion (6,7,8,9,10) and especially the highly sensitive real-time ultrasound equipment developed during recent years have proved most valuable (8,10,11).

Human placental lactogen (HPL) and pregnancy-specific beta-1-glyco-

Pregnancy Proteins in Animals

protein (SP_1) are both products of the trophoblast of the placenta (12,13) and are thus markers of the placental function. The purpose of the present study was to assess the value of HPL and SP_1 determination in the evaluation of pregnancies complicated by vaginal bleeding during the first 20 weeks of gestation.

Materials and Methods

The study comprised 173 women admitted to the Departments of Obstetrics and Gynecology, Gentofte and Herlev Hospitals, University of Copenhagen, Denmark, because of vaginal bleeding during the first 20 weeks of pregnancy. To enter the study the menstrual bleedings had to be regular prior to conceivement in order to achieve an accurate gestational age. At pelvic examination at admission the uterus had to be enlarged, the cervix intact and the os closed as well as the HCG pregnancy test should be positive. In case of abortion the products were examined histologically excluding cases of ectopic pregnancy and hydatidiform mole.

Blood samples for HPL and SP_1 determinations were collected at admission, twice weekly during the stay in hospital and 10 to 14 days after discharge in case of successful outcome. Serum was separated and stored at -20°C until assayed. The results of the analyses did not influence the treatment of the individual patient.
HPL was measured by radioimmunoassay (Radiochemical Centre, Amersham, England). The detection limit was 0.05 mg/l and the day-to-day analytical variation less than 10%. SP_1 was determined by immunoelectrophoresis using monospecific anti-SP_1 (lot 018B, Dakopatts, Copenhagen, Denmark) and calibrated with an SP_1 standard (lot 042405, Behringwerke AG, Marburg, FRG). The detection limit was 0.7 mg/l and the day-to-day analytical variation less than 10%.

Results

Ninety-three (54%) of the 173 women included in the study continued the pregnancy until at least the 28th gestational week and 80 (46%) aborted. HPL was determined in all 173 patients whereas SP_1 only

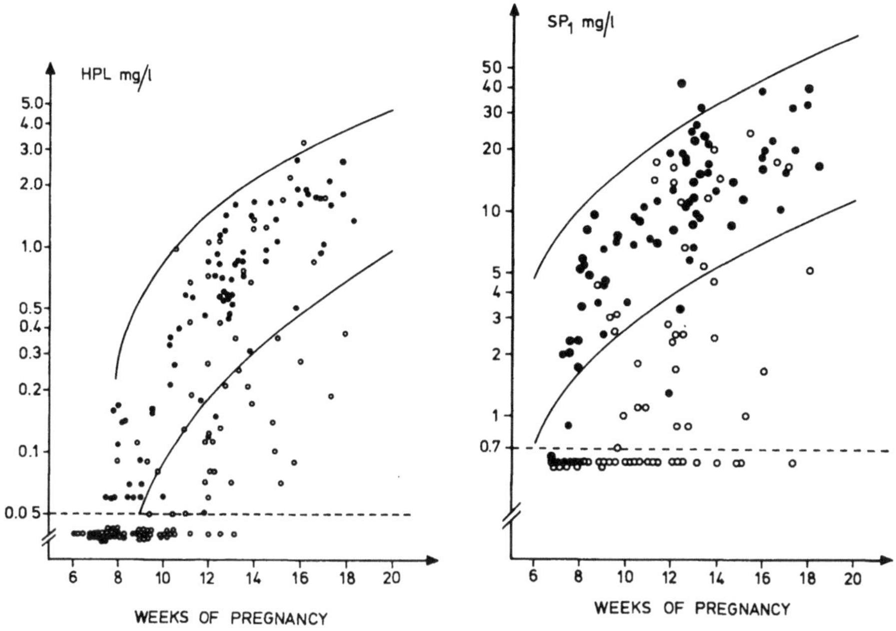

Fig. 1. The HPL and SP_1 values from the day of admission in 173 (HPL) and 141 (SP_1) patients with threatened abortion.
o = aborted , • = gave birth .

was measured in 141 patients. Based on values from patients continuing pregnancy until at least the 28th gestational week, and after logarithmic transformation of the data, the 95% reference range (mean $\pm$ 2SD) was calculated for each parameter. The HPL and SP_1 values from the day of admission are shown in Fig. 1. The reference ranges (solid lines) and the detection limits (dotted lines) are indicated.

The lower border of the reference range has been used as "the cut-off-level" in discriminating between the normal and abnormal test results. Followingly the sensitivity, specificity and the predictive values of abortion and successful outcome have been calculated for the parameters separately and in combination concerning the entire gestational period studied (Table 1). In Table 2 the predictive values of abortion and successful outcome have been calculated for separate gestational periods.

Table 1. The Sensitivity, Specificity and Predictive Values (%) of HPL and SP_1 Determinations in Patients with Threatened Abortion during the first 20 Weeks of Pregnancy*.

parameter	sensitivity	specificity	prediction of abortion	prediction of successful outcome
HPL	64	94	90 (100)	74 (90)
SP_1	74	87	82 (93)	81 (91)
HPL + SP_1	70	97	94	81

* Values are based on initial analysis from the day of admission and on serial analyses (in parentheses).

Table 2. The Predictive Values (%) of HPL and SP_1 Determinations in Patients with Threatened Abortion in Separate Gestational Periods*.

parameter	prediction of abortion		prediction of successful outcome	
	weeks		weeks	
	6 - 9	10 - 20	6 - 9	10 - 20
HPL	67 (100)	91 (100)	94 (100)	70 (87)
SP_1	50[1] (100)	95 (92)	100 (100)	76 (87)
HPL + SP_1	100	94	100	78

* Values are based on initial analysis from the day of admission and on serial analyses (in parentheses).

[1] 95% from the 8th gestational week on.

In Figs. 2 and 3 the regression lines between HPL and SP_1 are shown for the gestational periods weeks 6 to 20 and weeks 6 to 9, respectively. In both cases a highly significant positive correlation was found between HPL and SP_1 ($p<0.001$).

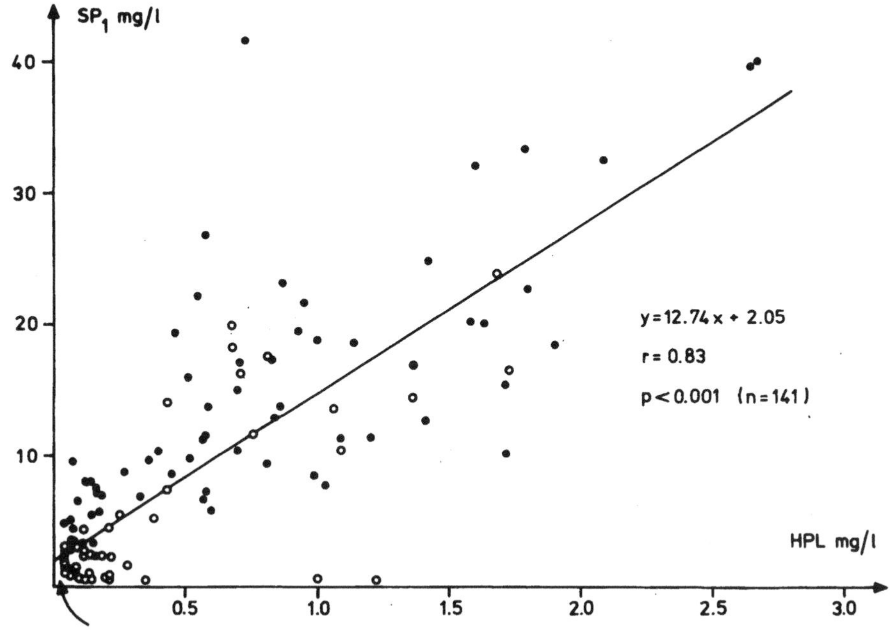

Fig. 2. The regression line between HPL and SP_1 in the gestational period weeks 6 to 20. Arrow indicates 18 women who aborted and 7 who gave birth. o = aborted , ● = gave birth.

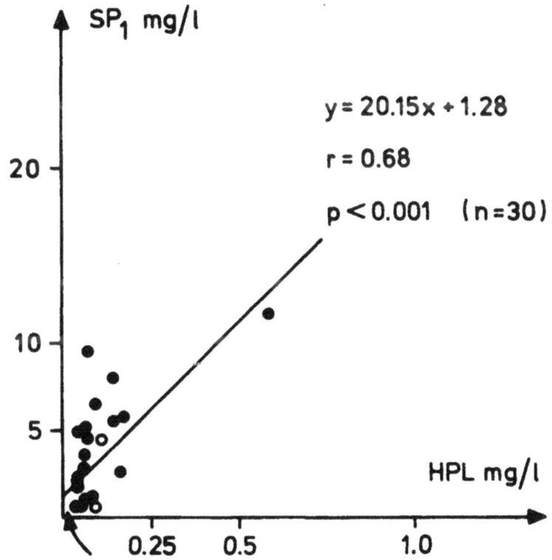

Fig. 3. The regression line between HPL and SP_1 in the gestational period weeks 6 to 9. Arrow indicates 6 women who aborted and 3 who gave birth. o = aborted , ● = gave birth.

Discussion

Both investigated parameters are products of the trophoblast and followingly a certain degree of placental development is to be expected before thewe two proteins become detectable in the maternal circulation. By means of a sensitive radioimmunoassay (14) and enzyme immunoassay (15) however, SP_1 has been detected as early as 6 to 14 days after ovulation, levels rising steadily thereafter, allowing the presence of SP_1 to be used in the early detection of pregnancy. The lower border of the reference range for SP_1 in this study exceeds the detection limit of the analysis before the 6th gestational week making the interpretation of the SP_1 values valid from this stage of pregnancy on (Fig. 1). Due to the minor amounts of HPL in the maternal circulation this protein hormone first becomes detectable at a later gestational age. With the present HPL analysis low or undetectable levels of HPL are only to be interpreted safely from the 9th week on when the lower border of the reference range exceeds the detection limit (Fig. 1).

The predictive values seem to improve after serial determinations (Table 1 and 2), but in the respect of identifying the failing gestations and followingly accelerate definitive treatment they do, however, waste time. The combination of parameters seems more convenient and the results are comparable (Table 1 and 2).

The prognostic value of HPL and SP_1 have been extensively studied, the results of some of the largest studies are summarized in Table 3. Beyond this Niven et al. (21) found a single HPL determination to be a sufficiently reliable index of the outcome of the threatened pregnancy from the 9th week, before this gestational age serial samples were needed. They did not calculate the predictive values. Gartside & Tindall (22) found a single reading of HPL between 9 and 19 weeks of pregnancy to give the correct immediate prognosis in 86% of threatened abortions, an equivocal prognosis in 11% of cases and a wrong prognosis in 3%.

Jouppila et al. (23) found the results of SP_1 estimations highly encouraging in the evaluation of the prognosis of vaginal bleeding in early pregnancy being "at least as useful as the measurements of HCG or HPL". They found the predictive value of abortion from low SP_1 values to be 96%. Jandial et al. (19) found serial measurements of SP_1 advisable before 10 weeks' gestation and the prognostic value of a single HPL measurement was found to be inferior to that of a single SP_1 measurement. We found a single low SP_1 value to be followed by abortion in 95% of the cases from the 8th week on (Table 2).

Table 3. The Predictive Values (%) of HPL and SP_1 Measurements in Maternal Serum or Plasma in Threatened Abortion.

reference	N° of patients studied	prediction of abortion	prediction of successful outcome
HPL			
Garoff & Seppälä (16)[1]	86	100	60
Kunz & Keller (17)	65	81	61
Gerhard & Runnebaum (18)[1]	84	92	45
Duff et al. (7)	69	91	52
Westergaard et al. (10)	108	83	77
Present study	173	90	74
SP_1-			
Jandial et al. (19)	64	87	75
Duff et al. (7)	69	72	83
Tamsen (20)	105	100	65
Westergaard et al. (10)	108	81	82
Present study	141	82	81

[1] The predictive values were not stated in these studies but they have been calculated from the reported results.

From Table 3 it appears that the prediction of abortion almost unanimously has been found to be higher than that of successful outcome. This is in some cases to be ascribed to the missed and anembryonic gestations where the trophoblastic protein production is continued. But also the miscarriages where an intact and live conceptus is expelled, as it has been confirmed by ultrasound (8), may be responsible for the lower predictive value of successful outcome.

The greater amount of SP_1 in maternal blood in very early pregnancy compared to the amount of HPL and the highly significant positive correlation between HPL and SP_1 found in this study (Figs. 2 and 3) indicate that SP_1 measurements might replace those of HPL in the evaluation of vaginal bleeding in early pregnancy. Since ultrasound examination with the detection of fetal life has proved fully reliable in the prediction of abortion from the 9th to 10th week on (6,8,11) the early gestational period from weeks 6 to 9 offers special attention. Unfortunately neither HPL nor SP_1 determinations resulted in convincing predictive values of abortion during this early stage of pregnancy, but when a low HPL and SP_1 value were combined abortion succeeded in every case (Table 2).

In conclusion, we have found HPL to be of value in the evaluation of threatened abortion from the 9th gestational week on and SP_1 of value from the 8th week on. The highly significant positive correlation between HPL and SP_1 suggests that SP_1 determinations might replace those of HPL in the evaluation of early threatened pregnancy. It must be underlined, however, that the total reliability of ultrasound examination in the prediction of abortion from the 9th to 10th week on has made ultrasound the diagnostic mainstay in the evaluation of threatened abortion. Formerly measurements of biochemical parameters only seemed to add valuable information to ultrasound in the diagnosis of impending abortion before the 10th week of pregnancy (6,8). Most recently, however, Westergaard et al. (10) have found low levels of the pregnancy associated plasma protein-A (PAPP-A) in all patients who subsequently aborted in the presence of a live fetus. By means of PAPP-A measurements it thus seems possible to identify these socalled "live-abortions" and followingly accelerate definitive treatment in these patients.

Acknowledgement

This study was supported by a grant from Carl og Ellen Hertz's Familielegat.

References

1. South, J., J. Naldrett. 1973. The effect of vaginal bleeding in early pregnancy on the infant born after the 28th week of pregnancy. J. Obstet. Gynaecol. Br. Commonw. 80, 48-52.

2. Stickle, G. 1968. Defective development and reproductive wastage in the United States. Am. J. Obstet. Gynecol. 100, 442-447.

3. Hertig, A.T., J. Rock, E.C. Adams, M.C. Menkin. 1959. Thirty-four fertilized human ova, good, bad and indifferent, recovered from 210 women of known fertility. Pediatrics. 23, 202-211.

4. Miller, J.F., E. Williamson, J. Glue, Y.B. Gordon, J.G. Grudzinskas, A. Sykes. 1980. Fetal loss after implantation. Lancet. II, 554-556.

5. Jouppila, P. 1980. Clinical and ultrasonic aspects in the diagnosis and follow-up of patients with early pregnancy failure. Acta Obstet. Gynecol. Scand. 59, 405-409.

6. Hertz, J.B. 1984. Diagnostic procedures in threatened abortion. Obstet. Gynecol. 64, 223-229.

7. Duff, G.B., J.J. Evans, M. Legge. 1980. A study of investigations used to predict outcome of pregnancy after threatened abortion. Br. J. Obstet. Gynaecol. 87, 194-198.

8. Jouppila, P., I. Huhtaniemi, J. Tapanainen. 1980. Early pregnancy failure: Study by ultrasonic and hormonal methods. Obstet. Gynecol. 55, 42-47.

9. Salem, H.T., S.A. Ghaneimah, M.M. Shaaban, T. Chard. 1984. Prognostic value of biochemical tests in the assessment of outcome in threatened abortion. Br. J. Obstet. Gynaecol. 91, 382-385.

10. Westergaard, J.G., B. Teisner, M.J. Sinosich, L.T. Madsen, J.G. Grudzinskas. 1985. Does ultrasound examination render biochemical tests obsolete in the prediction of early pregnancy failure? Br. J. Obstet. Gynaecol. 92, 77-83.

11. Anderson, S.G. 1980. Management of threatened abortion with real-time sonography. Obstet. Gynecol. 55, 259-262.

12. Sciarra, J.J., S.L. Kaplan, M.M. Grumbach. 1963. Localization of anti-human growth hormone serum within the human placenta: Evidence for a human chorionic "growth hormone-prolactin". Nature. 199, 1005-1006.

13. Horne, C.W.H, C.M. Towler, R.G.P. Pugh-Humphreys, A.W. Thomson, H. Bohn. 1976. Pregnancy specific beta-1-glycoprotein - a product of the syncytiotrophoblast. Experientia. 32, 1197-1199.

14. Grudzinskas, J.G., Y.B. Gordon, D. Jeffrey, T. Chard. 1977. Specific and sensitive determination of pregnancy-specific beta-1-glycoprotein by radioimmunoassay. Lancet. I, 333-335.

15. Ahmed, A.G., A. Klopper. 1983. Diagnosis of early pregnancy by assay of placental proteins. Br. J. Obstet. Gynaecol. 90, 604-611.

16. Garoff, L., M. Seppälä. 1975. Prediction of fetal outcome in threatened abortion by maternal serum placental lactogen and alpha fetoprotein. Am. J. Obstet. Gynecol. 121, 257-261.

17. Kunz, J., P.J. Keller. 1976. HCG, HPL, oestradiol, progesterone and AFP in serum in patients with threatened abortion. Br. J. Obstet. Gynaecol. 83, 640-644.

18. Gerhard, I., B. Runnebaum. 1978. Aussagewert von HCG-, HPL-, progesteron- und östriol-bestimmungen bei frauen mit drohender fehlgeburt. Geburtsh. u. Frauenheilk. 38, 785-799.

19. Jandial, J., C.M. Towler, C.H.M. Horne, D.R. Abramovich. 1978. Plasma pregnancy-specific beta-1-glycoprotein in complications of early pregnancy. Br. J. Obstet. Gynaecol. 85, 832-836.

20. Tamsen, L. 1984. Pregnancy-specific beta-1-glycoprotein (SP_1) levels measured by nephelometry in serum from women with vaginal bleeding in the first half of pregnancy. Acta Obstet. Gynecol. Scand. 63, 311-315.

21. Niven, P.A.R., J. Landon, T. Chard. 1972. Placental lactogen levels as guide to outcome of threatened abortion. Br. Med. J. 2, 799-801.

22. Gartside, M.W., V.R. Tindall. 1975. The prognostic value of human placental lactogen (HPL) levels in threatened abortion. Br. J. Obstet. Gynaecol. 82, 303-309.

23. Jouppila, P., M. Seppälä. 1980. Pregnancy-specific beta-1-glycoprotein in complications of early pregnancy. Lancet. I, 667-668.

CARBOHYDRATE MICROHETEROGENEITY OF MATERNAL SERUM AFP IN PREGNANCY.

K. Toftager-Larsen
Department of Clinical Chemistry, Sønderborg Hospital, Sønderborg and
Hormone Department, Statens Seruminstitut, Copenhagen, Denmark.

Introduction

In pregnancy alpha-fetoprotein (AFP) is produced in the fetal yolk sac and in the fetal liver (1). Investigations of the carbohydrate heterogeneity pattern indicates that amniotic fluid AFP in early pregnancy originates mainly from the fetal yolk sac, because the heterogeneity pattern of amniotic fluid AFP then closely resembles yolk sac AFP (2,3). With increasing gestation the pattern changes, and in the third trimester the amniotic fluid AFP resembles fetal liver AFP, which is different from yolk sac AFP with respect to the carbohydrate heterogeneity (2,3).
No AFP is produced by the mother in pregnancy, and it has been argued that maternal serum AFP originates mainly from amniotic fluid (4). In contradiction others find that maternal serum AFP originates from fetal liver via fetal serum (5). In the present study the carbohydrate heterogeneity pattern of maternal serum AFP was analysed, and the results indicate that the major part of maternal serum AFP is of fetal liver origin, but possibly with contributions of amniotic fluid AFP.

Material and methods

Maternal serum samples were obtained in a screening program for neural tube defects. The AFP concentration was determined by radioimmunoassay, and the following samples were selected for analysis:

A. Fifteen samples from women with a normal course of pregnancy. In three cases a neural tube defect of the fetus was detected. The 12 remaining cases continued normally, and normal infants were delivered.

B. Ten samples from twin pregnancies ending with delivery of normal infants.

C. Five samples from pregnancies with threatened abortion, but where the course of late pregnancy was normal and a normal infant was delivered.

D. Ten samples from pregnancies ending with a spontaneous or missed abortion.

Pregnancy Proteins in Animals

Only samples with an AFP concentration above 100 µg/l were selected for analysis (see below).
The median AFP concentration in the samples from patients with fetal neural tube defects was 360 µg/l, whereas the median concentrations in the other groups of patients were approximately 175 µg/l with no consistent differences between groups. The range of concentrations was from 110 to 750 µg/l.
All samples were obtained between 15 and 20 weeks of gestation.

The carbohydrate heterogeneity of maternal serum AFP was investigated by radio crossed affinity immunoelectrophoresis. Fifty microliter samples of maternal serum was electrophoresed in a 1% agarose gel (Indubiose A37, Industrie Biologique Francaise) containing 300 µg of concanavalin A (con A) (Pharmacia Fine Chemicals) per square cm. After electrophoresis at 5 V/cm until a Bromophenol Blue stained albumin marker had migrated 3.8 cm, the gel slabs were transferred to the lower end of a 10 x 20 cm glass plate with an upper gel containing anti-AFP antibody (DAKO Immunoglobulins Ltd.) above the gel slabs. Between the two gels a 3 mm wide line gel was cast containing a small amount of AFP. To increase the sensitivity of the crossed affinity immunoelectrophoresis the antibody content of the upper gel was lowered to 0.006 µl of anti-AFP per square cm, and a few nanograms of AFP labelled with 125-Iodine was added to the line gel for visualization by autoradiography.
Four gel slabs from the first dimension electrophoresis were run on each plate, three maternal serum samples and a standard of amniotic fluid. The con A non-reactive fraction of this sample was known to be 6.7%. After electrophoresis at 2 V/cm for at least 12 hours in the second dimension perpendicular to the first dimension electrophoresis the plates were dried and placed on X-ray film. After 24-48 hours the film was developed, and the fractions reactive and non-reactive with con A were evaluated by measuring the height of the precipitates outlined on the film.
The appearance of electrophoreses after development of the film is shown in Figure 1.
To evaluate the validity of the radio technique three amniotic fluid samples with con A non-reactive fractions of AFP of 15, 18, and 20% and the standard sample (6.7%) were diluted with AFP-free human serum to a concentration of approximately 150 µg/l. A plate was run by the radio technique, and values of 14, 17, 16, and 5.8% respectively were obtained.
The crossed affinity immunoelectrophoresis is described in more detail by Toftager-Larsen et al. (2,3,8), and the modification with incorporation of a radioactive tracer is described by Nørgaard-Pedersen et al. (6).

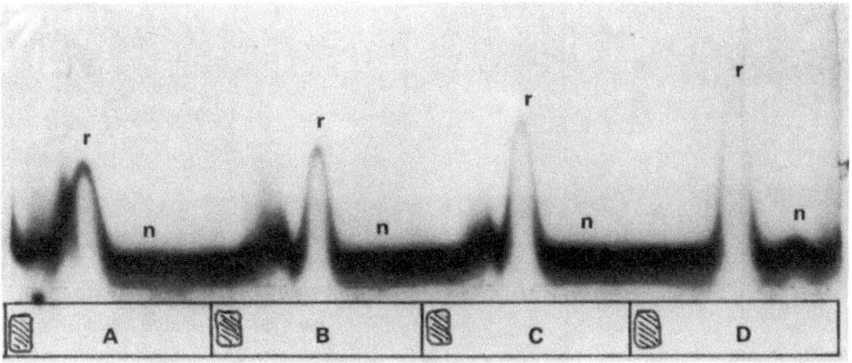

Figure 1. Radio crossed affinity immunoelectrophoresis of maternal serum AFP. Appearance of X-ray film after development. A - D indicates gel slabs from first dimension electrophoresis. A: Twin pregnancy. B and C: Missed abortion. D: Standard sample. See text for details. r: Reactive fraction. n: Non-reactive fraction.

In rocket immunoelectrophoresis the detection limit for a sample volume of 50 µl is a concentration in the sample of approximately 1 µg/l of AFP (6). Maternal serum AFP separates into two fractions, and the detection limit corresponds in the radio crossed affinity immunoelectrophoresis to a con A non-reactive fraction of at least 1% in a sample with a concentration of AFP of 100 µg/l. Precipitates obtained with lower concentrations of antigen and antibody are increasingly blurred and unreadable. For this reason only samples containing at least 100 µg/l of AFP were selected for analysis.

Results

The concentration of AFP in the samples investigated ranged from 110 µg/l to 750 µg/l. No consistent correlation between the concentration of AFP and the con A non-reactive fraction of AFP was found. The con A non-reactive fraction of AFP in the samples analysed was determined as the average of three measurements of the height of the precipitates. Results are shown in Figure 2. The median con A non-reactive fraction of AFP was approximately 2% in all groups, varying from 1.6% in pregnancies with threatened abortion to 2.3% in pregnancies with a fetal neural tube defect.
The results from twin pregnancies, normal pregnancies with a normal outcome, and pregnancies with spontaneous/missed abortion were compared. No statistically significant differences were found (Mann-Whitney rank sum test, $p>0.10$). The values obtained in the three cases of fetal neural tube defect were indistinguishable from the other groups.

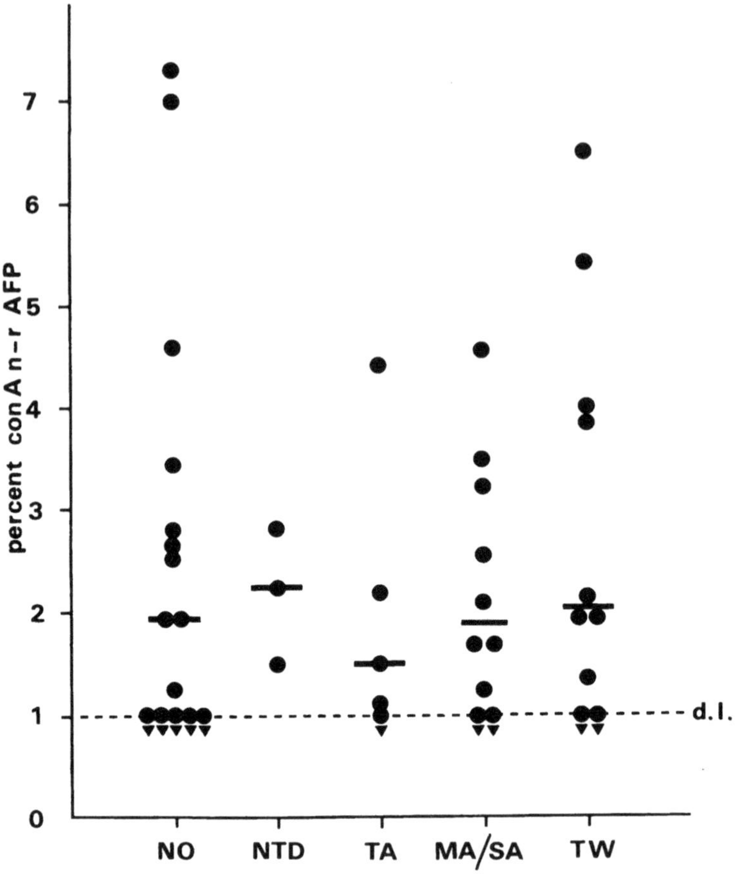

Figure 2. Fractions of maternal serum AFP non-reactive with con A. Bars indicate median values. d.l.: Detection limit. NO: Normal pregnancies with a normal outcome. NTD: Neural tube defect of the fetus. TA: Threatened abortion. MA/SA: Missed/spontaneous abortion. TW: Twin pregnancy.

Values of the con A non-reactive fraction of AFP above 4% were obtained in only seven of the 40 cases investigated, and no value exceeded 7.3%.

In six of the 12 cases with a normal outcome and a normal course of pregnancy amniocentesis was performed. Normal concentrations of amniotic fluid AFP were obtained in all cases.

The standard sample included in each second dimension electrophoretic run had a median con A non-reactive fraction of 7.2%. The coefficient of variation for this sample was 18%.

Discussion

Fetal yolk sac AFP contains approximately 50% of AFP non-reactive with con A, whereas fetal liver AFP contains less than 4% of this fraction (2). The composition of amniotic fluid AFP reflects that the AFP in early pregnancy originates from the yolk sac, but with the turnover of amniotic fluid by fetal swallowing and replacement by fetal urine, a change towards the heterogeneity pattern of fetal liver AFP is seen. At 15 weeks of gestation the mean con A non-reactive fraction of AFP in amniotic fluid is 17%, decreasing to 9% at 20 weeks (3). The findings of median values of 2% in maternal serum AFP in this study indicates that maternal serum AFP originates mainly from the fetal liver via fetal serum and the placenta. In spontaneous/missed abortion and possibly also in threatened abortion the integrity of the placental barrier is disturbed, and in these cases the fetal origin of the maternal serum AFP is obvious (7). However, in twin pregnancies the placental barrier is intact, but the mean concentration of AFP in maternal serum is almost exactly twice the mean concentration in normal singleton pregnancies. The median con A non-reactive fraction of AFP found in twin pregnancies in the present study was 2.0%, which is the same as the fraction obtained for fetal liver AFP (2), and much less than the findings for AFP in second trimester amniotic fluid (3). No significant difference between the values obtained in twin pregnancies and spontaneous/missed abortions was found , and the results from pregnancies with an elevation of the maternal serum AFP but with a normal course of pregnancy and a normal outcome are not discernible from the results obtained in the other groups of patients including the samples from pregnancies carrying a fetus with a neural tube defect. This seems to confirm fetal serum AFP as the major source of maternal serum AFP. This is what is expected, because in a normal pregnancy the concentration of AFP in fetal serum in the second trimester is about 100-fold higher than the concentration in amniotic fluid, and the diffusion gradient across the placenta is therefore comparatively higher than the gradient across the membranes. Furthermore, diffusion is facilitated by a considerable fetal and maternal blood flow through the placenta, and by the several-fold higher villous area in the placenta compared to the area of the membranes.

In seven of the 40 cases investigated values of con A non-reactive AFP of more than 4% were obtained. This indicates that probably some of the AFP in maternal serum originates from the amniotic fluid AFP, and certainly it is known that in cases of fetal malformations with high concentrations of AFP in amniotic fluid the maternal serum AFP concentration increases even though the placental barrier is intact. It therefore appears that some diffusion or transport of AFP across the membranes takes place also under normal circumstances.

Ishiguro et al. (5) did not find any con A non-reactive AFP in maternal serum throughout pregnancy with a method similar to the above described using peroxidase conjugated protein A for visualization instead of radiolabelled iodine. The sensitivity of their method is less than the technique described here, and small fractions of con A non-reactive AFP may have been missed. However, fractions of a size as found in amniotic fluid in the midtrimester should have been detected by their technique if present, and generally their conclusions are in agreement with the findings in the present study.

It has previously been shown that in pregnancy a strong correlation exists between the con A non-reactive fraction of AFP and the fraction of AFP weakly reactive with lentil lectin (8), and that the fraction of AFP not reacting with lentil lectin probably is of fetal liver origin (2). Ishiguro et al. (5) also analysed the affinity of AFP in maternal serum to lentil lectin. In conjunction with our results they found only the fraction of fetal liver origin in maternal serum. They concluded that maternal serum AFP originates from the fetal circulation via the placenta, but again, with the limitations due to sensitivity they may have missed minor fractions indicating that a small contribution comes from amniotic fluid.
Contrary to the results of Ishiguro et al. (5) is the report of Kerckaert & Bayard (9) who found that maternal serum also contains a considerable fraction of AFP reactive with lentil lectin decreasing with increasing gestational age. Whether this reflects changes in the amount of AFP originating from amniotic fluid is unclear, because their results are difficult to interpret. Their analyses revealed two fractions of AFP only with lentil lectin. It has been shown previously that AFP fractionates into three fractions provided the proper concentrations of lectin are used. Toftager-Larsen et al. (8) obtained three fractions of AFP with a concentration of lentil lectin six fold lower per ml of gel compared to the analyses of Kerckaert & Bayard (9).

Kelleher and Smith (4) found a good correlation between the con A non-reactive fraction of AFP in maternal serum and in amniotic fluid in eight pairs of samples from pregnancies with neural tube defects of fetuses and three pairs of samples from normal pregnancies. These results have not been confirmed in the present study. It is realised that the coefficient of variation was high due to to precipitates being more blurred than in conventional immunoelectrophoresis. However, the results obtained by the present method were reproducible and reliable.

In conclusion, it was found that the major part of maternal serum AFP originates from fetal liver AFP via fetal serum, but with a small contribution from amniotic fluid AFP. Because of this it is not believed that analysis of the lectin reactivity of maternal serum AFP will play any role in the context of prenatal diagnosis of fetal neural tube defects or other malformations or in any complications of pregnancy.

Acknowledgements

This study was supported by Fonden for Lægevidenskabelig Forskning ved Sygehusene i Ringkøbing, Ribe og Sønderjyllands Amter. The technical assistance of Mrs. Jette Rasmussen is gratefully acknowledged.

References

1. Gitlin, D., A. Pericelli, M.G. Gitlin. 1972.
   Cancer Res. 3,979-982.

2. Toftager-Larsen, K., P. Lund Petersen, B. Nørgaard-Pedersen. 1981.
   In: Lectins - Biology, Biochemistry, Clinical Biochemistry (T.C. Bøg-Hansen, ed.). W. de Gruyter, Berlin - New York pp. 283-292.

3. Toftager-Larsen, K., E. Kjærsgaard, J.C. Jacobsen, B. Nørgaard-Pedersen.1980.
   Clin. Chem. 26,1656-1659.

4. Kelleher, P.C., C.J.P. Smith. 1979.
   Lancet ii,1302-1303.

5. Ishiguro, T., H. Sakaguchi, M. Fukui, I. Sugitachi. 1985.
   Biol. Res. Pregnancy Perinatol.

6. Nørgaard-Pedersen, B., K. Toftager-Larsen, N.H. Axelsen. 1984.
   In: Handbook of Immunoprecipitation-in-gel Techniques (N.H. Axelsen, ed.). Blackwell, Oxford pp. 256-263.

7. Brock, D.J.H. 1976.
   Lancet ii,345-346.

8. Toftager-Larsen, K., E. Kjærsgaard, B. Nørgaard-Pedersen. 1983.
   Clin. Chem. 29,21-24.

9. Kerckaert, J. P., B. Bayard. 1981.
   In: Lectins - Biology, Biochemistry, Clinical Biochemistry (T.C. Bøg-Hansen, ed.). W. de Gruyter, Berlin - New York pp. 271-281.

PRENATAL DIAGNOSTIC SIGNIFICANCE OF A SPECIFIC QUANTITATIVE IMMUNOASSAY FOR ACETYLCHOLINESTERASE IN AMNIOTIC FLUID

A.G. Rasmussen, B. Nørgaard-Pedersen
Hormone Department, Statens Seruminstitut, DK-2300 Copenhagen S

J. Hangaard
Department of Clinical Chemistry, Sønderborg Hospital, DK-6400 Sønderborg

Introduction

Amniotic fluid alpha-fetoprotein (AFP) measurement is the main prenatal diagnostic test for open neural tube defects (NTD) (1). However, the test may yield both false positive and false negative results.

In order to confirm and eventually correct the results of the AFP-test different additional biochemical methods have been used including mainly detection of amniotic fluid acetylcholinesterase (AChE), but also determination of concanavalin A non reactive fraction of AFP and other confirmatory tests (2).

The diagnostic significance of a qualitative determination of AChE in amniotic fluid has been confirmed by several studies (3,4,5) especially by the polyacrylamide gel electrophoresis (PAGE) technique (6,7,8). The U.K. Collaborative Study (9) found that PAGE is very effective when used for amniotic fluid samples with a positive AFP-result. The qualitative gel test has substantially reduced the number of false-positive AFP-results with only a very small proportion of misdiagnoses of open neural tube defects. However, the PAGE test is very time-consuming and somewhat complicated to use as a screening test for all amniotic fluid samples. Furthermore, the number of AChE false-positive results is not known when the test is used for screening of all amniotic fluid samples.

Recently a new quantitative enzyme antigen immunoassay (EAIA) for AChE has been developed (10,11). This method utilize an antibody against AChE to immobilize AChE in an enzymatically active immunocomplex. The method has the advantages of higher throughput compared with PAGE and a potential for use in screening. The clinical signifi-

Pregnancy Proteins in Animals

cance of this method as a confirmatory test in screening for different fetal malformations is evaluated in this study.

Materials and Methods

Samples

Amniotic fluids were sampled from diagnostic amniocentesis at 12-26 weeks of gestation. All samples were centrifuged at 3000 G for 10 minutes and supernatants were stored at -20°C for as long as four years. Neither repeated freeze-thaw procedures nor prolonged frozen storage appeared to affect the AChE-activity in the amniotic fluid samples.

The AFP-concentration was determined as part of the routine screening procedure at the time of amniocentesis by "rocket" immunoelectrophoresis technique (12). The discrimination between normal and abnormal concentration of AFP was based on the 95-th percentile for gestational age as determined in our laboratory (12).

The samples were classified by their appearance in clear, unclear and blood-contaminated fluids. A clear sample is defined as a sample with no visible turbidity, an unclear sample is defined as a sample with visible turbidity and a blood-contaminated sample is defined as a sample with visible blood-staining. For blood-contaminated samples a further classification is carried out after degree and type of blood-contamination (fetal or adult). Furthermore, it is registrated whether the blood-contamination is fresh or old i.e. red or brown and whether the contamination is with intact red cells or haemolyzed cells. The samples used in this study were:

1) 1400 consecutively collected control samples from pregnancies with apparently normal outcome. All the samples were clear fluids with normal AFP values for gestational age.
2) 70 further samples, in which the AFP-concentration exceeded the 95-th percentile, but with apparently normal outcome of pregnancy. In 15 of these samples the total AFP-concentration exceeded the cut off value proposed by the U.K. Collaborative Study (1).
3) 160 samples from pregnancies with various fetal malformations. The different diagnoses were based on information from different clinical departments and pathologists.
4) 40 blood-contaminated samples, all from pregnancies with apparently normal outcome.

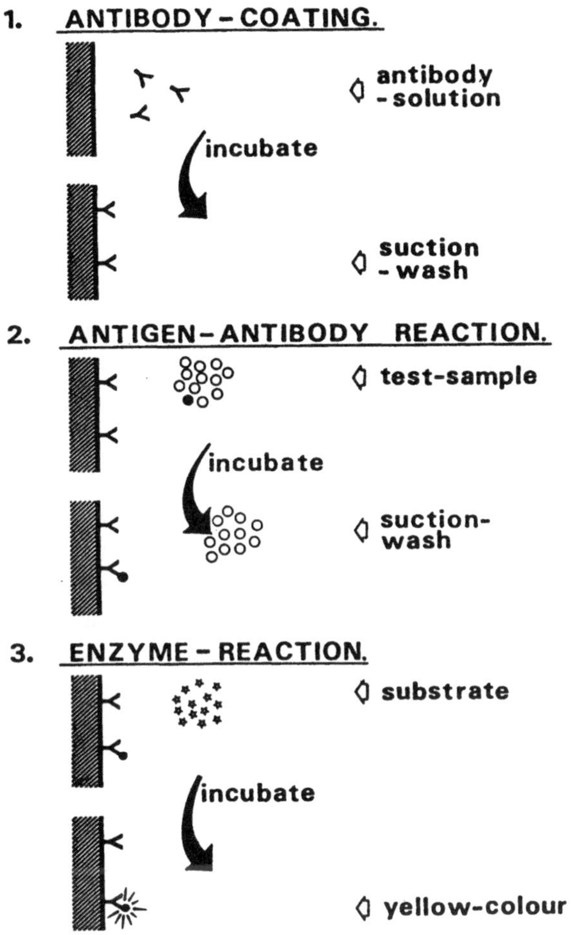

Figure 1. Schematic analytical procedure for the enzyme antigen immunoassay. Note that the demonstration of bound antigen is performed using the enzymatic activity of the antigen in the immunocomplex.

Procedure

Antibodies against AChE: The antibodies were raised by injecting rabbits with an affinity purified human erythrocyte AChE (11,13).

Procedure for enzyme antigen immunoassay: The technique was carried out in three main steps as shown in Figure 1.

1) Ad 100 µl of anti-AChE (IgG-fraction), diluted 1000-fold in sodium

carbonate buffer (50 mmol/l, pH 9.6), to each well of a 96-well microtiter plate (Nunc-Immuno Plate I). After overnight incubation at 37°C wash the plate three times with phosphate-buffered saline, pH 7.2 containing 0.05% Tween 20 (PBS-Tween).
2) Dispense 100 µl of test samples, controls and standards, respectively, into the wells, and incubate for 4 hours at 37°C. Then wash the wells three times with PBS-Tween.
3) Ad 100 µl of Ellman's substrate (14) to each well and incubate for 2 hours at 37°C. The substrate solution is made by dissolving 30 mg acetylthiocholine iodide (Sigma No. A 5751) and 10 mg of Ellman's reagent (5,5-dithio-bis(2-nitrobenzoic acid), DTNB; Sigma No. D 8130) in 100 ml of 50 mmol/l potassium-phosphate buffer, pH 7.0. Read the plates on a ELISA-photometer at 405 nm. With a multiscan photometer (Titertek Multiskan Spectrophotometer, FLOW Laboratories) the absorbance of all 96 wells can be recorded within 1 minute.
Polyacrylamide gel electrophoresis: For qualitative detection of AChE and cholinesterase (ChE) activity, electrophoresis was carried out as described by Smith et al. (6). For specific inhibition of AChE-activity we used BW 284 C 51 (1,5-bis(4-allyldimethylammonium)-phenyl-pentan-3-one dibromide, Sigma No. A 9013).

Standardization

As standard a 40-donor serum pool, defined to contain 1000 arbitrary AChE units/l (arb.U/l), was used in the EAIA. The standards were prepared by two-fold dilution of the primary standard in 9 g/l sodium chloride solution. The primary standard contained 308 International Units/l (IU/l) when the AChE-activity was measured as described by Chubb and Smith (15), i.e. 1 arb.unit = 0.308 IU.
Results from precision studies are shown in Table 1. Four different samples (control 1 to control 4) have been tested. The within-assay precision was determined in 10 experiments and the values given are the coefficients of variation (CV) within the individual plate. The between-assay CV was determined by measuring the samples on different days, with different lot-number of plates. In all cases an internal standard (the above mentioned 40-donor pool) was used.
The CV's in within- and between-assays were between 2.7% and 6.5% at levels from 15 to 350 arb.U/l.

TABLE 1. Precision Studies for EAIA.

	Mean AChE arb.U/l	Within-assay CV% (n)	Between-assay CV% (n)
Control 1	15	6.1 (10)	6.5 (10)
Control 2	70	2.8 (10)	3.5 (10)
Control 3	240	2.7 (10)	3.5 (10)
Control 4	350	4.4 (10)	4.4 (10)

Results and Discussion

AChE-levels in amniotic fluid samples from pregnancies with normal AFP-values and normal outcome have been analysed. Results from this are shown in Table 2 and Table 3.

TABLE 2. Amniotic Fluid AChE-concentrations and Gestational Age in 1400 Clear Amniotic Fluid Samples from Pregnancies with Normal Outcome.

Week	Number of samples	Mean AChE-content*	± 2 SD
12	2	25	18-32
13	8	19	4-33
14	71	22	6-38
15	487	23	9-36
16	461	22	7-35
17	174	21	6-35
18	73	23	10-35
19	53	23	8-37
20	35	26	12-38
21	18	23	5-40
22	6	25	1-47
23	4	26	4-47
24	2	25	14-36
25	4	26	5-49
26	2	25	19-31

* : In arb.U/l

No relationship was found between AChE-level and gestational age from week 12 to week 26; the mean value was 23 arb.U/l independent of gestational age. The AChE-values was in the range 5 to 70 arb.U/l. Based on the range and the mean value we have selected a cut off value on 75 arb.U/l.

Examination of the AChE-levels in 70 amniotic fluid samples from pregnancies with normal outcome but elevated AFP-values (false-positives) gave AChE-values below the cut off level in all cases (Table 3).

TABLE 3. AChE- and AFP-results from 1470 Clear Amniotic Fluid Samples from Pregnancies with Normal Outcome.

Number of samples	AFP $<$95-th**	AFP $\geq$95-th**	AFP $\geq$2.5 x NM+	AChE-results* $<$75	AChE-results* $\geq$75
1400	1400	0	0	1400	0
70	0	70	15	70	0

* : In arb.U/l
** : The 95-th percentile (12)
\+ : 2.5 times the normal median (16)

Among 40 blood-contaminated amniotic fluid samples 9 gave elevated AChE-values by the EAIA, but none were positive by PAGE. However, blood-contaminated samples will be detected before analyses and will therefore always be especially evaluated.

The amniotic fluid samples from the pregnancies with various fetal malformations are divided into three groups as shown in Table 4. Group one with neural tube defects (NTD) which includes anencephaly, spina bifida and encephalocele (77 samples), group two with abdominal wall defects (AWD) which includes omphalocele and gastroscisis (21 samples) and the third group with other fetal defects (62 samples).

In the NTD-group the EAIA detects all the anencephalies, all the encephaloceles and 96% of the spina bifidas (26 of 27 cases).

The EAIA does not detect all the AWD-pregnancies. Only 57% of the omphaloceles (8 of 14 cases) and 86% of the gastroscises (6 of 7 cases) shows elevated AChE-levels.

In the third group the AChE-levels for the chromosome defects and congenital nephrosis are below the cut off level in all cases, whereas the two cases of Meckel's syndrome, all cases with various multiple malformations and 83% (10 of 12 cases) of amniotic fluid samples from pregnancies with intrauterine fetal death (including cases which were not dead at the time of amniocentesis) had elevated AChE-levels. Only one case out of 6 cases of hydrocephalus had elevated AChE-level.

The main clinical significance of EAIA is in the NTD- and AWD-group. Detailed AChE-results from these groups including AFP-results and PAGE-results are found in Table 5.

TABLE 4. AChE-levels in 160 Amniotic Fluid Samples from Pregnancies with Various Fetal Malformations.

Diagnosis	Number of samples	Number of samples with elevated AChE-level*
Group one - NTD:		
Anencephaly	45	45
Spina bifida	27	26
Encephalocele	5	5
Group two - AWD:		
Omphalocele	14	8
Gastroscisis	7	6
Group three - Others:		
Chromosome defects	35	0
Intrauterine death	12	10
Congenital nephrosis	2	0
Hydrocephalus	6	1
Meckel's syndrome	2	2
Multiple malformations	5	5

NTD: Neural tube defects
AWD: Abdominal wall defects
* : Elevated AChE-level i.e. 75 arb.U/l

The positive AFP-results for the anencephaly- and encephalocele-pregnancies i.e. with values above 2.5 x the normal median were all confirmed by EAIA and PAGE, which gave corresponding positive results. For the spina bifida group 7 of the 27 samples were in the so-called gray zone i.e. with AFP-values between the 95-th percentile and 2.5 x the normal median. These cases represented a diagnostic dilemma before the AChE-detection was introduced. By the use of EAIA 6 of the 7 cases were AChE-positive and the seventh case gave a borderline value of 71 arb.U/l. The remaining 20 samples with AFP-values above 2.5 x the normal median all gave positive results in the EAIA.

TABLE 5. AFP-, AChE- and PAGE-results from NTD- and AWD-pregnancies.

Abnormalities	Total number	AFP-results $<$95-th**	$\geq$95-th**	$\geq$2.5xNM^{+}	AChE* $<$75	$\geq$75	PAGE pos***
Anencephaly	45	0	45	45	0	45	45
Spina bifida	27	0	27	20	1^{++}	26	26
Encephalocele	5	0	5	5	0	5	5
Omphalocele	14	2	12	11	6	8	8
Gastroscisis	7	0	7	7	1	6	6

* : In arb.U/l
** : The 95-th percentile (12)
+ : 2.5 times the normal median (16)
***: Number of polyacrylamide gel electrophoreses with positive AChE-results.
++ : 71 arb.U/l, but PAGE-positive

The PAGE also gave positive AChE-results in 26 out of the 27 samples. In the AWD-group nearly all the samples showed elevated AFP-values except for 2 omphalocele cases with normal values and 1 omphalocele case with a gray zone value. These 3 cases of omphalocele had normal AChE-levels. Although AChE-detection or determination gave additional information to the AFP-results in the AWD-group the diagnostic information was less significant than in the NTD-group.

In this study we have found similar results with the semiquantitative PAGE for AChE and the EAIA, but the EAIA is superior to PAGE in the following ways: 1) First of all it is possible to automize the EAIA and use it for routine screening. 2) Secondly there is no use of toxic substances in the EAIA. In PAGE both an acetylcholinesterase-inhibitor and acrylamide is used. 3) Furthermore, the sample throughput is much higher in the EAIA technique where a technician daily can perform 200 analyses compared with only 20 for PAGE. 4) Finally the EAIA gives quantitative results.

In conclusion the quantitative EAIA technique should replace the qualitative PAGE method as a routine complementary test to amnion AFP-determination in the prenatal diagnosis of neural tube defects. In our opinion all amniotic fluids should be examined for the AChE-concentration independent of the AFP-concentration.

Acknowledgement

This study was supported by a grant from Michaelsen Foundation.

References

1. Wald, N.J., H.S. Cuckle. 1979. Second Report of the U.K. Collaborative Study on Alpha-fetoprotein in Relation to Neural-Tube Defects. Amniotic-fluid alpha-fetoprotein measurement in antenatal diagnosis of anencephaly and open spina bifida in early pregnancy. Lancet ii, 651-662.

2. Wald, N.J., H.S. Cuckle. 1984. In: Antenatal and neonatal screening (N.J. Wald, ed.). Oxford University Press, Oxford, New York, Tokyo. pp. 25-73.

3. Dale, G., J.R. Bonham, P. Lowdon, D.F. Roberts. 1979. Amniotic-fluid acetylcholinesterase and neural-tube defects. Lancet i, 880-881.

4. Seller, M.J., K.J. Cole, A.H. Fensom, P.E. Polani. 1980. Amniotic fluid acetylcholinesterase and prenatal diagnosis. Br. J. Obstet. Gynaecol. 87, 501-505.

5. Hodgson, A.J., P.M. Pilowsky, E.F. Robertson, A.C. Pollard, I.W. Chubb. 1981. Combined analysis of acetylcholinesterase and alpha-fetoprotein improves the accuracy of antenatal diagnosis of neural-tube defects. Med. J. Aust. 1, 457-460.

6. Smith, A.D., N.J. Wald, H.S. Cuckle, G.M. Stirrat, M. Bobrow, H. Lagercrantz. 1979. Amniotic-fluid acetylcholinesterase as a possible diagnostic test for neural-tube defects in early pregnancy. Lancet i, 685-688.

7. Barlow, R.D., H.S. Cuckle, N.J. Wald. 1982. A simple method for amniotic fluid gel-acetylcholinesterase determination, suitable for routine use in the antenatal diagnosis of open neural tube defects. Clin. Chim. Acta 119, 137-142.

8. Nørgaard-Pedersen, B., S. Jensen, J. Wandrup. 1982. Identification of acetylcholinesterase and cholinesterase in amniotic fluid by immune absorption technique. Prenat. Diagn. 2, 257-264.

9. Wald, N.J., H.S. Cuckle. 1981. Report of the Collaborative Acetylcholinesterase Study. Amniotic fluid acetylcholinesterase electrophoresis as a secondary test in the diagnosis of anencephaly and open spina bifida in early pregnancy. Lancet ii, 321-327.

10. Nørgaard-Pedersen, B., J. Hangaard, O.J. Bjerrum. 1983. Quantitative Enzyme Antigen Immunoassay of Acetylcholinesterase in Amniotic Fluid. Clin. Chem. 29, 1061-1064.

11. Hangaard, J., K. Sørensen, U. Brodbeck, B. Nørgaard-Pedersen. 1984. Quantitative determination of acetylcholinesterase by enzyme antigen immunoassay: methodological aspects and clinical use. Scand. J. clin. Lab. Invest. 44, 717-724.

12. Nørgaard-Pedersen, B. 1976. Human alpha-fetoprotein. A review of recent methodological and clinical studies. Scand. J. Immunol. 5, Suppl. 4, 1-45.

13. Harboe, N.M.G., A. Ingild. 1983. Immunization, Isolation of Immunoglobulins and Antibody Titre Determination. Scand. J. Immunol. 17, Suppl. 10, 345-351.

14. Ellman, G.L., K.D. Courtney, V. Andres Jr., R.M. Featherstone. 1961. A new and rapid colorimetric determination of acetylcholinesterase activity. Biochem. Pharmacol. 7, 88-95.

15. Chubb, I.W., A.D. Smith. 1975. Isoenzyme of soluble and membrane-bound acetylcholinesterase in bovine splanchnic nerve and adrenal medulla. Proc. R. Soc. Lond. B. 191, 245-261.

16. Toftager-Larsen, K., E. Kjærsgaard, B. Nørgaard-Pedersen. 1983. Comparison of Amniotic Fluid Alpha-Fetoprotein Reactivity to Lens cullinaris Agglutinin and Concanavalin A in Crossed-Affinity Immunoelectrophoresis: Ancillary Tests in the Prenatal Diagnosis of Severe Fetal Malformations. Clin. Chem. 29, 21-24.

CHANGING PATTERNS OF MICROHETEROGENEITY OF HUMAN CHORIONIC GONADOTROPIN (hCG) DURING NORMAL PREGNANCY AND TROPHOBLASTIC DISEASE

D.L. Hay
Department of Pathology, Royal Women's Hospital,
Melbourne, 3053, Australia.

Introduction

Human chorionic gonadotropin (hCG) is a glycoprotein which contains approximately 30 per cent carbohydrate by weight and is particularly rich in sialic acid (1). The importance of the carbohydrate moiety to the function of hCG was first demonstrated when removal of its sialic acid residues resulted in the rapid clearance of asiolo-hCG from circulation (2). Consequently asiolo-hCG demonstrated little biological activity in vivo due to its preferential and rapid removal from circulation by hepatic uptake (3). Further studies in vitro demonstrated that asiolo-hCG bound to ovarian receptors but with reduced ability to activate adenylate cyclase (4). Also sequential removal of carbohydrate residues internal to sialic acid resulted in hCG species that demonstrated progressively decreasing ability to stimulate steroid activity in vitro (5). These findings indicate that modification of the carboydrate side chains can play a crucial role in the function of the hormone. The aim of the present study was to evaluate whether such changes occur in vivo to the carbohydrate moiety of hCG.

Highly purified hCG preparations which are homogenous by immunological analysis can display microheterogeneity by isoelectric focusing in gels (6,7). These microvariants reflect small changes in electrical charge of hCG species due to variations in their carbohydrate content. In this study, hCG microheterogeneity was determined by high pressure liquid chromoatography (HPLC) techniques using chromofocusing columns which elute proteins on the basis of their different individual charges (pI). The patterns of microheterogeneity were then compared between hCG isolated from serum and urine in patients of different gestations in normal pregnancy, or with

Pregnancy Proteins in Animals

trophoblastic disease. The aim was to establish whether patterns of hCG microheterogeneity change systematically during pregnancy or in malignancy. Any significant changes in these patterns would indicate that modifications occur in vivo to the carbohydrate moieties of hCG. Such changes may represent an important physiological mechanism which regulates the functions of hCG during different stages of pregnancy.

Materials and Methods

Specimens

Paired sera and early morning urine samples (n=20) were collected from 10 healthy pregnant women of various gestational ages (7 to 38 weeks) as confirmed by ultrasonography. Also sera and urine samples were collected from patients with benign hydatidiform moles or choriocarcinoma at the start of their treatment cycle. The diagnosis of trophoblastic neoplasia was by histopathologic examination.

Reagents and Apparatus

Antisera to hCG suitable for affinity chromatography was supplied by Silenus Australia, Melbourne, Australia. Polybuffer 74 and CN-BR activiated Sepharose 4B were purchased from Pharmacia Fine Chemicals, Uppsala, Sweden. Reagents for buffers including Bis-Tris (Cat. No. B9754), Glycine (G2879), Mono Sodium Phosphate (S0751), Disodium Phosphate (S0876), Sodium Chloride (S9625) were supplied by Sigma Chemical Co., St. Louis, 63178, U.S.A. The reagents for the hCG immunoradiometric assay (IRMA) were supplied as a kit called "hCG Maiaclone" by Serono Diagnostic Ltd., Woking, U.K.

Centriflo membrane cones (Code No. CF25) were purchased from Amicon Membrane Filtration Industries, Danvers, 01923, U.S.A. Filter units (0.22 µm) were supplied by Millipore Co., Bedford, 01730, U.S.A. Chromofocusing column MonoP H5/20, and a FRAC-100 fraction collector were purchased from Pharmacia. The HPLC was performed with a Waters Model M45 pump and a U6K injection system purchased from Millipore Waters, Milford, 01757, U.S.A. The pH in each fraction was determined by a pH M62 Standard meter from Radiometer, Copenhagen, Denmark.

Extraction

HCG was extracted from sera or urine samples by affinity chromatography. The recommended protocol supplied by Pharmacia was used to couple anti-hCG to CN-BR activated Sepharose. Sera or urine samples were first filtered and then applied to anti-hCG sepharose beads packed into an 0.8 x 12 cm column. The unbound fractions were removed by pumping 0.1 M phosphate buffer (pH 7.5) at 10 ml/min for 20 mins. The bound hCG fractions were eluted with 2 M NaCl in Glycine/HCl buffer (pH 2.3). The aliquots containing hCG were desalted and concentrated by centrifugation in Centriflo cones at 1000g for 10 mins. These cones retained molecules greater than 25,000 Daltons. The hCG extracts were then washed and equilibrated by twice resuspending the concentrates with 0.02 M Bis-Tris/HCl buffer (pH 7.1) and repeating the centrifugation step.

Chromofocusing

The Mono-P chromofocusing column had a resolution power of 25,000 plates per metre. It was first equilibrated with start buffer Bis-Tris/HCl (pH 7.1). The focusing procedure commenced by pumping a small volume (5 mls) of eluting buffer Polybuffer 74/HCl (pH 3.5) onto the column. Samples of hCG extract (25 µl) containing 2.5 to 10 IU of activity were applied, and the eluting buffer pumped through the column at a flow rate of 0.4 ml/min. Fractions were collected at 20 sec. intervals and the pH of each fraction determined by a glass electrode.

Immunoreactivity

The levels of hCG in each fraction were measured by an IRMA procedure which utilized 3 highly specific monoclonal antibodies in a 3 tiered sandwich procedure. Before each fraction was assayed, it was buffered with 200 µl of 0.5 M phosphate buffer (pH 7.5). Assay standards were based on the first IRP (75/537). The IRMA assay demonstrated no cross-reactions with free alpha or free beta subunits, while the cross-reactions observed with LH, FSH, and TSH were 1.1%, 1.3% and 0.5% respectively. The sensitivity of the IRMA assay was 1 mIU/L; while its intra-assay and inter-assay precision (CV%) were 2.9% and 4.4% respectively. All fractions from the same specimen were measured in a single assay procedure to reduce assay variation. The concentrations of each fraction (mIU) were plotted in relation to the pH of each fraction and against fraction number. A spike or peak in hCG levels was considered significant if rises were greater than 6 per cent of the previous fraction. Usually peaks were sustained over 2 to 3 fractions. The patterns obtained on the same sample were reproducible as demonstrated in rerunning 5 samples, however individual spikes were not refocused to confirm their pI value. The average

The IEF patterns of urinary hCG are similar to the serum IEF patterns and demonstrate a decreasing number of variants from 11 to 5 with increasing gestation. Table 2 lists the urine IEF variants with increasing gestation age. Compared to paired serum IEF variants, there is a higher proportion of urine microvariants in the more basic pH zone of 7 - 6. This is most evident in samples collected from early gestation.

Table 2 Urine IEF Variants of hCG

Gestation weeks	No. of variants	Percentage Distribution of hCG Variants over four pH zones 7 - 6	6 - 5	5 - 4	4 - 3
7	11	7	45	40	8
8	11	8	46	38	8
10	11	8	48	38	6
12	10	10	50	36	4
15	10	11	52	33	4
20	9	13	54	29	4
25	8	12	59	25	4
35	7	14	60	22	4
37	5	14	62	20	4
38	5	15	58	23	4

The patterns of microvariants from hCG isolated from patients with trophoblastic disease are more complex than those observed in normal pregnancy as illustrated in Figure 2. The range of IEF variants in these patients are detailed in table 3. Compared to normal early pregnancy, hCG from trophoblastic disease demonstrates increased number of variants with a particular increase in acidic variants. No characteristic microvariants appear in patients with trophoblastic disease. Rather there is a more complex IEF pattern which varies between individuals, and this perhaps reflects their different histologies.

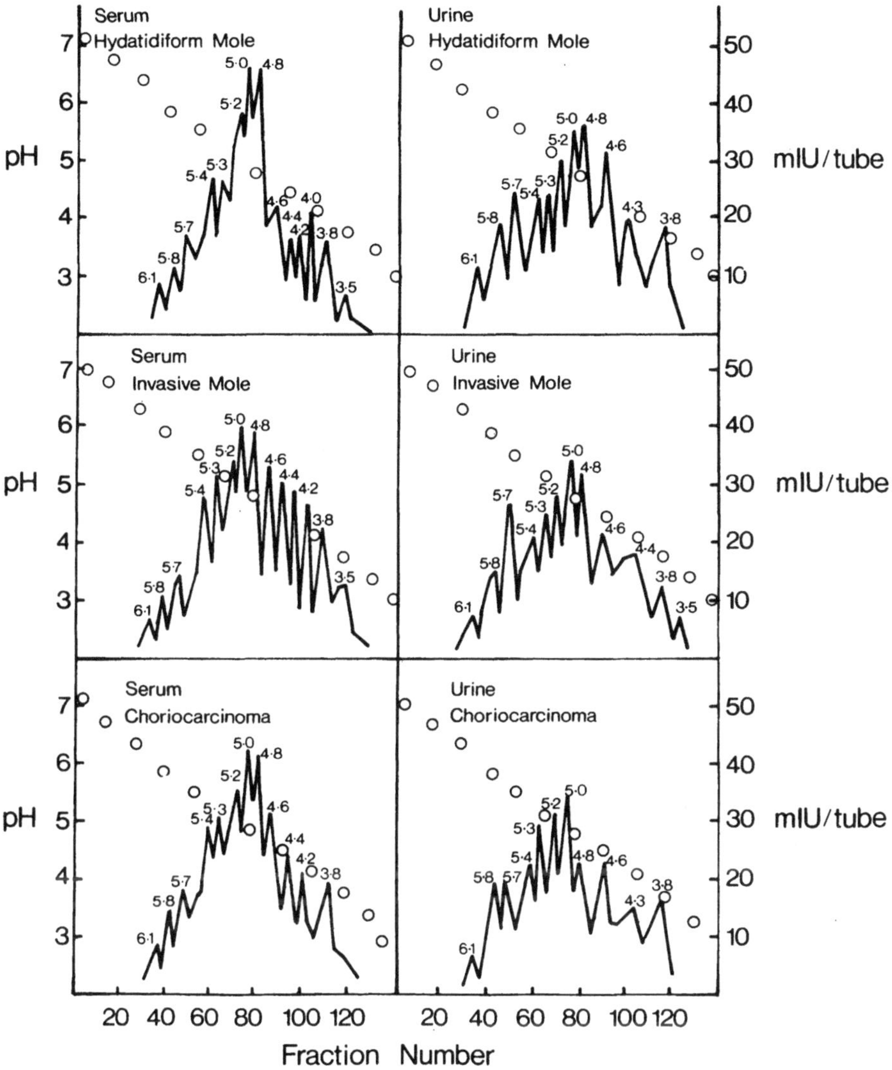

Figure 2 The microvariants of hCG on paired serum and urine samples from 3 patients with trophoblastic disease.

Table 3 IEF Variants of hCG in 4 patients with Trophoblastic Disease

	Classification	Serum or urine	No. of Variants	Percenage Distribution of hCG Variants over four pH zones 7 - 6	6 - 5	5 - 4	4 - 3
1.	Hydatidiform mole	Serum	14	3	43	41	13
2.	" "	Urine	11	6	43	41	10
3.	" "	Serum		4	42	41	13
4.	Invasive mole	Serum	14	3	39	44	14
5.	" "	Urine	12	4	49	38	9
6.	Choriocarcinoma	Serum	12	4	46	37	13
7.	"	Urine	11	4	50	35	11

Discussion

A number of groups have studied IEF microheterogeneity of hCG in gels. Merz et al (8) observed 6 microvariants with pIs ranging from 2.2 to 7.3. Graesslin et al (9) isolated 6 bands with pI ranging from 4.0 to 5.2. Yazaki et al (10) reported seven peaks of hCG with pI ranging from 3.9 to 6.7 and also observed an additional 3 acidic microvariants from a patient with choriocarinoma. Nwokoro et al (11) observed 11 variants with pI ranging from 4.4 to 6.3. In the present preliminary study, the improved resolution of IEF microvariants by HPLC has demonstrated that the number of microvariants that can be distinguished decreases with increasing gestational age. Furthermore the average distribution of hCG microvariants over four pH zones shifts progressively to become more basic with increasing gestational age.

Changes in the carbohydrate content of hCG during pregnancy has been reported by Fein et al (12). This group measured the elution volume (Ve) of hCG by gel chromatography and demonstrated that changes in Ve reflects variations in the carbohydrate content of hCG. Also hCG from first trimester sera was observed to be more glycosylated than third trimester sera, and serum hCG was more glycosylated than urinary hCG. The findings of the present study, suggest that changes in Ve may be due to a decrease in the average number of hCG microvariants rather than reduced glycosylation of a single hCG species.

Changes in the average pI of hCG observed in this study may be analagous to changes in pI observed in the related glycoprotein hormone FSH, as reported by Wide (13). In that study the average pI of FSH from men and postmenopausal women was found to be more acidic (pI 4.1) than FSH from pre-menopausal women (pI 4.4). Furthermore treating male patients with large doses of oestrogens changed the average pI of FSH to the more basic form which contained less sialic acid. Similarly the origin of microheterogeneity of hCG has been attributed to its sialic acid content, and the IEF patterns of hCG in gels can be reduced to a single band by treatment with neuramidase (9). Other glycoproteins such as α_1 acid glycoprotein have demonstrated oestrogen related changes to its carbohydrate moiety during pregnancy (14). These findings indicate that rising oestrogen levels may reduce the activity of sialyl transferases associated with glycosylation of glycoproteins. Consequently progressively increasing oestrogen levels during pregnancy may serve as a controlling mechanism to reduce the sialic acid contents of hCG species. Further studies however are required to determine whether decreasing sialic acid content with increasing gestational age causes the changes in the number and distribution of hCG microvariants observed in this study.

The carbohydrate content of hCG isolated from patients with trophoblastic disease has been reported by Choy et al (15). In their study, hCG purified from hydatidiform mole contained less N-acetylglucosamine and mannose than hCG from normal pregnancies. In contrast the microheterogeneity of hCG in the present study reveals a more complex mixture of variants than observed in normal pregnancy rather than a single species with unusual glycosylation. These changes in patterns of microheterogeneity probably reflect the release of partially glycosylated intermediates alongside the hCG pattern for early gestation. Similar findings of the release of incompletely sialylated hCG species has been reported by Braunstein et al (16) who observed ectopic hCG production by ovarian carcinoma cell lines. It was not possible in this study to confirm the observations of Yazaki et al (10) that patients with choriocarcinoma secrete characteristic hCG microvariants.

The significance of changes in the glycosylation of hCG remains to be determined but it is likely to be important in determining the biological activity of hCG. Although the primary role of hCG is luteotropic, its continued secretion implies that it serves other functions during pregnancy. HCG also appears important to placental steroidogenesis, adrenal steroidgenesis (17), lipogenesis (18), prostaglandin levels (19), and relaxin levels (20). It also demonstrates secondary activity such as thyroid stimulating activity (TSA) which can be increased by removing the carbohydrates on the carboxyl terminus of the beta subunit of hCG (21). This present study shows that the carbohydrate content of hCG changes during pregnancy. Thus it is possible that factors regulating the glycosylation steps of hCG may modify its biological activity to enhance one role while reducing another role.

REFERENCES

(1) Morgan, F.J., S. Birken, R.E. Canfield. 1975. J. Biol. Chem. 250, 5247

(2) Van Hall, E.V., J.L. Vaitukaitis, G.T. Ross, J.W. Hickman, G. Ashwell. 1971. Endocrinology 88, 456.

(3) Ashwell, G., A.G. Morell. 1974. Adv. Enzymol. 41, 99.

(4) Brand, E.C., J. Odink, G. Klok, E.V. Van Hall. 1981. Acta Endocrinol. (Kbh) 97, 270.

(5) Kalyan, N.K., H.A. Lippes, O.P. Bahl. 1982. J. Biol. Chem. 257, 12624.

(6) Bell; J.J., R.E. Canfield, J.J. Sciarra. 1969. Endocrinology. 84, 298.

(7) Puett, D.A., A. Kenner, R. Benveniste, D. Rabubiwutz. 1978. Acta Endocrinol. (KbH). 89, 612.

(8) Merz, W.E., V. Hilgenfeldt, M. Dorner, R. Brossmer. 1974. Hoppe-Seylers Z. Physio. Chem. 355, 1035.

(9) Graesslin, D., H.C. Weise, M. Rick. 1976. Anal. Biochem. 71, 492.

(10) Yazaki, K., C. Yazaki, K. Wakabayashi, M. Igarashi. 1980. Am. J. Obstet. Gynecol. 138, 189.

(11) Nwokoro, N., H-C. Chen, A. Chrambach. 1981. Endocrinology. 108, 291.

(12) Fein, H.G., S.W. Rosen, B.D. Weintraub. 1980. J. Clin. Endocrinol. Metab. 50, 1111.

(13) Wide, L. 1981. J. Clin. Endocrinol. Metab. 55, 682.

(14) Wells, C., T.C. Bog-Hansen, E.H. Cooper, M.R. Glass. 1981. Clin. Chem. Acta. 109, 59.

(15) Choy, Y.M., K.M. Lau, P.H. Ma, C.Y. Lee. 1978. Clin. Chem. Acta. 85, 7.

(16) Braunstein, G.D., V.V.Kamdar,, J.Kanabus, T.Rasor, 1978. J. Clin. Endrocrinol. Metab. 47, 326.

(17) Seron-Ferre, M., C.C. Lawrence, R.B. Jaffe. 1978. J. Clin. Endocrinol. Metab. 46, 834.

(18) Hwang, J., K.M. Menon. 1983. J. Biol. Chem. 258, 7322.

(19) Inazu, N., H. Kogo, Y. Alzaway. 1983. Biochem. Biophys. Acta. 750, 98.

(20) Quagliarello, J., N. Szlachter, J.S. Niselbaum, et al. 1981. Fertil Steril. 36, 399.

(21) Carayon, P., B. Nisula, S. Lissitzky. 1981. Endocrimology. 108, 1891.

USE OF ONCODEVELOPMENTAL PROTEINS IN DIAGNOSIS AND MONITORING OF TESTICULAR CANCERS.

D R Pledger, A Belfield, J Mabon

Department of Biochemistry, Royal Maternity Hospital, Rottenrow, Glasgow, Scotland

Introduction

Testicular tumours are a relatively common form of cancer in males. Teratomas comprise about 1% of all malignant tumours in males, with seminomas being the commonest malignant tumour of the testes comprising about 40% of testicular germ cell tumours.

The incidence of teratomas appears to be increasing and until recently they ranked as the most common fatal cancer in men aged 24-34[1]. However, new chemotherapeutic regimes have greatly improved the outlook for the patient with this tumour and the availability of tumour marker assays is regarded as an essential component of staging and follow-up of these patients[2]. Moreover multivariate analysis has shown that the initial serum concentration of alpha-foetoprotein (AFP) and human chorionic gonadotrophin (hCG) identifies prognostic groups[1,3].

No similar markers exist for seminoma although about 10% of patients exhibit raised hCG concentrations[2]. There is a need for a reliable marker in patients with seminoma and carcinoplacental alkaline phosphatase (CPAP) has been suggested as a possible candidate[4,5].

We have investigated the use of three oncodevelopmental proteins, pregnancy specific β_1 glycoprotein (SP_1), pregnancy associated plasma protein A (PAPP-A), and carcinoplacental alkaline phosphatase (CPAP) in patients with testicular teratoma and seminoma.

Pregnancy Proteins in Animals

Materials and Methods

Serum concentrations of SP_1, PAPP-A, CPAP, hCG and AFP were measured in preoperative samples obtained from 25 patients in whom the diagnosis of testicular teratoma was established. Preoperative blood specimens were also obtained from 13 patients with seminoma. In addition to the preoperative measurements, serial postoperative measurements were made where possible. Twenty-five normal male subjects comprised the control group.

Serum concentrations of SP_1 and PAPP-A were measured by enzyme linked immunoassay (ELISA) [6,7] and CPAP by immunoenzymatic assay[8]. Chorionic gonadotrophin and AFP were measured by radio-immunoassay (RIA).

Results

Preoperatively 72% of patients with teratoma had elevated serum AFP concentrations and 68% had elevated serum hCG concentrations. 60% of patients exhibited an elevation of both markers. By contrast only 40% of patients exhibited elevated serum SP_1 and 30% had a raised serum PAPP-A (Figure 1). One patient with testicular teratoma showed elevated CPAP.

Serial monitoring of 3 patients with testicular teratoma is shown in Figures 2-4.

23% of patients with seminoma had elevated hCG, 30% had raised PAPP-A and 69% showed elevated SP_1 concentrations preoperatively. However, of this 69% many were marginal elevations and only 23% of patients were judged to have marked SP_1 elevations. 38% of patients had elevated CPAP concentrations (Figure 5).

Serial measurements of markers in patients are shown in Figures 6-8.

Detectable concentrations of SP_1 and PAPP-A were found respectively in 12% and 46% of sera from healthy males.

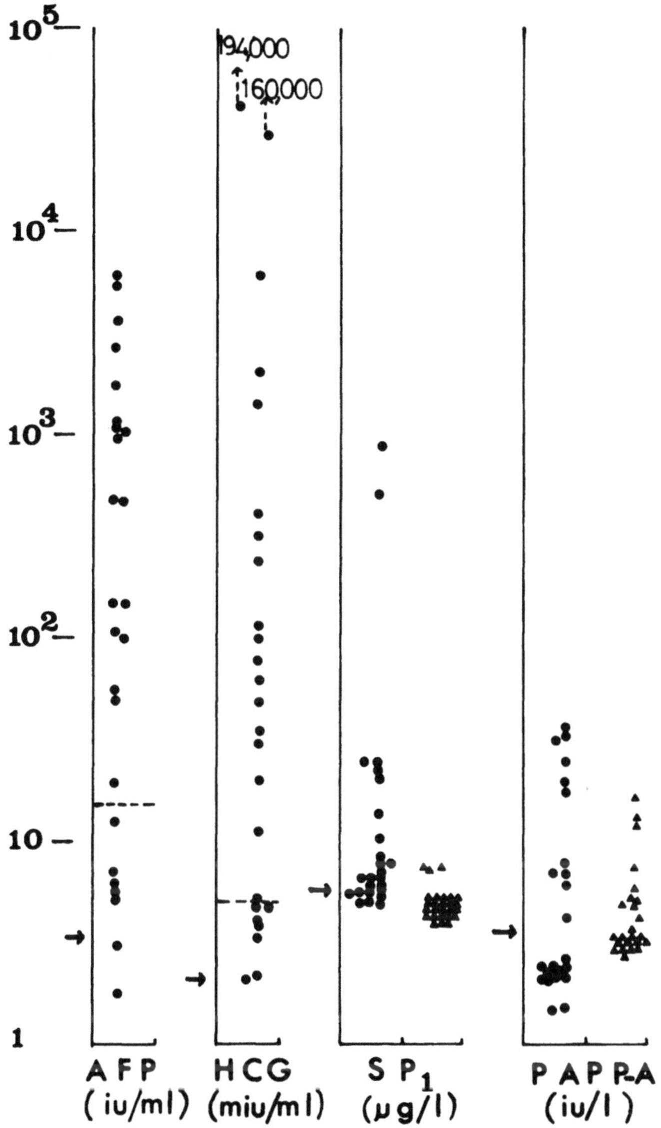

Figure 1

Serum concentrations of various markers in patients with testicular teratoma (●). (—>) represents the limit of detection of the assay and (- - -) represents the upper limit of normal. Where a reference range could not be assigned, comparison was made with results from a group of normal male sera (▲).

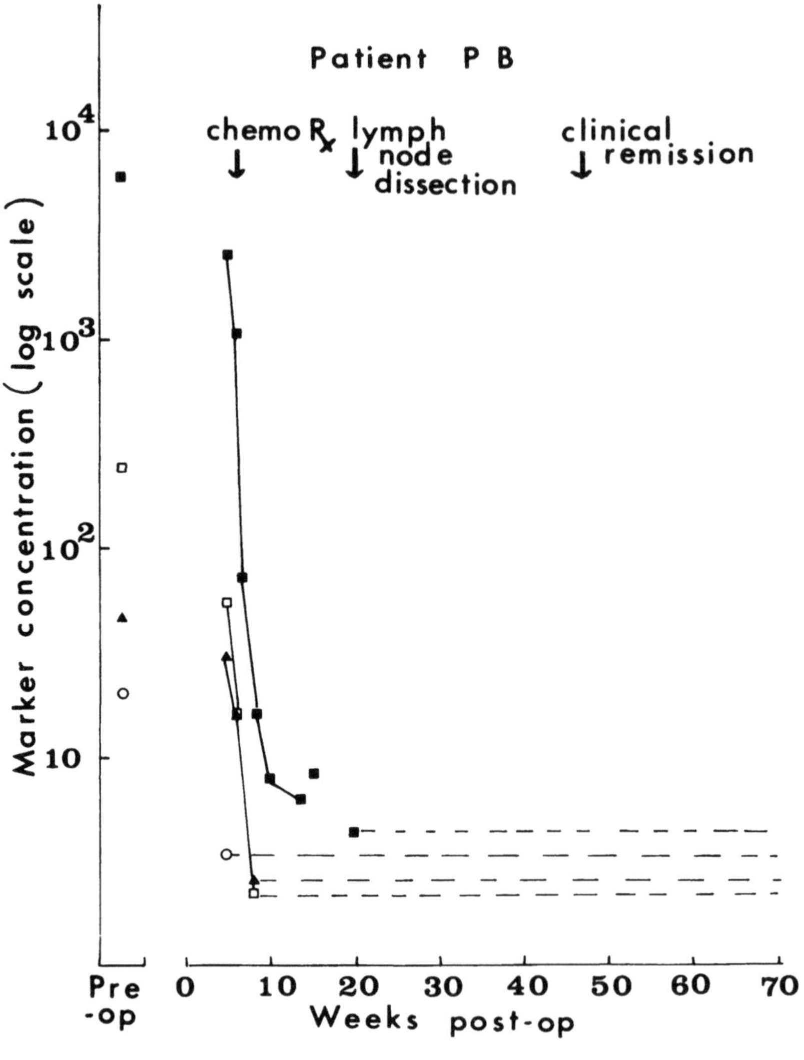

Figure 2

Patient PB was a 21 year old with teratoma of the (R) testis. Following orchidectomy he commenced chemotherapy. He was found to have retroperitoneal metastases and underwent lymph node dissection. He is now in clinical remission and remains marker free.

(■) AFP (□) hCG (▲) PAPP-A (O) SP_1

Broken lines indicate results remaining within normal range to present.

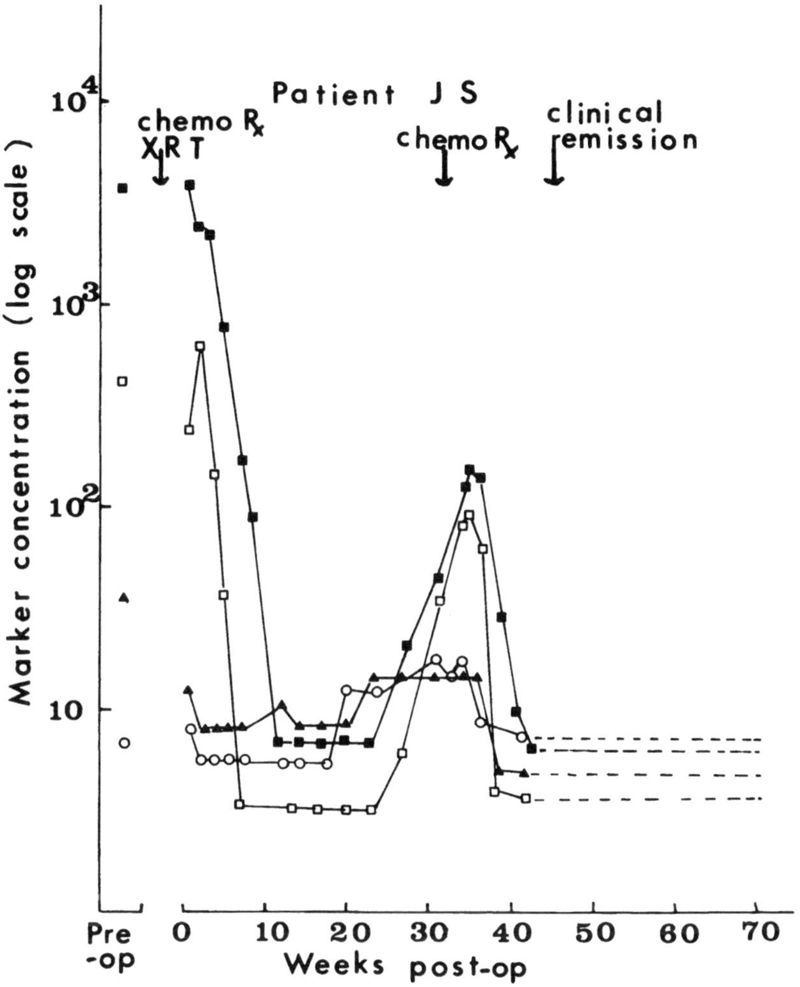

Figure 3

Patient JS was a 32 year old with teratoma of the (R) testis and lymphatic spread. Post orchidectomy he underwent chemotherapy and XRT, followed by a further course of chemotherapy. Clinical remission was achieved and the patient remains alive and well.
(■) AFP (□) hCG (▲) PAPP-A (O) SP_1

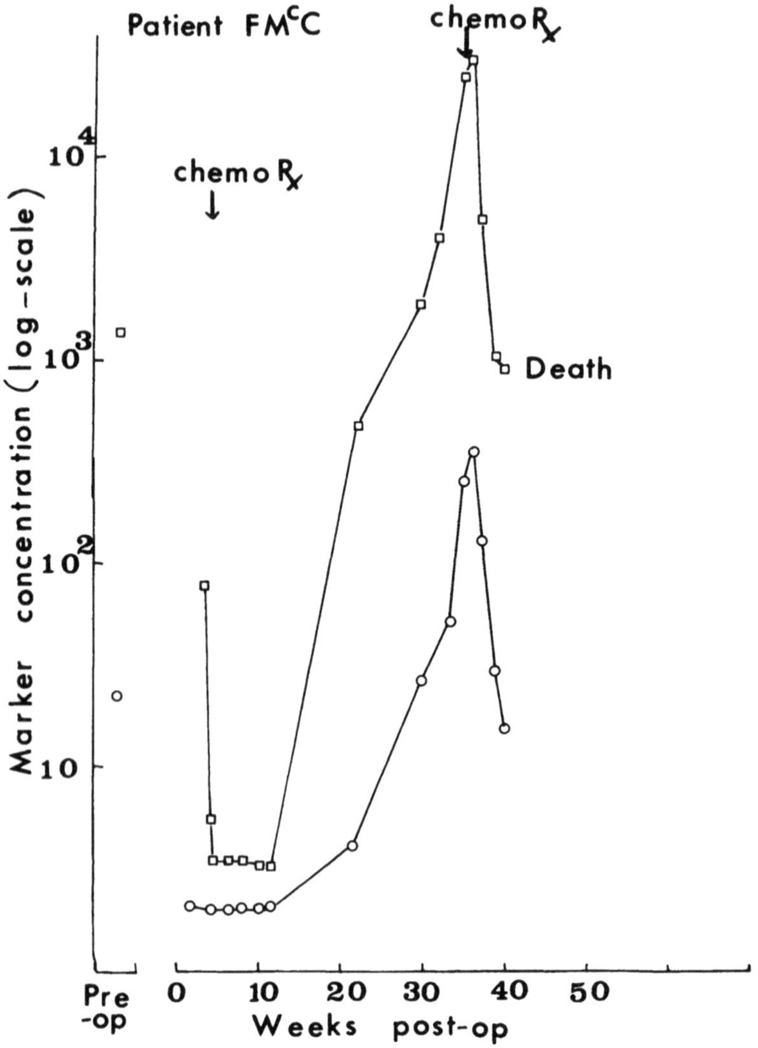

Figure 4

Patient FMcC was a 19 year old with teratoma of the (R) testis. Following a course of chemotherapy, there was evidence of recurrence and a second course of chemotherapy was commenced. However, the patient's condition deteriorated and he died 40 weeks post orchidectomy. At post-mortem examination evidence of pneumococcal meningitis was found together with lung, liver and paraaortic node metastases.

(□) hCG (O) SP_1

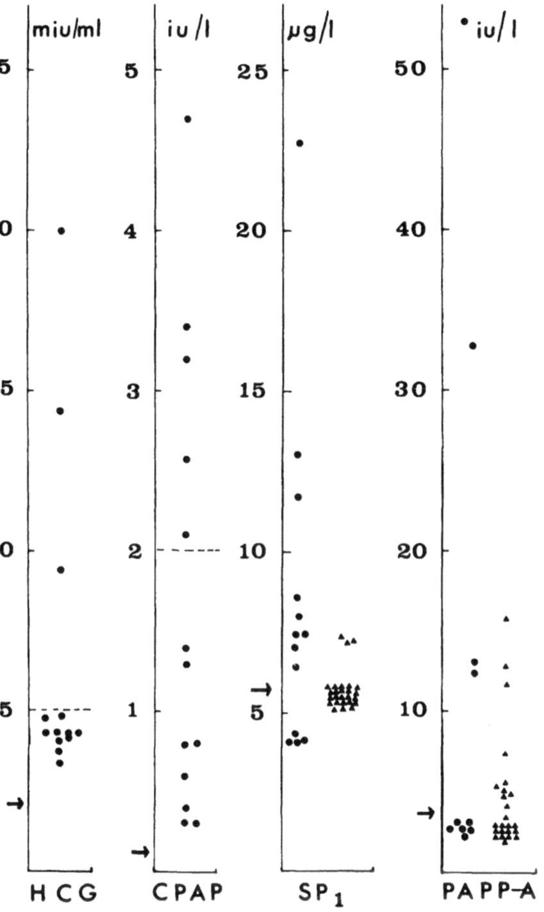

Figure 5

Serum concentrations of various markers in patients with seminoma (●). (—>) represents the limit of detection of the assay and (- - -) represents the upper limit of normal. Where a reference range could not be assigned, comparison was made with results from a group of normal male sera (▲).

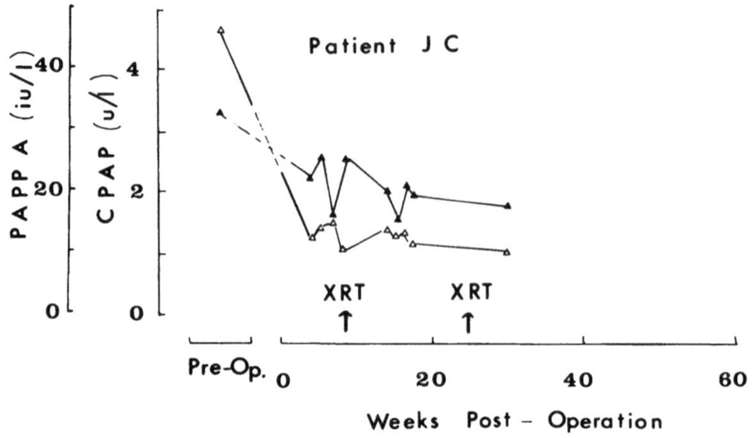

Figure 6

Patient JC was a 38 year old man who presented with swelling of the right testis. He underwent right orchidectomy and was found to have seminoma infiltrating the spermatic cord. Involvement of the para-aortic nodes was also suspected. The patient underwent two courses of radiotherapy to para-aortic and iliac nodes and remains alive and well.
(Δ) CPAP (▲) PAPP-A

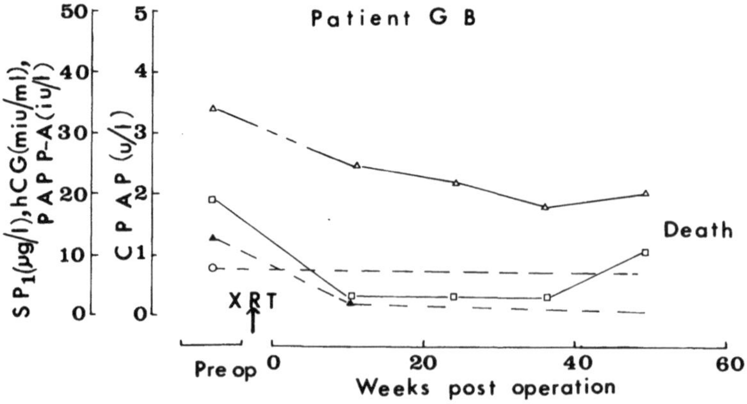

Figuro 7

Patient GB was a 43 year old man who presented with seminoma in an undescended testes and underwent orchidectomy. Disease appeared localised and the patient had no evidence of mediastinal disease at any time. He received irradiation of the para-aortic nodes and initially appeared well. However about 12 months after surgery he presented with signs of spinal cord compression and died. Post-mortem examination revealed a secondary tumour deposit in the spine. (Δ) CPAP (▲) PAPP-A (O) SP_1 ( ) hCG

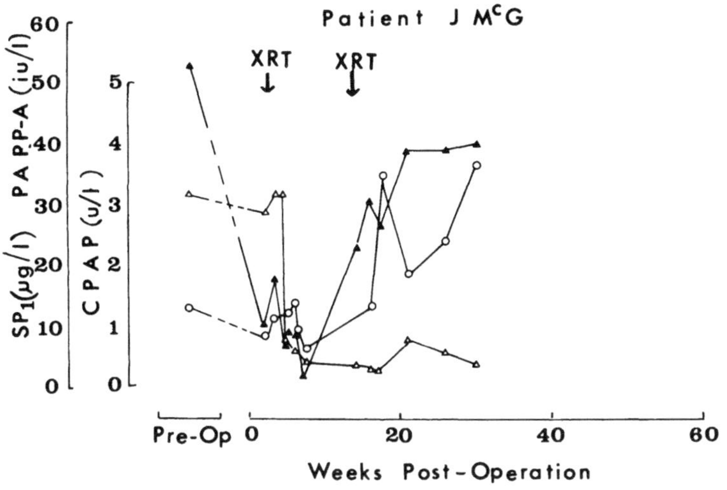

Figure 8

Patient JMcG was a 65 year old man with a previous history of right orchidopexy at 12 years of age. At operation he was found to have a spermocytic seminoma of the right testis, with extensive metastases. The patient underwent a righ orchidectomy followed by 2 courses of radiotherapy and remains alive and well.
(▲) PAPP-A (Δ) CPAP (O) SP_1

Discussion

The detection of oncodevelopmental proteins in the sera of normal subjects is largely due to improved assay sensitivity. Whilst the presence of AFP, hCG and CPAP in the sera of normal subjects is generally accepted[9,10,11] the occurrence of other oncodevelopmental proteins has been disputed[12,13]. The results obtained here confirm the presence of SP_1 and PAPP-A in the serum of healthy males. It is not possible therefore to use the presence of these proteins in serum to diagnose cancer; rather we must use the extent to which the marker concentration exceeds the upper limit of the reference range.

Initial observations on patients with testicular teratoma suggest that SP_1 and PAPP-A are raised more frequently in the serum of patients with widespread disease compared with patients having restricted disease. More extensive investigations are needed to confirm this point. Overall, hCG and AFP are raised more often than SP_1 and PAPP-A in preoperative patients, so there would seem to be no place for SP_1 and PAPP-A in diagnosis.

The potential of serum markers in the management of patients with teratoma is shown by the serial measurements made in 3 patients (Figures 2-4).

Successful eradication of disease results in decline of the conventional markers AFP and hCG to within the normal range (patients PB, Fig 2), and they remain normal unless there is a recurrence of disease as with patients JS and FMcC (Figures 3 & 4). On the grounds of rising hCG concentrations (there was no clinical evidence of recurrence at that time), JS received further chemotherapy and remission was achieved, the patient remaining alive and well. There is little doubt that hCG and AFP are effective monitors of this condition[1,3]. The question remaining is whether alternative markers will be more useful?

Concordant elevation of hCG and SP_1 in patients with testicular teratoma has been reported previously[14,15]. The latter group also reported that SP_1 is raised infrequently in patients without elevation of hCG and AFP. Only two patients in the present study showed such rises.

Would SP_1 or PAPP-A estimation allow an increased lead-in time in patients who had initially elevated levels? Only occasionally (patient JS) would a rising SP_1 have allowed earlier commencement of treatment. Even with later intervention clinical remission was still achieved and it appears the use of SP_1 or PAPP-A would offer little therapeutic advantage although there has been some dissent from this view[14,15]. The conventional markers are likely to remain the mainstay of diagnosis, staging, prognosis and monitoring the therapy of testicular teratomas.

In contrast to the use of hCG and AFP in the investigation of cases of testicular teratoma there are no established markers for seminoma which are widely used. The use of CPAP in this context has been suggested[4] and we have presented encouraging data on this use elsewhere[16].

PAPP-A concentrations were raised in 4 of the patients, 3 of whom also had raised CPAP. In patient JC, PAPP-A remained above the detection limit, but in patient GB, PAPP-A fell to below the assay detection limit and remained there despite clinical evidence of progression of disease. Conversely in patient JMcG PAPP-A levels declined initially but then became elevated again. Despite this finding the patient showed no sign of progression of disease.

Of the 3 patients who showed significantly elevated SP_1 concentration, 2 also had elevated CPAP. In patient JMcG, SP_1 levels show a similar trend to PAPP-A and do not reflect the patient's clinical condition. No further specimens were received on the second patient with combined elevation of markers. The patient who had only elevated SP_1 remains alive and well with no sign of residual disease, his SP_1 concentration having declined to normal post-operatively and remained thus.

Lange et al[15] in a study of patients with seminoma found elevated SP_1 in some cases initially with corresponding elevation of hCG. Szymendera[17] presented similar results but neither author clarified the usefulness of this test.

We have too little data at present to indicate whether serum hCG reflects the clinical course more accurately than CPAP. However, it appears that since both SP_1 and PAPP-A concentrations are usually elevated in combination with CPAP, that the former proteins are unlikely to be of use in monitoring the course of disease.

In the investigation of testicular tumours we find little evidence to commend the use of the more recently discovered oncodevelopmental proteins. HCG and AFP have advantages over the use of SP_1 and PAPP-A in monitoring testicular teratoma, whilst for seminoma only CPAP appears promising. The initial enthusiasm with which the "new" placental proteins were greeted as possible tumour markers must be tempered.

References

1. Newlands, E.E.S., Rustin, G.J.S., Begent, R.J.H., Parker, D and Bagshawe, K.D. 1983. Further advances in the management of malignant teratomas of the testes and other sites. Lancet, April 30th, 948-951.

2. Hainsworth, J.D. and Greco, F.A. 1983. Testicular Germ Cells Neoplasms. Am. J. Med. 78, 817-832.

3. MRC Working Party on Testicular Tumours 1985. Prognostic Factors in Advanced Non-Seminomatous Germ Cell Testicular Tumours: Results of a Multicentre Study. Lancet January 5th, 1985.

4. Lange, P.H., Millan, J.L., Stigbrand, T., Vessella, R.L., Ruislahti, E. and Fishman, W.H. 1982. Placental alkaline phosphatase as a tumour marker for seminoma. Cancer Res. 42, 3244-3247.

5. Paiva, J., Damjanov, I., Lange, P.H. and Harris, H. 1983. Immunohistochemical localisation of placental like alkaline phosphatase in testes and germ cell tumours using monoclonal antibodies. Am. J. Pathol. 111, 156-165.

6. Macdonald, D.J., Belfield, A., Steele, C.J., Mack, D.S. and Shah, M.M. 1979. The Quantitation of Pregnancy Specific β_1 Glycoprotein by Enzyme Linked Immunoassay. Clin. Chim. Acta. 94, 41-49.

7. Pledger, D.R. and Belfield, A. 1983. An ELISA for pregnancy associated plasma protein A. Ann. Clin. Chem. 20, 236-240.

8. Mabon, J., Pledger, D.R. and Belfield, A. 1984. Immunoenzymatic assay of carcinoplacental alkaline phosphatase. Med. Lab. Sci. 41, 29-34.

9. Ruislahti, E. and Seppala, M. 1971. Studies on carcino-fetal proteins III Development of a radioimmunoassay for AFP. Demonstration of AFP in serum of healthy human adults. Int. J. Cancer, 8, 374-383.

10. Borkowski, A and Muquardt, C. 1979. Human chorionic gonadotrophin in the plasma of normal non-pregnant subjects. N. Engl. J. Med. 301, 298-302.

11. Haye, W.G., Meerwaldt, J.H., Talerman, A., Kuipers, T., Baggerman, L., Teeuw, A.H., Van der Pompe, W.B. and Van Driel, J. 1979. The value of a sensitive assay of carcino-placental alkaline phosphates (CPAP) in the follow-up of Gynaecological cancers. Int. J. Cancer. 24, 288-293.

12. Bischof, P., Duberg, S., Herrmann, W. and Sizanenko, P.C. 1981. Pregnancy Associated Plasma Protein-A (PAPP-A) and hCG in Early Pregnancy. Brit. J. Obstet. Gynaecol. 88, 973-975.

13. Sørensen, S. and Trentemøller, S. 1984. Non-specific serum/ plasma response in radioimmunoassay of pregnancy-specific β_1 glycoprotein (SP_1) in healthy, non-pregnant subjects. Clin. Chim. Acta. 140, 101-108.

14. Blacker, C., Feinstein, M.C., Rodger, M and Scholler, R. 1981. Evaluation of $PS\beta_1G$ and other tumour markers in germ cell tumours of the testes. Andrologia 13, 458-467.

15. Lange, P.H., Bremner, R.D., Horne, C.H.W., Vessella, R.L. and Fraley, E.E. 1980. Is SP_1 a marker for testicular cancer? Urology 15, 251-255.

16. Pledger, D.R., Mabon, J. and Belfield, A. 1985. Preliminary Observations on the Application of Carcinoplacental alkaline phosphatase to the Investigation of Patients with Seminoma of the testes. Clin. Biochem. Accepted for publication.

17. Szymendera, J.J., Zborzil, L., Lenko, J., Kaminska, J.A. and Cadek, A. 1983. Evaluation of Five Tumour Markers (AFP, CEA, hCG, hPL and SP_1) in Monitoring Therapy and Follow-up of patients with Testicular Germ Cell Tumours. Oncology 40, 1-10.

ESTROGEN, GROWTH HORMONE AND PREGNANCY-ASSOCIATED PROTEINS

Nils Fröhlander and Bo von Schoultz
Department of Obstetrics and Gynecology, University Hospital, S-901 85 Umeå, Sweden

Introduction

Exogenous sex steroids greatly affect the serum concentrations of the human pregnancy-associated proteins sex-hormone binding globulin (SHBG, SP_2) and pregnancy zone protein (PZP, α_2-PAG, SP_3). It is believed that the liver synthesis of these proteins during hormonal therapy is regulated by the estrogen/androgen balance. The levels of SHBG and PZP are increased in a strictly dose dependent manner following oral estrogen administration and are usually decreased following administration of androgens, progestrogens and their analogues and both proteins have been suggested as markers of "estrogenicity" (1, 2). However, several observations are at variance with the concept that endogenous sex steroids regulate SHBG and PZP synthesis. In a recent study of men with prostatic cancer protein levels were unchanged after orchidectomy despite the fact that testosterone concentrations were reduced to less than 5 % of the pre-treatment values (3). In postmenopausal women oral micronized estradiol-17β caused a prompt increase in circulating SHBG and PZP, whereas percutaneous application of the same estrogen, yielding the same serum estradiol-17β levels as during oral administration, did not result in any change in protein levels (4). Certainly, factors other than the absolute concentration of the steroids must be operative.

In rats the sexual differences in hepatic steroid metabolism were shown to be regulated by the secretory pattern of growth hormone (GH), a relatively high base level with frequent small spikes being the "feminizing factor". Estrogens were shown to induce a female mode of hepatic steroid metabolism via the hypothalamo-pituitary system (5). In order to elucidate whether growth hormone might be involved in the regulation of SHBG or PZP synthesis, we have studied the effects of oral estrogen therapy and of tamoxifen, an antiestrogen, on the serum GH level. In addition, as a pilot study, the effects of short-term parenteral administration of GH on the serum levels of SHBG and PZP were studied.

Pregnancy Proteins in Animals

Material and methods

Subjects. A total of 38 women aged 32-62 years (mean age 50 years) attending the clinic for typical symptoms of estrogen deficiency were recruited for the study of the effects of estrogen on GH. All women had their last period at least six months earlier and all had postmenopausal FSH levels. They were all apparently healthy and without medication for the last three months. Seven women, aged 27-54 years (mean age 42 years), from the hospital staff were recruited for the pilot study of the effects of GH administration. They were all having regular periods and were apparently healthy taking no estrogens or other drugs.
Hormonal treatment. Twentyfour women, mean age 46 years (range 32-60) received ethinyl estradiol in a dose of 30 µg daily. Treatment was prescribed cyclically for three weeks with an interval of one week, and continued for three cycles. Fourteen women, mean age 53 years (range 44-62), received 10 µg of ethinyl estradiol daily. After three cycles of unopposed estrogen treatment, tamoxifen in a dose of 10 mg twice daily was added during the last ten days of each cycle. This combined treatment was continued for three cycles.
Venous blood samples were drawn before therapy was started and at clinical examinations during the last days of the third cycle. In the ethinyl estradiol-tamoxifen group additional samples were collected after the fourth and sixth cycles. Serum was stored at -20°C until assayed.
Seven women, mean age 42 years (range 27-54), received intramuscular injections of GH, (Crescormon^{R}, KabiVitrum, Stockholm, Sweden), 4 IU twice daily for four days (day 1-4). Injections were given at 8.00 h and 16.00 h each day. The mean SHBG and PZP levels of two pre-treatment blood samples, one fasting and one non-fasting, were used as pre-treatment values. Additional blood samples were collected between 9.00 h and 11.00 h on days 2, 3, 4, 5 and 9. Additional samples were obtained from four subjects on days 12, 16 and 23. Serum was stored at -20°C until assayed.
Analyses. Growth hormone (GH) was determined by radioimmunoassay using a commercial kit (Phadebas hGH PRIST^{R}, Pharmacia Diagnostics, Uppsala, Sweden). Pregnancy zone protein (PZP) was determined by radioimmunoassay according to Damber et al (6) using [^{125}I] labelled PZP, a mono-specific rabbit anti-serum and the purified protein as standard. Sex-hormone binding globulin (SHBG) binding capacity was determined using a two-phase equilibrium partition technique as previously described (7). In the study of the effects of parenteral GH administration SHBG was determined by an immunoradiometric technique using a commercial kit obtained from Farmos Diagnostica OY, Turku, Finland.
Within and between-assay variation was 5 and 11 % respectively for SHBG, 10 and

15 % for PZP and 8 and 11 % for GH.

Statistical analyses. The results were evaluated using Wilcoxon's signed rank test.

Results

The serum levels of GH, SHBG and PZP before and during treatment with ethinyl estradiol only and in combination with tamoxifen are given in Table I.

Table I. Serum levels (mean ± SEM) of GH, SHBG and PZP in A) 24 postmenopausal women before and during replacement therapy with 30 µg ethinyl estradiol daily and B) 14 postmenopausal women before and during replacement therapy with 10 µg ethinyl estradiol daily with and without addition of tamoxifen 10 mg twice daily. GH (mU/l), SHBG (nM), PZP (mg/l).

		Before treatment	Estrogen only third cycle	Tamoxifen, addition first cycle	Tamoxifen, addition third cycle
A)	GH	2.6 ± 0.49	9.1 ± 2.02***		
	SHBG	93.9 ± 7.87	220.9 ± 7.65***		
	PZP	18.9 ± 3.78	192.8 ± 22.02***		
B)	GH	2.8 ± 0.96	6.5 ± 1.36*	2.3 ± 0.48+	2.5 ± 0.78
	SHBG	104.9 ± 9.76	194.6 ± 11.14***	192.7 ± 10.32	166.0 ± 11.47++
	PZP	14.1 ± 2.00	155.4 ± 25.07***	82.0 ± 13.52	41.4 ± 6.63++

*p < 0.05 compared to pre-treatment values, ***p < 0.001 compared to pre-treatment values. +p < 0.02 compared to unopposed estrogen therapy, ++p < 0.01 compared to unopposed estrogen therapy.

GH, SHBG and PZP all showed a highly significant increase in serum concentration following treatment with 30 µg of ethinyl estradiol ($p < 0.001$). During treatment with the lower dose of 10 µg of ethinyl estradiol the serum levels of GH were significantly increased ($p < 0.05$) whereas highly significant increases were recorded for SHBG and PZP ($p < 0.001$). Already after the first cycle of tamoxifen addition the serum concentrations of GH were reduced to pre-treatment values ($p < 0.02$). The serum levels of SHBG and PZP were significantly decreased

at the end of the third cycle of tamoxifen addition ($p < 0.01$).
Following parenteral administration of GH, 4 IU twice daily, the serum SHBG levels rose sharply in four of the investigated women while there was no apparent effect in the remaining three women (Fig 1).

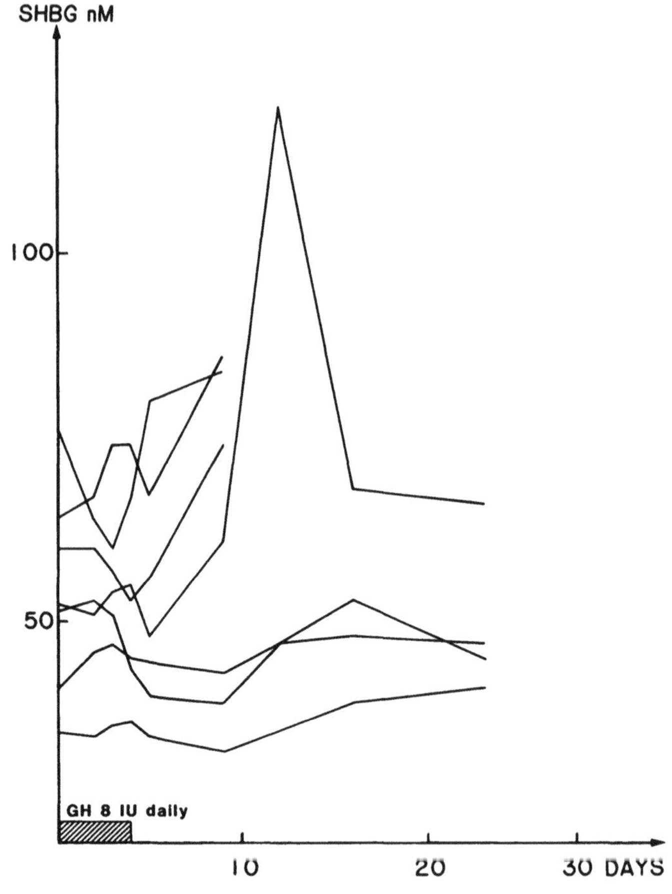

Fig.1. Individual serum concentrations of SHBG (nM) in seven women following short-term intramuscular administration of growth hormone (4 IU twice daily for four days.

When the results from days 9 and 12 were pooled the mean SHBG concentration was significantly increased as compared to the pre-treatment level ($p < 0.05$). Three women showed increased PZP levels on day 9 or 12 but the mean serum level was virtually unchanged.

Discussion

Despite the clearcut effects excerted by exogenous sex hormones upon the serum concentrations of SHBG and PZP very little is known about the physiological regulation of these pregnancy-associated proteins. The levels of both proteins increase considerably during pregnancy, but there is no apparent correlation between the levels of estrogen and proteins in individual pregnant women (8, 9). Furthermore, neither SHBG nor PZP show any significant variation during the menstrual cycle and their concentrations are not affected by the menopause (8, 10, 11). This indicates that other factors than the steroid levels per se are involved in the regulation of their synthesis.

The present data confirm previous reports that exogenous estrogens induce the synthesis of SHBG and PZP in a dose dependent manner. Furthermore the serum GH levels showed a quite similar and apparently dose dependent increase following estrogen administration. The stimulatory effect of 10 µg of ethinyl estradiol on GH levels was eliminated by the addition of tamoxifen, 10 mg twice daily, already during the first cycle of combined treatment. Tamoxifen is generally regarded as a pure antiestrogen without androgenic side-effects and when added during replacement therapy it strongly counteracts the theraputic effects of the estrogen. Our results indicate that estrogens have a direct effect on the secretion or synthesis of GH and that this effect is mediated by a specific receptor mechanism.

Whether growth hormone is involved in the regulation of estrogen induced protein synthesis is unknown, but there are some data to support such a hypothesis. The liver is a major target organ for GH and estrogen treatment has previously been shown to increase pituitary GH secretion (12). The presence of GH was found to be necessary for steroid induced protein synthesis in liver slices (13). Recent results from animal experiments indicate the presence of a pituitary factor, likely to be GH, that is estrogen sensitive and which influences liver metabolism (5). The serum levels of the murine PZP-analogue, PAMP-1 (pregnancy-associated mouse protein), decrease after oophorectomi to be restord after estrogen substitution, which seemingly indicates a sex steroid dependent synthesis. Yet, following hypophysectomy PAMP-1 levels rapidly decrease, in spite of estrogen substitution to physiological levels (14). The results from the pilot study of the effects of short-term parenteral GH administration should be considered with caution and the interpretation is complicated by the wellknown intra- and inter-individual variability of GH serum levels. While a stimulatory effect of GH on protein levels was recorded in some women the total results were inconclusive. The influence of the interaction between growth hormone and sex steroids on hepatic enzyme activity has recently been elucidated in the rat (5) and the mode of GH administration was found to be extremely important for the effect. Based on current knowledge it is tempting to speculate that growth hormone secretion is important for the regulation of "steroid sensitive" protein synthesis by the liver. However, a more thorough study of this subject requires detailed investigation of the secretory patterns of GH, including its pulsatile nature and dynamic responses to various stimuli. The existence of a murine analogue to the human PZP, PAMP-1 (14), provides an animal model for further studies on the regulation of pregnancy-associated proteins. Investigations are in progress and preliminary results indicate that continous subcutaneous administration of GH induces the synthesis of PAMP-1 in hypophysectomized mice.

Acknowledgements

This work was supported by the Swedish Medical Research Council (5982). Skilful technical assistance was provided by Mrs M Isaksson and Mrs M Wallén and excellent typing by Miss D Clemedtson.

References

1. Damber, M-G., Sandström, B., von Schoultz, B. and Stigbrand, T. 1979: A new sensitive method for quantifying and comparing the biological potency of various estrogens in man. Acta Obstet Gynecol Scand 58, 527-530.

2. Laurell, C-B. and Rannevik, G. 1979: A comparison of plasma protein changes induced by danzol, pregnancy and estrogens. J Clin Endocrinol Metab 49, 719-725.

3. Carlström, K., Eriksson, A., Gustafsson, S.A., Henriksson, P., Pousette, Å., Stege, R. and von Schoultz, B. 1985: Influence of orchidectomy or estrogen treatment on serum levels of pregnancy-associated α_2-glycoprotein and sex hormone binding globulin in patients with prostatic cancer. Int J Androl 8, 21-27.

4. Holst, J., Cajander, S., Carlström, K., Damber, M-G. and von Schoultz, B. 1983: A comparison of liverprotein induction in postmenopausal women during

oral and percutaneous estrogen replacement therapy. Br J Obstet Gynecol 90, 355-360.

5. Mode A 1983: Growth hormone - a regulator of the sexually differentiated steroid metabolism in rat liver. Thesis Karolinska Institutet, Sthlm, Sweden.

6. Damber, M-G., von Schoultz, B., Stigbrand, T. and Carlström, K. 1976: A radioimmunoassay for the pregnancy zone protein. Clin Chim Acta 66, 85-88.

7. Shanbhag, V.P., Södergård, R., Carstensen, H. and Albertsson, P.A. 1983: A new method för the determination of the binding capacity of testosterone-estradiol-binding globulin in human plasma. J Steroid Biochim 4, 537-550.

8. Anderson, D.C. 1974: Sex hormone binding globulin. Clin Endocrinol 3, 69-96.

9. Westergaard, J.G., Bach, A., Teisner, B., Hau, J. and Grudzinskas, J.G. 1982: Circulating human pregnancy zone protein and estradiol in twin pregnancies. J Reprod Fertil 66, 695-698.

10. Damber. M-G., von Schoultz, B., Stigbrand, T. and Carlström, K. 1976: Serum levels of the pregnancy zone protein during the normal menstrual cycle. Acta Obstet Gynecol Scand 55, 467-468.

11. Damber, M-G., von Schoultz, B., Stigbrand, T., Carlström, K. and Furuhjelm, M. 1977: Basic levels of the immunosuppressive serum protein PZ in women - influence of age, parity, gonadotrophins and estrogens. Acta Obstet Gynecol Scand 56, 95-99.

12. Wiedemann, E., Schwartz, E. and Frantz, A.G. 1976: Acute and chronic estrogen effects upon serum somatomedin activity, growth hormone and prolactin in man. J Glin Endocrinol Metab 42, 942-952.

13. Griffin, E.E and Miller, L.L. 1974: Effects of hypophysectomy of liver donors on net synthesis of specifik plasma proteins by the isolated perfused rat liver. J Biol Chem 249, 5062-5069.

14. Hau, J., Gidley-Baird, A.A., Teisner, B., Svendsen, P. and Westergaard, J.G. 1983: The influence of pituitary and gonadal hormones on serum levels of pregnancy-associated murine protein - 1. Acta Endocrinol 104, 23-26.

STUDIES ON THE STRUCTURE AND FUNCTION OF HUMAN PREGNANCY ZONE PROTEIN, A CLOSE HOMOLOG OF HUMAN α_2-MACROGLOBULIN

Lars Sottrup-Jensen, Ole Sand

Department of Molecular Biology, University of Aarhus, 8000 Aarhus C, Denmark

Jørgen Folkersen

Institute of Medical Microbiology, University of Odense, 5000 Odense, Denmark

Introduction

Among the pregnancy associated plasma proteins PZP is quantitatively the most important (1). It is also known as Pregnancy Associated α_2-Glycoprotein, Pregnancy Associated Globulin, α_2-Acute Phase Glycoprotein, α_2-Pregnoglobulin, Pregnancy Associated α-Macroglobulin, Schwangerschaftsprotein-3, Xh-antigen and Pa-1 (2). In the plasma of healthy non-pregnant donors PZP is a trace protein (women: 10-30 mg/l and men: < 10 mg/l (3)). At term the plasma level of PZP may increase to 1000-1400 mg/l (3,4). PZP is a glycoprotein with α_2-mobility (5-7) containing approx. 10 % carbohydrate (8-10). Under non-reducing denaturing conditions a 360 kDa molecular species is observed along with aggregates of high molecular weight (9-11). After reduction subunits of 180 kDa are found indicating that native PZP is a 360 kDa dimer (9-11) or perhaps a 720 kDa tetramer (8,11), i.e. a dimer of dimers just like the tetrameric human α_2M (4 x 180 kDa) (12-14). Apart from the similar size of the subunits of PZP and α_2M, and the similar amino acid composition (8,10,14,15), the apparent existence of common antigenic determinants (16,17) indicate that PZP and α_2M might be related proteins. We have shown recently by sequencing a random selection of tryptic peptides from PZP that PZP and α_2M indeed are strongly homologous (18). This result indicates that the function of PZP in pregnancy, still largely unknown, may be analogous to that of α_2M. In particular, PZP can be expected to function as a proteinase binding protein.

Pregnancy Proteins in Animals

Here we present a procedure for the large scale isolation of native and immunologically pure PZP from late pregnancy plasma or serum. Furthermore we show that this preparation has functionally intact β-cysteinyl-γ-glutamyl thiol esters and will form complexes with a variety of proteinases. Finally we discuss the structural and functional aspects of PZP, which indicate that PZP may be the human equivalent to the acute phase α-macroglobulins known in other species (19-21).

Material and Methods

Analytical procedures

Non-denaturing PAGE, SDS-PAGE, automated sequence determination and identification of phenylthiohydantoin amino acids were carried out as described previously (22,23). Evaluation of the extent of binding of $[^{125}I]$-labelled proteinases to PZP was done essentially as before (24), this time by gel chromatography on a column of TSK G-4000 SW (7.5 x 600 mm). Rocket- or crossed IEP were performed as described (25) using antibodies from Dako.

Preparative procedures

An anticontaminant column was prepared by coupling anti-(human serum) antibodies to CNBr-activated Sepharose 4B. Iminodiacetic acid and L-lysine-substituted Sepharose were prepared as described earlier (26-28). CNBr fragments of PZP were separated essentially as described for α_2M by combinations of gel chromatography and high performance liquid chromatography (22).

Purification of native PZP

The procedure described below has been found to result in a preparation of native PZP in about 20-30% yield and is suitable for processing 250-600 ml portions of pooled late pregnancy plasma or serum.

After depletion of plasminogen by adsorption to L-lysine Sepharose (28) the proteins precipitating between 6 and 16% (w/v) PEG were collected. This precipitate contains about 90% of the PZP and α_2M present in plasma. After dialysis for 16 h against running tap-water (11 °C) to precipitate euglobulins, the pH of the supernatant containing the pseudoglobulins was adjusted to 7.0 and the solution subjected to DEAE-Sephacel ion-exchange chromatography (Fig.1).

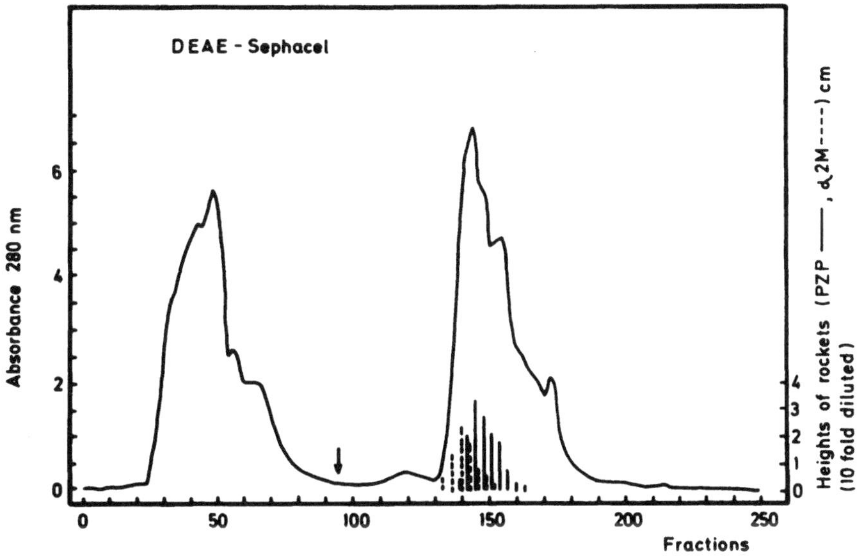

Fig.1. Separation of the pseudoglobulin fraction obtained from 370 ml late pregnancy plasma on a column of DEAE-Sephacel (5 x 21 cm). The column was equilibrated with 20 mM Na-phosphate, 50 mM NaCl, pH 7.0 and eluted with this buffer followed by a linear gradient (↓) in NaCl (50 mM → 500 mM, 1000 ml + 1000 ml). The flow rate was 100 ml/h and 10 ml fractions were collected. The elution position for PZP and α_2M were determined by rocket IEP. The fractions 142-151 were pooled and further separated on a column of Zn^{2+}-iminodiacetic acid Sepharose.

After locating the PZP and α_2M containing fraction by IEP and SDS-PAGE a pool containing about 70% of the PZP was loaded on a column of Zn^{2+}-iminodiacetic acid Sepharose. Using gradient elution (pH 7.0-5.0) PZP was separated from α_2M (Fig.2).

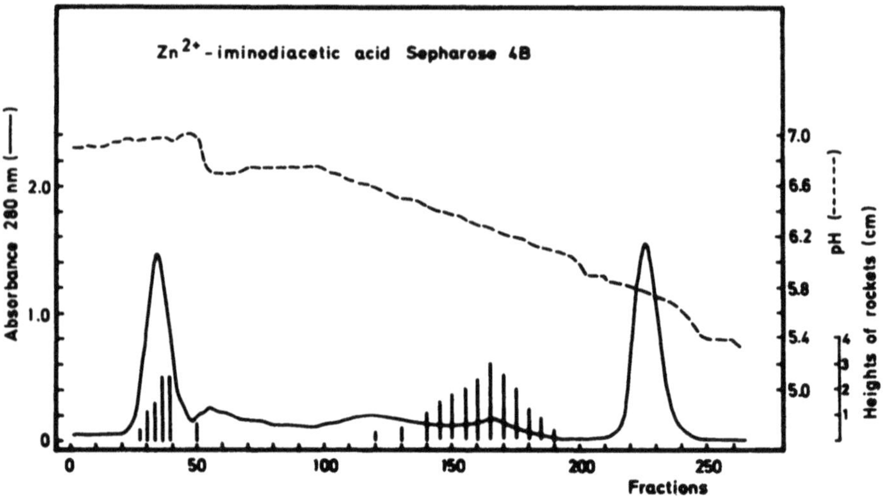

Fig.2. Separation of the PZP containing fraction from DEAE-Sephacel chromatography on a column of Zn^{2+}-iminodiacetic acid-Sepharose (5 x 24 cm). The column was equilibrated with 20 mM Na-phosphate, 50 mM NaCl, pH 7.0 and eluted at a flow rate of 100 ml/h; 10 ml fractions were collected. After sample application the column was eluted with 20 mM Na-phosphate, 1 M NaCl, pH 7.0, and then from fraction 50 the column was eluted with a pH-gradient. This was formed from 850 ml 20 mM Na-phosphate, 1 M NaCl, pH 7.0 and 850 ml 20 mM Na-phosphate, 1 M NaCl pH 5.0. Finally, the column was eluted with the pH 5.0 buffer. PZP was located in the effluent by rocket IEP. α_2M eluted exclusively in fractions 215-245.

Approximately 20% of the applied amount of PZP appeared in the break through volume, while the main portion eluted between pH 6.5 and 6.1, well separated from α_2M (pH 5.8-5.3). As judged from SDS-PAGE the material eluting in fractions 165-190 contained approx. 95% PZP, while the material eluting in fractions 120-164 only contained 25-50% PZP. After pooling fractions 165-190 this material, which contained about 20-30% of the original amount of PZP in plasma, was dialyzed against 0.1 M Na-phosphate pH 8.0 and passed through an anti-contaminant column to remove trace contaminants. The final preparation was concentrated to 2-3 mg/ml, filtered through 0.45 μm membrane filters and kept at -20 °C until use. As demonstrated in Fig.3 no contaminants could be detected upon crossed IEP.

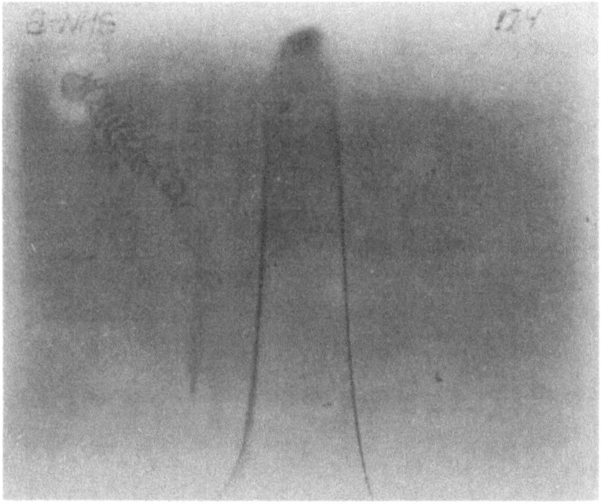

Fig.3. Crossed IEP of 29 μg purified PZP. The antibody containing gel contained antibodies against PZP and normal human serum proteins.

The material eluting in fractions 120-164 (Fig.2) could be combined with corresponding fractions from other preparations to yield a second crop of PZP. This procedure involved an additional step of concanavalin A-Sepharose chromatography, followed by negative affinity chromatography as described above (not shown).

Results

Comments on the purification scheme

Although several procedures for purification of PZP from late pregnancy serum or plasma or from placental extracts have appeared earlier (7-11,29,30) the recent demonstration that the sequences of PZP and α_2M are extensively homologous (18) has

prompted us to develop a new large scale procedure, which will yield native PZP of high purity. In the past it has been difficult to avoid contamination by $\alpha_2 M$, a fact which is now understandable in view of the very similar physico-chemical properties of PZP and $\alpha_2 M$. Even the use of immuno adsorbent procedures has not in general led to preparations of PZP, which are not contaminated by $\alpha_2 M$ (18,30), due to the existence of common antigenic determinants in PZP and $\alpha_2 M$ (16,17). Moreover, the use of acidic chaotropic solvents for elution of PZP from antibody columns (18,29,30) might be expected by analogy with $\alpha_2 M$ to lead to profound changes in the tertiary structure of PZP, resulting in cleavage of its putative internal β-cysteinyl-γ-glutamyl thiol esters (18,31). Such preparations, although essentially intact, would not be expected to bind proteinases.
The use of Zn^{2+}-chelate affinity chromatography is now the method of choice for preparation of highly active $\alpha_2 M$ (27). Pilot experiments showed that the binding of PZP was negligible at pH 6.2, while $\alpha_2 M$ bound quantitatively at this pH, thus indicating the feasability of using Zn^{2+}-chelate columns for separating PZP and $\alpha_2 M$. Using pH gradient elution it has been demonstrated earlier that the closely related rat $\alpha_1 M$ and $\alpha_2 M$ will separate on these columns (21). From this and the recent observation that PZP binds zinc ions less strongly than $\alpha_2 M$ does (32), we have now shown that these columns in fact can be used to separate PZP from $\alpha_2 M$ (Fig.2). In order to decrease the protein load on the Zn^{2+}-chelate column PEG precipitation and euglobulin precipitation followed by DEAE-Sephacel ion-exchange chromatography was used to separate PZP and $\alpha_2 M$ from most other plasma proteins (Fig.1). Following a single step of negative affinity chromatography the resultant preparation of PZP (obtained in a yield of 20-30%) was immunologically pure (Fig.3).

The reason why about 20% of the applied PZP appears in the break-through volume of the Zn^{2+}-chelate column (Fig.2) is not entirely clear. However, by raising the pH of the equilibrium buffer of this column to 8.0 and by loading the column at this pH the amount of PZP eluting here was very low. Apparently, the weak affinity of PZP for Zn^{2+} ions (32) is increased at relatively high pH-values. In contrast to α_2M, which elutes as a sharp peak, the main portion of PZP elutes as a broad peak, indicating an underlying heterogeneity. This may be due to the existence of a readily reversible dimer-tetramer equilibrium, or it may be due to the presence of slightly denatured forms of PZP (see below). In this regard we have found that highly purified preparations of PZP should not be concentrated to more than approx. 3 mg/ml, since upon thawing from -20 °C concentrated solutions frequently develop a precipitate, which in some cases may contain as much as one third of the PZP present. PZP prepared from serum appears to be identical to PZP prepared from fresh plasma with regard to SDS-PAGE, PAGE and crossed IEP patterns. Likewise the potential for appearance of SH-groups and the ability to bind proteinases seems to be the same.

Characterization of native PZP

From the result of amino acid analysis based on 131, 87, 65 and 54 residues of Asx, Gly, Phe and Arg, respectively, per 180 kDa subunit (18) the specific extinction coefficent was estimated at 8.2±0.2 ($E_{1\,cm}^{1\%}$, 280 nm), slightly lower than 8.9 determined earlier for α_2M (33). The SDS-PAGE pattern (reducing conditions) for PZP and α_2M is shown in Fig.4.

These results show that native PZP like α_2M (23,34,35) generates characteristic heat cleavage fragments of 120 kDa and 60 kDa from its 180 kDa subunits. In contrast, prior incubation with methylamine suppresses the formation of these fragments, thus clearly demonstrating that PZP contains an internal β-cysteinyl-γ-glutamyl thiol ester (see also below). The heat fragmentation could not be demonstrated in PZP purified by immuno adsorbent techniques (18).

In discontinuous non-denaturing PAGE two forms of α_2M can be readily distinguished (the "slow" form (native α_2M) and the "fast" form (methylamine or proteinase treated α_2M)(23,36,37).

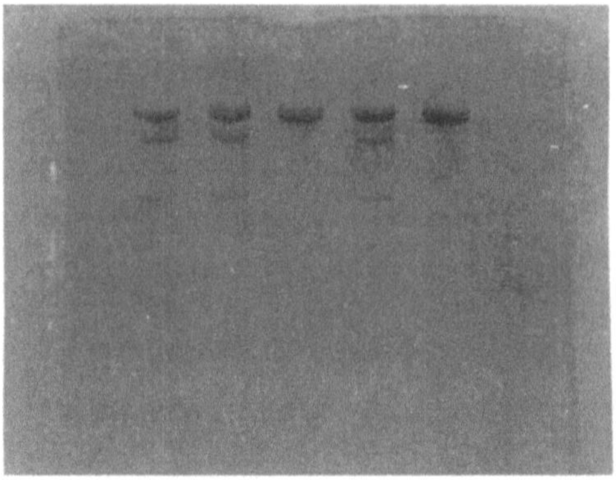

Fig.4. SDS-PAGE pattern (reduced) of preparations of PZP and α_2M. PZP or α_2M solutions (2.5 mg/ml in 0.1 M Na-phosphate pH 8.0) were denatured by 2% SDS containing 20 mM dithioerythritol for 5 min at 100 °C. Lane 1 represents freshly prepared PZP, lane 2 represents PZP which has been thawed and frozen 10 times and lane 4 represents α_2M. These preparations show besides the 180 kDa subunits the characteristic heat cleavage fragments of 120 and 60 kDa. These fragments are absent in methylamine treated PZP and α_2M (0.1 M, 45 min at RT)(lanes 3 and 5, respectively).

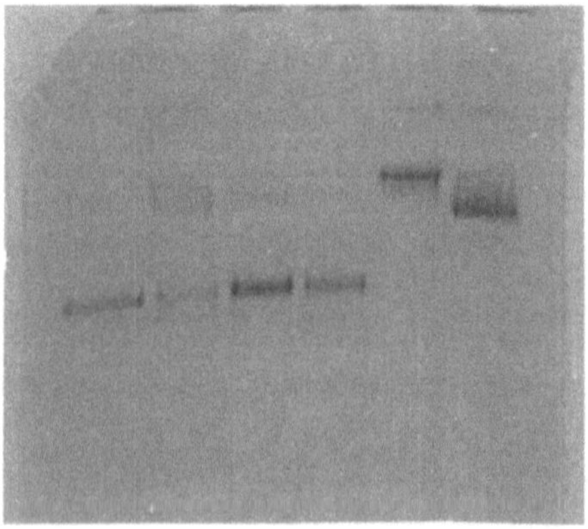

Fig.5. Non-denaturing PAGE of PZP and α_2M preparations. Lanes 1-4 contain PZP and lanes 5 and 6 contain α_2M. PZP and α_2M had been treated with methylamine for 45 min prior to electrophoresis (lanes 3, 4 and 6, respectively). Lane 1 represents freshly prepared PZP while lane 2 represents thawed and frozen PZP. Native α_2M is present in lane 5. Although the main portion of the PZP preparations migrates as a dimer tetrameric forms are also present, particularly in aged PZP.

As shown in Fig.5, (lane 1) native freshly prepared PZP and methylamine treated PZP (lanes 3 and 4, respectively) migrate as a 360 kDa dimer, consistent with earlier estimates (9-11). However, in a preparation of PZP, which had been thawed and frozen 10 times during a period of 30 days, large amounts of a molecular form of 720 kDa and aggregates were apparent (Fig.5, lane 2), indicating that the native structure of PZP is quite labile. Both slow and fast form $\alpha_2 M$ migrate as a tetramer (Fig.5, lanes 5 and 6, respectively), along with minor amounts of aggregates. These aggregates of $\alpha_2 M$ are frequently found as minor constituents of "aged" preparations of $\alpha_2 M$, and probably represent polymerization products resulting from a low degree of cleavage of the thiol esters. Likewise, when $\alpha_2 M$, C3 and C4 become proteolytically activated some degree of internal cross-linking between the thiol esterified Glx-residues and suitably oriented nucleophilic residues occurs (23,31,38,39). When assayed for its potential for methylamine dependent SH-appearance (see below) the aged PZP preparation (Fig.5, lane 2) was only 60-70% active.

Although PZP seems to be a dimer of 360 kDa as judged from PAGE these dimers readily associate reversibly to form tetramers as seen from gel chromatography experiments (Figs 6A, B and C). Freshly prepared PZP elutes as a closely spaced doublet (Fig. 6A), and aged PZP gives rise to a set of molecular species, probably representing dimers, tetrames and aggregates (Fig. 6B). Methylamine treated PZP also gives two closely spaced peaks, eluting in the same positions as those of native PZP (after 33 and 35 min, respectively)(Fig.6C). In contrast, the two main peaks of aged PZP elute after 32 and 36 min, respectively, indicating that this preparation contains molecular species with a changed conformation. Both native and methylamine treated $\alpha_2 M$, which migrate with slightly different rates in PAGE (Fig.5, lanes 5 and 6), elute as a single peak after 33 min (Fig.6D). The characteristic doublet pattern seen with native PZP was not changed, when EDTA was included in the buffer, indicating that the equilibrium between dimer and tetramer is not influenced by metal ions (results not shown).

When freshly obtained pregnancy plasma or serum was gel chromatographed, and PZP and α_2M located in the effluent by rocket IEP, the elution position for the PZP doublet and the α_2M singlet was unchanged. However, the distribution between dimeric and tetrameric PZP in plasma and serum was approx. 2:1, indicating that purified PZP has undergone a subtle conformational change, resulting in a higher proportion of tetramer (results not shown).

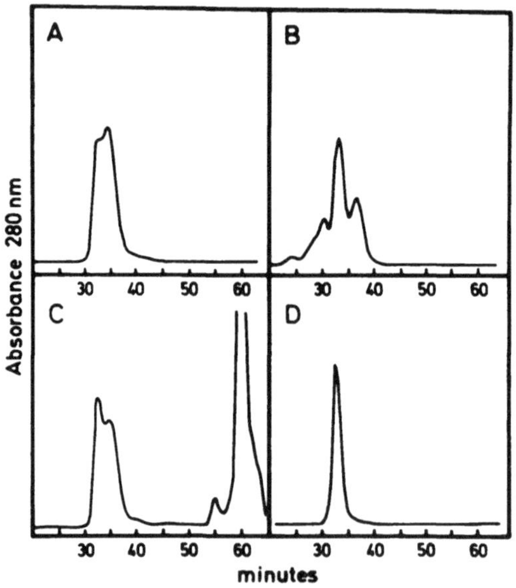

Fig.6. Gel chromatography of preparations of PZP and α_2M. A TSK G-4000 SW column (7.5 x 600 mm) was equilibrated and eluted with 0.1 M Na-phosphate pH 7.0 at a flow rate of 0.5 ml/min at RT. A: Elution profile of 125 µg freshly prepared PZP; B: elution profile of 125 µg aged PZP, thawed and frozen 10 times; C: elution profile of 100 µg methylamine treated PZP, containing DTNB (the same material as shown in Fig.5, lane 4); D: elution profile of 300 µg α_2M. The two closely spaced peaks seen in A and C appear after 33 and 35 min, respectively, and probably represents tetrameric and dimeric PZP. Tetrameric α_2M (D) elutes as a sharp peak after 33 min. The presumed dimeric and tetrameric species of conformationally changed, aged PZP elute after 36 and 32 min, preceded by aggregates eluting after 25-30 min (B).

Methylamine Dependent Appearance of SH-Groups and Sequence Location of the Reactive β-Cysteinyl-γ-Glutamyl Thiol Ester in PZP

As indicated by its characteristic pattern of heat cleavage, indistinguishable from that of α_2M (Fig.4) PZP was judged to contain an internal β-cysteinyl-γ-glutamyl thiol ester in each of its subunits. Fig.7 shows the appearance of DTNB titrable SH-groups (31) upon incubation of PZP and α_2M with methylamine.

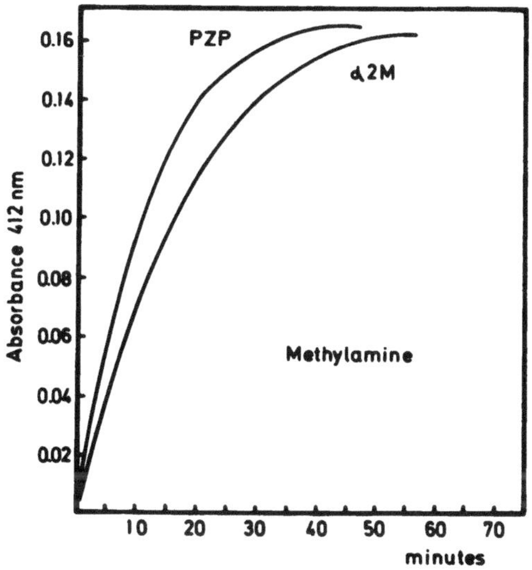

Fig.7. DTNB titration of the SH-groups that appear in PZP and α_2M by treatment with methylamine. After mixing 552 µl PZP and α_2M (2.7 mg/ml in 0.1 M Na-phosphate pH 8.0), respectively, with 18 µl 10 mM DTNB the reaction was initiated by the addition of 30 µl 2 M methylamine pH 8.0. The change in absorbance with time at 412 nm was recorded in a Zeiss spectrophotometer.

For both proteins the maximal appearance is 0.85-0.90 SH-groups per 180 kDa subunit. The initial rate of cleavage of the PZP thiol ester with methylamine is approx. 25% higher than that of the α_2M thiol ester. In order to locate the thiol ester in the PZP sequence PZP was incubated with methylamine in the presence of $[^{14}C]$-JCH_2CONH_2 essentially as done for α_2M (40). After separating the modified protein from excess reagents PZP was degraded with CNBr and the resulting fragments separated on a column of Sephacryl S-200 in urea containing buffers. A fragment of low molecular weight containing the ^{14}C-label was further purifed by HPLC (not shown) and subjected to sequence analysis. Fig.8 shows the sequence of the nonapeptide from PZP, containing the thiol ester site. Evidently the sequence of this peptide is identical to the corresponding stretch in α_2M (41,42) and strongly homologous to the sequences around the thiol ester sites in complement components C3 and C4 (43-45).

PZP:	Pro-Tyr-Gly-Cys-Gly-Glu-Glx-Asn-Met
α_2M:	Pro-Tyr-Gly-Cys-Gly-Glu-Glx-Asn-Met
C3:	Pro-Ser-Gly-Cys-Gly-Glu-Glx-Asn-Met
C4:	Pro-Arg-Gly-Cys-Gly-Glu-Glx-Thr-Met
	S ——— C=O

Fig.8. Alignment of the nonapeptide stretches from PZP, α_2M, C3 and C4, containing the reactive β-cysteinyl-γ-glutamyl thiol ester site. Identically located residues in all four proteins are boxed. The thiol ester site is comprised of Cys-949 and Glx-952 in α_2M (41).

Proteinase-induced Cleavage of the Thiol Esters in PZP and Formation of PZP-Proteinase Complexes

When PZP was incubated with a two-fold molar excess (based on the tetramer) of trypsin or chymotrypsin DTNB titrable SH groups appeared slowly (trypsin) or rapidly (chymotrypsin) (Fig.9). The corresponding appearance of SH-groups in α_2M is shown for comparison.

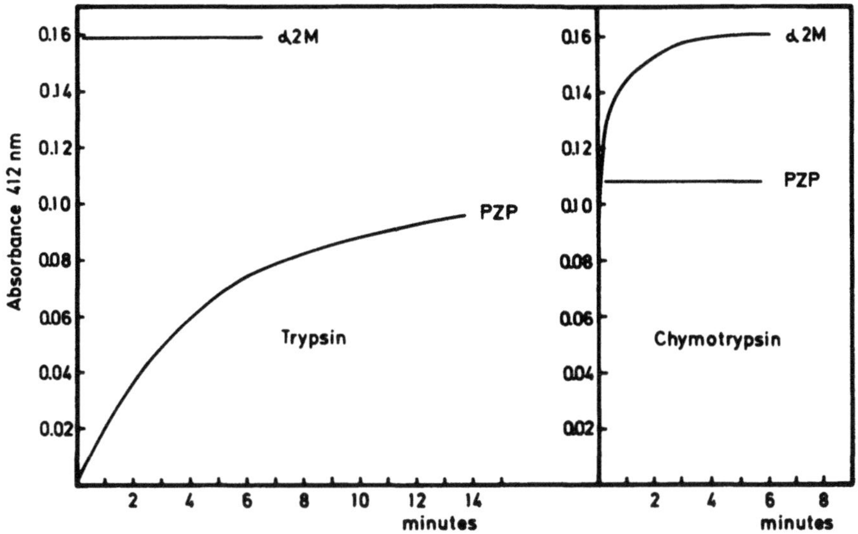

Fig.9. DTNB titration of SH-groups appearing in PZP and α_2M as a result of incubation with trypsin (left panel) or chymotrypsin (right panel). The stock solution of α_2M and PZP in 0.1 M Na-phosphate, pH 8.0 were 2.7 and 2.1 mg/ml, respectively. The molar ratio between proteinase and α_2M or PZP was 2:1. After mixing 552 µl α_2M or PZP and 18 µl 10 mM DTNB the reaction was initiated by the addition of 30 µl proteinase. The change in absorbance at 412 nm was recorded in a Zeiss spectrophotometer. The appearance of SH-groups in PZP-trypsin was complete after approx. 35 min.

The mechanism of α_2M-proteinase complex formation is now known in some detail (23,24,46). In brief, following proteolytic cleavage in the "bait" region of α_2M (residues 681-686 (41)) the α_2M tetramer undergoes a conformational change resulting in the generation of tight-fitting binding sites for the proteinase (maximally two per tetramer). When bound the proteinase is still active (47,48). The conformational change further results in the activation of the internal thiol esters, eventually leading to formation of covalent α_2M-proteinase complexes, engaging the thiol esterified Glx-residue and ε-lysyl amino groups of the proteinase being complexed (23,49). These complexes are rapidly cleared from the circulation (50-54). α_2M can form complexes with a wide variety of proteinases differing in their substrate specificity, and the overall rate of complex formation may differ by several orders of magnitude, reflecting the quality of the bait region sequence as a substrate for a particular proteinase (46,55). The results shown in Fig.9 indicate that PZP reacts slowly with trypsin, but very rapidly with chymotrypsin, thus probably excluding trypsin-like proteinases as targets for PZP. At a molar ratio of chymotrypsin and PZP (tetrameric) of less than 1.0 the rapid appearance of SH-groups is stoichiometrically related to the chymotrypsin: PZP ratio, similar to that observed earlier for trypsin and α_2M (31) (results not shown). Fig.10 shows the results of incubating PZP with increasing amounts of chymotrypsin or elastase and analyzing the cleavage pattern of PZP by reducing SDS-PAGE.

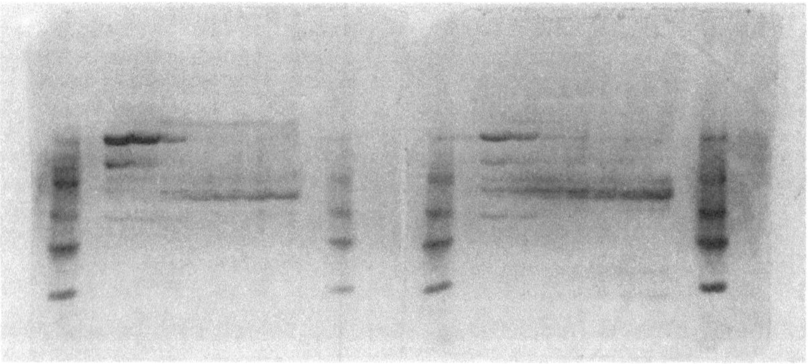

Fig.10. Reducing SDS-PAGE analysis of the cleavage pattern of PZP as a result of incubating PZP with increasing amounts of chymotrypsin (left gel) or elastase (right gel) for 2 min. Lanes 1 and 9 contain marker proteins (α_2M, 180 kDa; plasminogen, 92 kDa; serum albumin, 68 kDa; ovalbumin, 43 kDa, carbonic anhydrase, 29 kDa and ribonuclease, 14 kDa). Lanes 2-8 represent PZP incubated with the enzymes in the following molar enzyme/PZP ratios: 0.12, 0.25, 0.49, 0.98, 1.48, 1.97 and 2.46, respectively.

The pattern is analogous to that found in α_2M (13,14,23) and shows that PZP contains a bait region, which is cleaved in a stoichiometric fashion to produce a major approx. 85 kDa fragment as also observed earlier (18). Complete cleavage was found at a molar ratio of enzyme and PZP close to 1.0. Incubating PZP with chymotrypsin or elastase led to a characteristic doublet pattern in gel chromatography, similar to that seen with aged PZP preparations (Fig.11A,B).

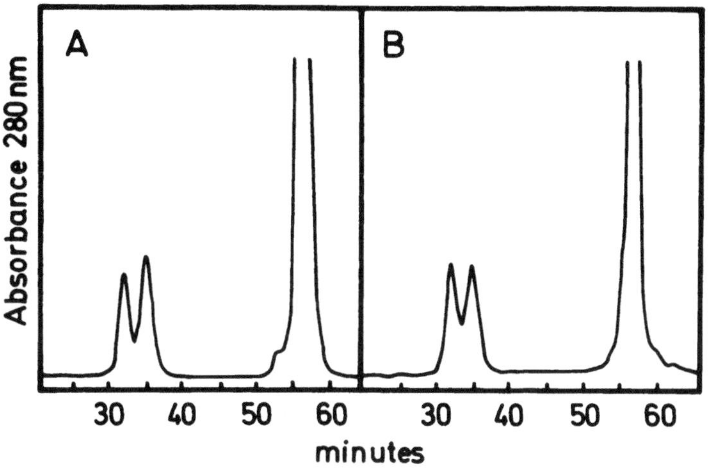

Fig.11. Gel chromatography of PZP-chymotrypsin (A) and PZP-elastase (B) on a column of TSK G-4000 SW, equilibrated and eluted as described in the legend to Fig.6. The molar ratio between proteinase and PZP was 2.0 for chymotrypsin and 1.78 for elastase. The two peaks, representing tetrameric and dimeric PZP-proteinase complexes elute after 32 and 35 min, respectively.

The pattern of peaks indicates that the incubation of PZP with both enzymes results in conformational changes in PZP. The formation of PZP-proteinase complexes was evident from gel chromatography experiments in which PZP was incubated with a variety of $[^{125}I]$-labelled proteinases. Like α_2M these complexes contained a fraction of covalently bound proteinase. These experiments revealed that the labelled proteinases were bound to both dimeric and tetrameric PZP. As demonstrated earlier for α_2M (31) no complex formation would take place between methylamine treated PZP and proteinases. Furthermore, the bound proteinase in PZP-proteinase complexes is enzymatically active (results not shown).

Discussion

The plasma of several animal species contains two related α-macroglobulins, e.g. rat α_1M and α_2M (19-21), rabbit α_1M and α_2M (56,57), dog α_1M and α_2M (58) and pig "slow" and "fast" macroglobulin (59). These proteins are both structurally and functionally related to human α_2M. Although the resemblance between PZP and α_2M has been considered briefly (11,60), the possible relation between these proteins has not been recognized neither in the α_2M literature nor in the literature on pregnancy associated plasma proteins. In order to investigate whether PZP and α_2M would be structurally related we prepared a random set of tryptic peptides from PZP for comparison with the α_2M structure. From the partial or complete sequence analysis of 38 peptides, covering 685 residues of the expected approx. 1450 residues of the PZP subunit we obtained definitive evidence that PZP and α_2M indeed are strongly related in terms of primary structure (18). The degree of identically placed residues in the sequenced stretches was 68%, indicating a close evolutionary relationship between PZP and α_2M. Thus PZP is a member of the class of α_2M related plasma proteins, which also includes the complement components C3 and C4 (61). From these results we suggested that PZP most likely would contain an internal reactive β-cysteinyl-γ-glutamyl thiol ester and would be able to react with and bind proteinases (18).

Although PZP has been studied intensely, no clear indication of its role in pregnancy has yet emerged. Apparently, low levels of PZP will result in an increased risk of spontaneous abortion (62-65), and it has been speculated that PZP might be involved in the maintenance of the immunologically priviledged status of the fetus. In this respect PZP apparently has immunosuppressive properties (65-68) like α_2M (69-70). Since plasma levels of PZP seem to be elevated in a number of malignant conditions (71-73), PZP resembles an acute phase reactant. Of the two αM's found in rats and rabbits rat α_2M and rabbit α_1M are typical acute phase reactants, since their plasma levels are greatly elevated in experimental inflammation (19-21,74-77). PZP could possibly be the human equivalent to these acute phase αM's (18).

Since PZP prepared earlier may have been contaminated with α_2M, or worse, has been subjected to acidic chaotropic conditions the interpretation of many experiments dealing with the interaction between PZP and cellular systems is doubtful. Evidently a reexamination is required using immunologically pure and native PZP, from which well defined molecular forms can be generated, following complex formation with proteinases, treatment with methylamine, etc. For these reasons we have now devised a procedure for the purification of native PZP from late pregnancy plasma or serum. With these preparations we have shown here that PZP as suggested earlier contains an internal β-cysteinyl-γ-glutamyl thiol ester and an exposed activation region (a bait region) in each 180 kDa subunit like α_2M. Native PZP does form complexes with a wide spectrum of proteinases, in which the proteinase is still active. Although the proteinase binding properties of PZP and α_2M are similar, it is likely that the inhibitory spectrum of PZP and α_2M will turn out to be slightly different. Thus, PZP does not seem to be important for complexing proteinases with trypsin-like specificity. The finding that the dimeric units of PZP, in contrast to α_2M, appear to be only loosely associated is a distinct feature of PZP and could point towards a functional difference for PZP and α_2M, perhaps with regard to interaction with cellular receptors.

The functional significance of having two human α-macroglobulins both of which can bind proteinases is presently unclear. However, the fact that the synthesis of α_2M seems to be constitutive (78) while the synthesis of PZP seems to be under hormonal control, possibly by estrogens during pregnancy (79, 80) or perhaps by other hormones during the acute phase response (81) suggests distinct physiological roles. For defining the role of PZP in pregnancy and in the acute phase response it will be important to identify not only the particular proteinases, which are complexed to PZP *in vivo*, but also the cells and tissues, which will interact with these complexes. This may eventually lead to an understanding of the potential of PZP (and perhaps also α_2M) for the control of those cellular events in the immune response and the acute phase response, which depend on proteinases.

Abbreviations: PZP, Human Pregnancy Zone Protein; α_2M: Human α_2-Macroglobulin; PEG, polyethylene glycol (4000 Da in average); SDS, sodium dodecyl sulfate; PAGE, polyacrylamide gel electrophoresis; DTNB, 5,5'dithiobis-2-nitrobenzoic acid; IEP, immuno electrophoresis.

Acknowledgements

Excellent technical assistance was given by Lene Kristensen and Bente J.Christensen. Birthe Hother Nielsen typed the manuscript and Else S.Christensen prepared the figures. We thank Drs. Karsten Rasmussen and K.Brogaard Hansen, Aarhus University Hospital, for supplying us with pooled late pregnancy serum. This work was supported by grants from the Harboe Foundation and The Novo Foundation.

REFERENCES

1. Pregnancy Proteins: Biology, Chemistry and Clinical Applications (1982) (Gruzinskas J.G., Teisner, B. and Seppala, M. eds), Academic Press, Sydney, Australia.

2. Berne, B.H., Bohn, H., Hofmann, R., Klausch, B., Straube, W., Horne, C.H.W., Kasukawa, R., Rittner, C., Stimson, W.H. and Than, G.N. 1975. Lancet i 367-368.

3. Folkersen, J., Teisner, B., Grunnet, N., Gruzinskas, J. G., Westergaard J.G. and Hindersson, P. 1984. Clin.Chim. Acta 110, 139-145.

4. Von Schoultz, B. 1974. Am.J.Obstet.Gynecol. 119, 792-797.

5. Smithies, O. 1959. Adv.Prot.Chem. 14, 65-113.

6. Cooper, D.W. 1963. Nature 200, 892.

7. Bohn, H. 1971. Arch.Gynak. 210, 440-457.

8. Stimson, W.H. and Eubank-Scott, L. 1972. FEBS Lett. 23, 298-302.

9. Bohn, H. and Winckler, W. 1976. Blut 33, 377-388.

10. Von Schoultz, B. and Stigbrand, T. 1974. Biochim.Biophys. Acta 359, 303-310.

11. Stimson, W.H. and Farquharson, D.M. 1978. Int.J.Biochem. 9, 839-843.

12. Jones, J.M., Creeth, J.M. and Kekwick, R.A. 1972. Biochem.J. 127, 187-197.

13. Harpel, P.C. 1973. J.Exp.Med. 138, 508-521.

14. Swenson, R.P. and Howard, J.B. 1979. J.Biol.Chem. 254, 4452-4456.

15. Sottrup-Jensen, L., Stepanik, T.M., Jones, C.M., Petersen, T.E. and Magnusson, S. 1979. In: The Physiological Inhibitors of Coagulation and Fibrinolysis (Collen, D., Wiman, B. and Verstraete, M., eds) Elsevier/North Holland pp.255-271.

16. Westrom, B.R., Karlsson, B.W. and Ohlsson, K. 1982. Hoppe Seyler's Z.Physiol.Chem. 364, 375-381.

17. Carlsson, L., Folkersen, J. and Stigbrand, T.(1985. Molec.Immunol. (in press).

18. Sottrup-Jensen, L., Folkersen, J., Kristensen, T. and Tack, B.F. 1984. Proc.Natl.Acad.Sci.USA 81, 7353-7357.

19. Ganrot, K. 1973. Biochim.Biophys.Acta 295, 245-251.

20. Gordon, A.H. 1976. Biochem.J. 159, 643-650.

21. Nelles, L.P. and Schnebli, H.P. 1982. Hoppe Seyler's Z. Physiol.Chem. 363, 677-682.

22. Sottrup-Jensen, L., Stepanik, J.M., Jones, C.M., Lønblad, P.B., Kristensen, T. and Wierzbicki, D.M. 1984. J.Biol. Chem. 259, 8293-8303.

23. Sottrup-Jensen, L., Hansen, H.F. and Christensen, U. 1983. Ann.New York Acad. Sci. 421, 188-208.

24. Sottrup-Jensen, L., Petersen, T.E. and Magnusson, S. 1981. FEBS Lett. 128, 127-132.

25. Axelsen, N.H., Krøll, J. and Weeke, B. 1973. Scand.J.Immunol. 2, suppl. 1, 1-169.

26. Porath, J., Carlsson, J., Ohlsson, I. and Belfrage, G. (1975) Nature 258, 598-599.

27. Sottrup-Jensen, L., Stepanik, T.M., Wierzbicki, D.M., Jones, C.M., Lønblad, P.B., Kristensen, T., Mortensen, S.B., Petersen, T.E. and Magnusson, S. 1983. Ann.New York Acad.Sci. 421, 41-60.

28. Deutsch, D.G. and Mertz, E.T. 1970. Science 170, 1095-1096.

29. Stigbrand, T., Damber, M.-G. and Von Schoultz, B. 1978. Acta Chem.Scand. B32, 717-719.

30. Folkersen, J., Teisner, B., Ahrons, S. and Svehag, S.-E. 1978. J.Immunol.Meth. 23, 117-125.

31. Sottrup-Jensen, L., Petersen, T.E. and Magnusson, S. 1980 FEBS Lett. 121, 275-279.

32. Sinosich, M.J., Davey, M.W., Teisner, B. and Grudzinskas, J.G. 1983. Biochem.Internat. 7, 33-42.

33. Hall, P.K. and Roberts, R.C. 1978. Biochem.J. 171, 27-38.

34. Harpel, P.C., Hayes, M.B. and Hugli, T.E. 1979. J.Biol. Chem. 254, 8669-8678.

35. Howard, J.B., Vermeulen, M. and Swenson, R.P. 1980. J. Biol.Chem. 255, 3820-3823.

36. Van Leuven, F., Cassiman, J.-J. and Van Den Berghe, H. 1981. J.Biol.Chem. 256, 9016-9022.

37. Barrett, A.J., Brown, M.A. and Sayers, C.A. 1979. Biochem.J. 181, 401-418.

38. Arnaout, M.A., Melamed, J., Tack, B.F. and Colten, H.R. 1981. J.Immunol. 127, 1348-1354.

39. Campbell, R.D., Dodds, A.W. and Porter, R.R. 1980. Biochem.J. 189, 67-80.

40. Sottrup-Jensen, L., Hansen, H.F., Mortensen, S.B., Petersen, T.E. and Magnusson, S. 1981. FEBS Lett. 123, 145-148.

41. Sottrup-Jensen, L., Stepanik, T.M., Kristensen, T., Wierzbicki, D.M., Jones, C.M., Lønblad, P.B., Magnusson, S. and Petersen, T.E. 1984. J.Biol.Chem. 259, 8318-8327.

42. Swenson, R.P. and Howard, J.B. 1980. J.Biol.Chem. 255, 8087-8091.

43. Tack, B.F., Harrison, R.A., Janatova, J. Thomas, M.L. and Prahl, J.W. 1980. Proc.Natl.Acad.Sci.USA 77, 5764-5768.

44. Campbell, R.D., Gagnon, J. and Porter, R.R. 1981. Biochem.J. 199, 359-370.

45. Harrison, R.A., Thomas, M.L. and Tack, B.F. 1981. Proc. Natl.Acad.Sci.USA 78, 7388-7392.

46. Christensen, U. and Sottrup-Jensen, L. 1985. Biochemistry 23, 6619-6626.

47. Mehl, J.W., O'Connell, W. and Degroot, J. 1964. Science 145, 821-822.

48. Ganrot, P.O. 1966. Clin.Chim.Acta 14, 493-501.

49. Sottrup-Jensen, L. and Hansen, H.F. 1982. Biochem.Biophys.Res.Commun. 107, 93-100.

50. Debanne, M.T., Bell, R. and Dolovich, J. 1975. Biochim. Biophys.Acta 411, 295-304.

51. Van Leuven, F., Cassiman, J.-J. and Van den Berghe, H. 1979. J.Biol.Chem. 254, 5155-5160.

52. Kaplan, J. and Nielsen, M.L. 1979. J.Biol.Chem. 154, 7323-7328.

53. Ohlsson, K. 1971. Acta Physicol.Scand. 81, 269-272.

54. Imber, M.T. and Pizzo, S.V. 1981. J.Biol.Chem. 256, 8134-8139.

55. Howell, J.B., Beck, T., Bates, B. and Hunter, M.J. 1983. Arch.Biochem.Biophys. 221, 261-270.

56. Lebreton de Vonne, T. and Mouray, M.H. 1968. C.R. Acad. Sci.Paris Ser.D. 266, 1076-1079.

57. Bloth, B., Chesebro, B. and Svehag, S.-E. 1968. J.Exp. Med. 127, 749-756.

58. Ohlsson, K. 1971. Biochim.Biophys.Acta 236, 84-91.

59. Westrom, B. 1979. Hoppe Seyler's Z.Physiol.Chem. 360, 1869-1878.

60. Starkey, P.M. and Barrett, A.J. 1977. In: Proteinases in Mammalian Cells and Tissues (Barrett, A.J., ed.). Elsevier/North Holland, Biomedical Press, pp.663-696.

61. Sottrup-Jensen, L., Stepanik, T.M., Kristensen, T., Lønblad, P.B., Jones, C.M., Wierzbicki, D.M., Magnusson, S., Domdey, H., Wetsel, R.A., Lundwall, Å., Tack, B.F. and Fey, G.H. 1985. Proc.Natl.Acad.Sci.USA 82, 9-13.

62. Berne, B.H. 1976. Prot.Biol.Fluids 24, 165-179.

63. Beckman, G., Beckman, L., Magnusson, S.S. and Von Schoultz, B. 1974. Acta Obstet.Gynecol.Scand. 53: 177-180.

64. Damber, M.-G., Von Schoultz, B., Solheim, F., Stigbrand, T. and Carlstrom, K. 1978. Obstet.Gynecol. 51, 677-681.

65. Than, G.N., Csaba, I.F., Karg, N.J., Szabo, D.O. and Novak, F. 1975. I.R.C.S.Med.Sci. 3, 94.

66. Stimson, W.H. 1972. Lancet i, 684.

67. Von Schoultz, B., Stigbrand, T. and Tarnvik, A. 1973. FEBS Lett. 38, 23-26.

68. Stimson, W.H. 1976. Clin.Exp.Immunol. 25, 199-206.

69. James, K. 1980. Trends Biochem.Sci. (February) 43-47.

70. Hubbard, W.J. 1978. Cell.Immunol. 39, 388-398.

71. Than, G.N., Csabo, I.F., Szabo, D.G., Karg, N.J. and Novak, P.F. 1975. Arch.Gynak. 218, 125-130.

72. Gropp, C., Lehmann, F.G., Bauer, H.W. and Haveman, K. 1977. Oncology 34, 267-272.

73. Anderson, J.M., Stimson, W.H., Gettinby, G., Jhunjkunwala, S.K. and Burt, R.W. 1979. Eur.J.Cancer 15, 709-714.

74. Morelis, P., Ambrosioni, J.-C., Got, R. and Fontanges, R. 1969. C.R.Acad.Paris Ser.D 269, 1453-1454.

75. Beaton, G.H., Selby, A.E., Veen, M.J. and Wright, A.M. 1961. J.Biol.Chem. 236, 2005-2008.

76. Heim, W.G. 1962. Nature 193, 941.

77. Weimer, H.E. and Benjamin, D.C. 1965. Amer.J.Physiol. 209, 736-744.

78. Ganrot, P.O. and Bjerre, B. 1967. Obst.Gynecol.Scand. 46, 126-137.

79. Stimson, W.H. and Blackstock, T.L. 1975. Experientia 31, 371-373.

80. Lundgren, E., Damber, M.-G., Roos, G., Von Schoultz, B., Stigbrand, T., Nilsson, K. and Alexander, T.T. 1979. Int. J.Cancer 24, 45-48.

81. Kushner, I. 1982. Ann.New York Acad.Sci. 389, 39-48.

IMMUNOCHEMICAL RELATIONSHIP BETWEEN PREGNANCY-ASSOCIATED α_2-GLYCOPROTEIN AND α_2-MACROGLOBULIN AS DEFINED BY POLY- AND MONOCLONAL ANTIBODIES.

L. Carlsson, T. Stigbrand
Department of Physiological Chemistry, University of Umeå, S-901 87 Umeå, Sweden

J. Folkersen
Institute of Medical Microbiology, University of Odense, Odense, Denmark

Introduction

Serum from most higher animals contain high molecular weight α_2-mobile serum-proteins, known to exert protease inhibiting activity. In several animals like the pig, dog, rat and rabbit two immunochemically distinct α-macroglobulins have been described with slightly different properties. In man only one protease inhibiting α_2-macroglobulin have been identified (1). During human pregnancy, however, two high molecular weight pregnancy-associated α_2-globulins different from human α_2-macroglobulin (α_2M) have been identified in the maternal serum, pregnancy-associated α_2-glycoprotein (PAα_2G) and pregnancy-associated plasma protein A (PAPP-A). PAα_2G is quantitatively dominating (2,3).

In normal non-pregnant females and normal males PAα_2G has been reported to be present in low amounts, below 30 mg/l but increases during pregnancy to approx. 1000 mg/l (4,5). The physiological significance of the 2 high molecular weight α_2-globulins associated with pregnancy is still unknown.

PAα_2G is a glycoprotein, containing approx. 10 % carbohydrate (6) which after reduction of the disulfide bonds gives rise to 180 kDa subunits (3). It has also been suggested that PAα_2G is a tetrameric protein, a dimer of dimers built up by these 180 kDa subunits (7). These data are very similar to those of α_2M. Purification of PAα_2G by conventional or immunosorbent techniques (6,8) have indicated that these molecules also share immunochemical properties since purification of one of these proteins often is crosscontaminated by the other.

Pregnancy Proteins in Animals

Most higher animals (ape, monkey, pig, cow, dog) have proteins immunologically crossreactive with monospecific polyclonal antisera against α_2M and PAα_2G. Recently, porcine α_2M has been shown to crossreact not only with human, dog and rat α_2M but also with human PAα_2G (9).

In lower animals (rat, mouse) the analogous to these pregnancy-associated proteins have been difficult to identify, although pregnancy-associated murine protein 1 (PAMP-1) has been postulated to be an analogue to PAα_2G (10).

The present study was performed to analyze and characterize the immunochemical relationship between PAα_2G and α_2M, and some animal analogues by using combinations of polyclonal and monoclonal antibody technology.

Material and methods

Poly- and monoclonal immunoglobulins. Rabbit polyclonal antisera against PAα_2G were obtained both from DAKO, Copenhagen, Denmark (antihuman pregnancy serum absorbed with male serum) and obtained by immunizing rabbits with purified human PAα_2G (8). Antisera against α_2M were likewise obtained from DAKO.

Monoclonal antibodies were raised by fusing spleen cells from BALB/c mice, immunized with affinity-purified PAα_2G (8) with cells of the X63-Ag8-P635 myeloma line (11) using 50 % polyethyleneglycol (for 1 min) as fusing agent. The mice were immunized with 400 µg of PAα_2G in complete Freund's adjuvant intraperitoneally and were given booster i.v. injections 6 weeks later during 3 consecutive days with 100 µg of antigen. The spleen was removed on the fourth day and prepared for fusion (12).

The identification of specific antibodies from the growing clones after HAT selection were made by and ELISA technique. Coating was performed in 96-well microtiter plates with 1 µg/ml PAα_2G or α_2M in 0.1 M $NaHCO_3$. After washing supernatants (100 µl) were added followed by rabbit anti-mouse immunoglobulin coupled to horseradish peroxidase (HRP) (DAKO) (dilution 1:10,000). As substrates o-phenylenediamine and H_2O_2 in citrate-phosphate buffer, pH 5.0, were used. The absorbance at 450 nm was recorded by a Titertek Multiskan spectrophotometer. The epitope specificities of these monoclonal antibodies have been described elsewhere (13).

Immunoelectrophoretic techniques. Crossed immunoelectrophoresis (14) was performed in 1 % (w/v) agarose in 0.02 M Tris-barbital buffer, pH 8.6. The same buffer was used as electrophoresis buffer. Electrophoresis in the first dimension was performed as a tandem modification (15) at 10 V/cm until a bromphenol blue stained albumin marker had migrated 5.0 cm. An intermediate gel (15) with or without antibodies to α_2M was interposed between the first- and second-dimension gels, the latter which contained absorbed or non-absorbed antibodies to $PA\alpha_2G$. Second-dimension electrophoresis was performed overnight at 2.5 V/cm. Immunoprecipitates were stained with Coomassie Brilliant Blue.

Line immunoelectrophoresis (14) was performed in 1 % (w/v) agarose in 0.02 M Tris-barbital buffer, pH 8.6. The same buffer was used as the electrophoresis buffer. Electrophoresis was performed overnight at 2.5 V/cm. Coomassie Brilliant Blue was used for staining.

Biological samples. Normal and pregnancy sera were obtained from Department of Obstetrics and Gynaecology, University Hospital, Umeå. Term normal and pregnant rhesus monkey sera were obtained from the primate laboratory, University of Uppsala, Sweden. Normal and pregnant rat sera were obtained from Department of Anatomy, University of Umeå, Sweden.

Affinity-purified $PA\alpha_2G$ was prepared as described earlier (8). To eliminate any possible contamination of α_2M the purified $PA\alpha_2G$ was passed through an anti-α_2M column (16). The saturation of the anti-α_2M column was monitored by frontal analysis using rocket immunoelectrophoresis as the test system. Fractions with detectable α_2M (above 2µg/ml) were reperfused until no α_2M was present. Purified α_2M, prepared according to Sottrup-Jensen et al (17), was kindly given to us by Ragnar Österberg, University of Gothenburg, Sweden and purified PAMP-1 (18) by Jan Hau, Copenhagen.

Results

Figure 1 shows crossed-immunoelectrophoresis experiments. In Fig. 1 a only one main precipitate is formed between $PA\alpha_2G$ (from well 2) and anti-$PA\alpha_2G$ including a trailing (arrow) to the left not due to the presence of α_2M (from well 1). Replacement of the monospecific anti-$PA\alpha_2G$ with the unabsorbed crude antiserum raised against approx. 96 % $PA\alpha_2G$ and 2-4 % α_2M (plus a few other contaminants) (Fig. 1b) revealed that the "trailing precipitate" in Fig 1a was changed in morphology due to the contemporaneous presence of the α_2M/anti-α_2M precipitate (double arrow). Figure 1 further shows a complete fusion between the trailing part of the PAα_2G precipitate and the α_2M precipitate with the disappearance of the right foot of the latter precipitate. Inclusion of monospecific anti-α_2M antiserum in the lower intermediate gels of Fig 1c) and d) resulted in a downward movement of the PAα_2G precipitate and formation of a pronounced "merging" precipitate (encircled) with the α_2M precipitate. This was concluded to represent a significant number of crossreactive epitopes on PAα_2G and α_2M.

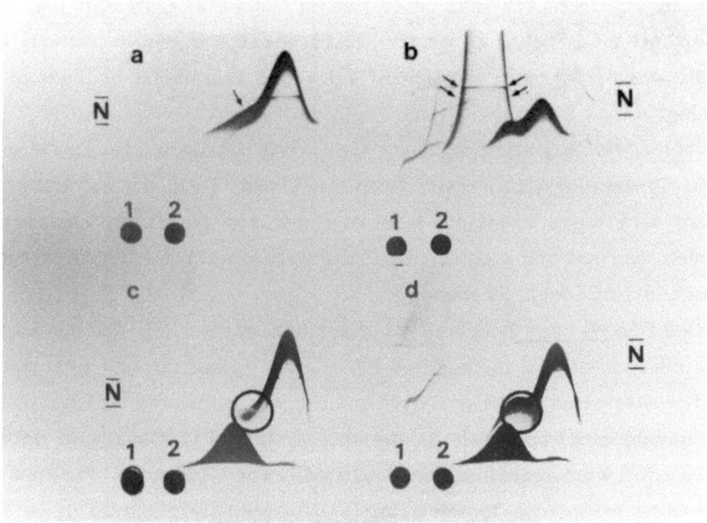

Fig. 1. Crossed-tandem immunoelectrophoresis of 4 µl normal human male serum (without detectable PAα_2G in double-diffusion test) in well 1, and 4 µl eluate from an anti-PAα_2G affinity chromatography in well 2. In a) and c) monospecific (absorbed) anti-PAα_2G was used in the upper part of the gel. In c) and d) anti-α_2M was included in the lower intermediate gels. In b) and c) unabsorbed anti-PAα_2G was used in the upper gels. N denotes neutral gel with no antibodies.

<u>Monoclonal antibodies to PAα_2G.</u> Supernatants from hybridoma cultures were screened for binding by the ELISA technique to PAα_2G, α_2M and PAMP-1 which were used for coating. Supernatants were added under saturating conditions. Altogheter 9 positive clones were obtained from 3 fusions and relevant clones were recloned. Figure 2 summarizes the reactivity of the monoclonal antibodies. Eight of them were reactive in the ELISA with PAα_2G (G5, H3, G3, G6, D7, BB4, B7 and F2), 3 with α_2M

(G6, D7 and B4) of which 1 is only reactive with α_2M (B4). None was demonstrated to show any binding to PAMP-1.

The chain constituents of the monoclonal antibodies were determined by double diffusion. Most of them were IgG_1, κ, i.e. G5, H3, G3, G6, BB4, B7 and F2 except D7 (IgM_κ) and B4 (IgG_{2a}, κ).

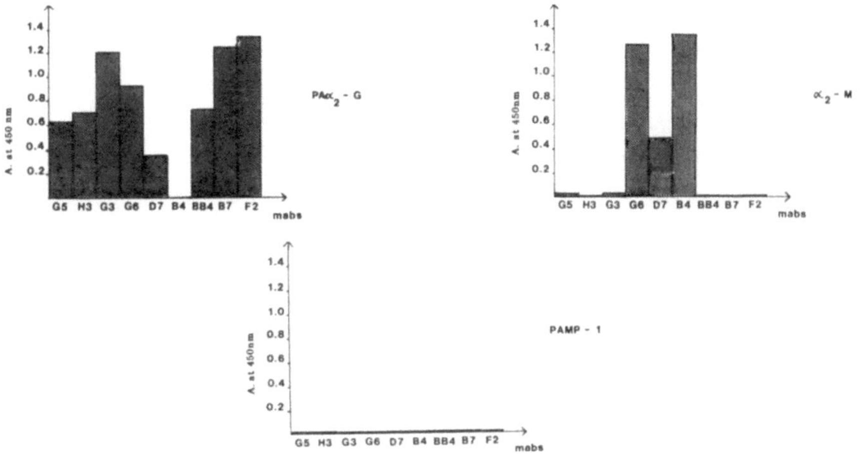

Fig. 2. Reactivity of the 9 monoclonal antibodies to highly purified preparations of $PA\alpha_2G$, α_2M and PAMP-1. Each antigen was used for coating at 1 μ g/ml. Supernatants from the hybridoma cultures were added under saturating conditions. HRP-rabbit anti-mouse immunoglobulin was used in the final step. The plate was developed by the use of H_2O_2 and o-phenylenediamine.

<u>Some animal analogues to $PA\alpha_2G$ and α_2M.</u> Crossreactivity between some animal analogues to $PA\alpha_2G$ and α_2M was tested in line immunoelectrophoresis experiments. Sera from pregnant and non-pregnant rhesus monkey, rat and mouse was tested against rabbit anti-$PA\alpha_2G$ and α_2M. Fig. 3a shows that both purified $PA\alpha_2G$ (wells 5 and 6) and rhesus monkey normal and pregnancy sera (wells 7 and 8) and human pregnancy sera (well 9) gave significant elevation of the baseline. Sera from normal and pregnant mice and rat did not cause an elevation of the baseline indicating that no analogues to this protein are present in these sera (wells 1-4). The same type of experiment is shown in Fig. 3b employing anti-α_2M in the upper gel and a baseline

formed by α_2M in normal male serum and the antiserum. The mouse sera and normal rat serum gave no elevation of the baseline but pregnant rat serum contains 1 component giving a clear precipitate with anti-human α_2M serum (well 4). All human and rhesus monkey sera gave, as expected, clear precipitates with this antiserum.

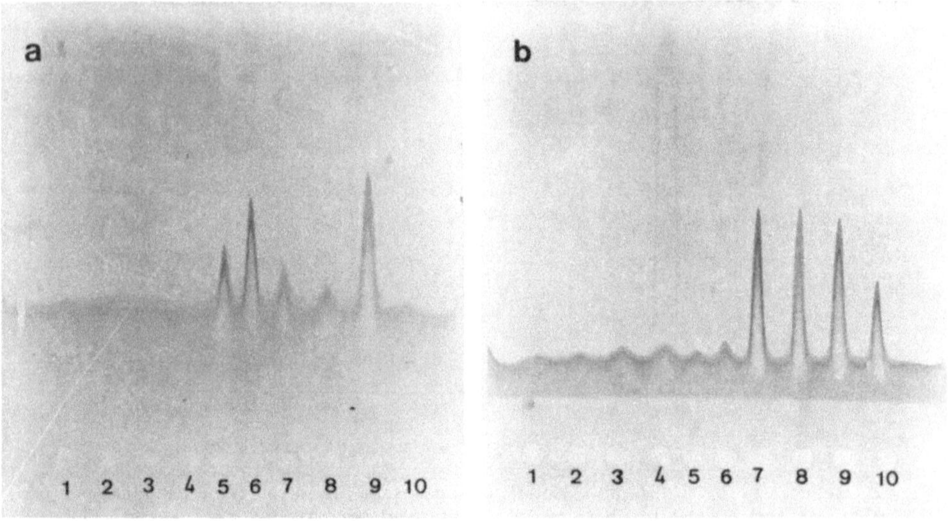

Fig. 3. Line immunoelectrophoresis with different human and animal sera. In a) human pregnancy serum was used in the lower part of the gel and anti-PAα_2G in the upper part of the gel. The sera used were: (1) mouse normal serum, (2) mouse pregnancy serum, (3) rat normal serum, (4) rat pregnancy serum, (5) highly purified PAα_2G (0.6 mg/ml), (6) purified PAα_2G (0.9 mg/ml), (7) rhesus monkey serum, (8) rhesus monkey pregnancy serum, (9) human pregnancy serum, and (10) human male serum. In b) normal human male serum was used in the lower part of the gel and anti-α_2M used in the upper part of the gel.

Discussion

Several observations have indicated evolutionary relatedness between some of the macroglobulins occurring in sera from normal and pregnant mammaliens, although the analogue to PAα_2G in lower animals has been difficult to establish. Usually, in order to obtain monospecific antisera against pregnancy-associated proteins like PAα_2G, absorption with normal male serum is performed. This procedure yields antisera mainly reactive with PAα_2G-specific determinants, but not with determinants common with α_2M.

To further confirm the existence of similar epitopes on α_2M and PAα_2G attempts were made to exclude the possible error induced by the presence of unrelated contaminating antibodies against α_2M in the anti-PAα_2G preparation and unrelated antibodies against PAα_2G in the anti-α_2M preparation. This was done by including a monospecific antibody preparation against α_2M in the lower intermediate gels of Fig. 1c) and d). If there were no common epitopes between α_2M and PAα_2G this arrangement should just move the anti-α_2M precipitate down in the lower intermediate gel without affecting the PAα_2G precipitate. As can be seen from Fig. 1c) and d) this was not the case. The PAα_2G precipitates were moved downwards and a very pronounced "merging" precipitate (encircled in Fig. 1c) formed. From a theoretical point of view the downward movement of the PAα_2G precipitates may be due to both contaminating anti-PAα_2G antibodies in the anti-α_2M precipitate and the presence of cross-reacting antibodies. However, the formation of the "merging" precipitate most likely reflects the existence of a great number of common epitopes on the PAα_2G and α_2M molecule.

The generation of 9 monoclonal antibodies to human PAα_2G confirms the existence of common determinants on both these antigens since 2 out of 9 monoclonal antibodies are reactive with both PAα_2G and α_2M. Five of the others were mainly reactive with PAα_2G alone and 1 only reactive with α_2M.

Recently, the primary structure of PAα_2G partially has been determinated and compared with α_2M, showing a close homology with α_2M (19).

This immunological and sequence homology indicates a close evolutionary relationship between these two macroglobulins.

The potential analogue in the mouse (PAMP-1) to human PAα_2G cannot be identified by immunochemical techniques using either poly- or monoclonal antibodies, which is contrary to earlier observations (10). Negative results were obtained both with the screening ELISA method using purified PAMP-1 for coating, and in double-diffusion tests. Interestingly, however, pregnancy in rats seems to be accompanied by an increase in another α_2-glycoprotein which is clearly crossreactive with antisera against human α_2M. This rat α_2M-glycoprotein has been described as an α_2-acute-phase macroglobulin (20) which has also been suggested for PAα_2G (21). Since the biological function of $PA\alpha_2G$ in man is unknown, the observation in rat of a pregnancy-induced related protein might add significant information to the biological role of $PA\alpha_2G$. It is tempting to postulate that the rat analogue to $PA\alpha_2G$ is α_2-acute-phase macroglobulin which is immunologically closely related to human α_2M. This protein (α_2M) furthermore is conserved during evolution to a much higher degree than $PA\alpha_2G$.

Acknowledgement

The present investigation was supported by grants from the Swedish Cancer Society, the Swedish Medical Research council and the Medical Faculty, University of Umeå.

References

1. Starkey, P.M. and A.J. Barrett. 1977. In: Proteases in Mammalian Cells and Tissues (Ed. A.J. Barrett) pp. 663-696, Vol. 2, North Hollan, N.Y.

2. Stimson, W.H. and L. Eubank-Scott. 1972. FEBS Lett. 23, 298-302.

3. von Schoultz, B. and T. Stigbrand. 1974. Biochim. Biophys. Acta 359, 303.

4. Damber, M.G., von Schoultz, B., Stigbrand, T. and K. Carlström. 1976. Clin. Chim. Acta 66, 85-88.

5. Folkersen, J., Teisner, B., Grunnet, N., Grudzinskas, J.G., Westergaard, J.G. and P. Hinderson. 1981. Clin. Chim. Acta 110, 139-145.

6. von Schoultz, B. and T. Stigbrand. 1973. Acta Obstet. Gynecol. Scand. 52, 51-54.

7. Harpel, P.C., Hayes, M.B. and T.E. Hugli. 1979. J. Biol. Chem. 254, 8665-8678.

8. Folkersen, J., Teisner, B., Ahrons, S. and S.-E. Svehag. 1978. J. Immunol. Meth. 23. 117-125.

9. Weström, B.J., Karlsson, B.W. and K. Ohlsson. 1983. Hoppe-Seyler's Z. Physiol. Chem. 364, 375-381.

10. Hau, J., Westergaard, J.G., Svendsen, P., Bach, A. and B. Teisner. 1981. J. Reprod. Immunol. 3, 341-349.

11. Kearny, J.F., Radbruch, A., Liesegang, B. and K. Rajewsky. 1979. J. Immunol. 123, 1548-1550.

12. Uotila, M., Engvall, E. and E. Ruoslahti. 1980. Mol. Immunol. 17, 791-794.

13. Carlsson, L., Folkersen, J. and Stigbrand, T. 1985. Mol. Immunol. (In press).

14. Laurell, C.B. 1965. Anal. Biochem. 10, 358-361.

15. Axelsen, N.H., Krøll, J. and B. Weeke (Eds) 1973. A Manual of Quantitative Immunoelectrophoresis. Scand. J. Immunol. 2, suppl. No. 1.

16. Folkersen, J., Teisner, B., Hinderson, P. and S.E. Svehag. 1980. J. Immunol. Meth. 32, 327-338.

17. Sottrup-Jensen, L., Petersen, T.E. and S. Magnusson. 1980. FEBS Lett. 121, 275-279.

18. Hau, J., Svendsen, B., Teisner, B. and S.-E. Svehag. 1978. J. Reprod. Fert. 54, 239-243.

19. Sottrup-Jensen, L., Folkersen, J., Kristensen, T. and B.F. Tack. 1984. Proc. natl. Acad. Sci., 81, 7353-7357.

20. Ganrot, K. 1973. Biochim. Biophys. Acta. 295, 245-251.

21. Bohn, H. 1973. Blut 26, 205-209.

THE MAJOR SOURCE OF PLACENTAL PROTEIN 12 IS DECIDUA/ENDOMETRIUM

E-M. Rutanen, M. Menabawey, T. Chard, J.G. Grudzinskas
Joint Academic Unit of Obstetrics, Gynaecology and Reproductive Physiology, The London Hospital and St. Bartholomew's Hospital Medical Colleges, London, United Kingdom

M. Seppälä
Department 1 of Obstetrics and Gynecology, University of Helsinki, Finland

H. Bohn
Behringwerke AG, Marburg, W. Germany

Introduction

Placental protein 12 (PP12) was originally isolated and purified from soluble extracts of term human placenta and membranes (1). Molecular weight estimates of PP12 range from 25,000 to 51,000 (1); the carbohydrate content is 4.3% and the electrophoretic mobility is that of an alpha-1-globulin. Immunohistochemical studies have shown PP12 in placental syncytiotrophoblast, decidua, and chorion and amnion (2,3). PP12 is also found in the non-pregnant endometrium and ovary (4,5), whereas all other adult tissues are negative. Here, we present data on the content of PP12 in endometrial, decidual and trophoblastic tissues, and demonstrate PP12 synthesis by early pregnancy decidua in tissue culture.

Material and methods

Endometrial samples were obtained from 50 non-pregnant women, aged 35 to 49 years, who underwent curettage or hysterectomy for benign pelvic disease. Part of each tissue was fixed in buffered formalin for histologic dating of the endometrium. Specimens of decidua and trophoblast were collected from 16 women who had elective termination by dilatation and curettage, and

Pregnancy Proteins in Animals

one pair of decidual and trophoblastic tissue samples was obtained from a spontaneous abortion at 17 weeks gestation. Fetal membranes, decidual and trophoblastic tissues were obtained from 15 normal term pregnancies after spontaneous delivery or elective Caesarean section. In the latter cases, the amnion was first separated from the chorion by peeling, and the decidua was then scraped from the maternal surface of chorion with a scalpel. Tissues were rinsed immediately in cold PBS, pH 7.4, frozen in liquid nitrogen, and stored at -20°C until used for protein and PP12 measurements. A piece of each tissue was fixed in buffered formalin for histologic identification.

PP12 synthesis

Synthesis and release of PP12 was studied in tissue culture using explants of decidua and trophoblast from six early pregnancies (weeks 6 to 11). Part of the tissue was fixed in formalin for histologic examination and part was frozen in liquid nitrogen until assay. Tissues for culture were rinsed three times with culture medium and minced into 2 x 2 mm explants. Each experiment included six parallel cultures in polystyrene multiwell plates; each well contained approximately 50 mg of explants and 5 ml Ham's F10 medium supplemented with 10% fetal calf serum, glutamine (290µg/ml) and antibiotics. In two experiments, three of the six wells contained a protein synthesis inhibitor (cycloheximide, 50 µM). After 28-h incubation at 37°C in 5% CO_2-95% air the media and tissues were processed as described elsewhere (6). PP12 concentrations were measured by radioimmunoassay (7,8). The sensitivity of the assay was 7.8 to 15.6 ng/ml. No cross-reaction was observed with hCG, human placental lactogen, pregnancy specific beta-1-glycoprotein (SP1), placental protein 5 or human pituitary prolactin.

<u>De novo</u> synthesis of PP12 was assessed by incorporation of ^{35}S methionine. After 8 h preincubation in Ham's F10 the medium was changed to minimal essential medium (MEM) (80% without methionine plus 20% with methionine) containing 15 µCi/ml ^{35}S methionine. Incubation was continued for an additional 20 h. Media and tissues were then processed as described previously (6). [^{35}S]PP12 was determined by immunoprecipitation. A sample of medium (0.4 ml)

was incubated with 20µl anti-PP12(A) antiserum (Behringwerke,AG, FRG) or 20µl normal rabbit serum in the presence of 0.5% Triton X-100 at room temperature overnight. 100 µl 20% (w/v) Protein-A Sepharose was then added and incubated for an additional 2 h. After washing twice in PBS with 0.5% Triton X and twice in 0.05M Tris-HCl, pH 6.8, the pellet was boiled in sodium dodecyl sulfate (SDS) buffer at 100°C for 2 min and analyzed in SDS polyacrylamide gel electrophoresis (9) . The migration mobility of newly synthesized PP12 was compared to that of ^{14}C-methylated marker proteins. The gel was examined by fluorography (10).

Results

Tissue content of PP12

PP12 was not detected in proliferative phase endometrium. In the secretory phase, 22 of 35 tissues examined contained detectable amounts of PP12; both the frequency of positive results and the concentration of PP12 increased towards the end of the cycle (Table 1).

Table 1 Concentrations of PP12 in Proliferative and Secretory Phase Endometria

Phase	No	PP12 detected no	%	PP12 (µg/g protein) (range)
Proliferative	15	0		
Early secretory	10	3	30	2.0 - 9.4
Mid secretory	12	8	67	1.1 - 21
Late secretory	13	11	85	5.3 - 900

Decidua from early pregnancy contained 80 times more PP12 than late secretory phase endometrium. PP12 content was higher in decidua than in trophoblast throughout pregnancy (Fig. 1). In early pregnancy, the ratio of decidual to trophoblastic PP12 was 50 (±12SE) and in term pregnancy it was 400 (±102SE) the difference being due to the lower PP12 concentration in late pregnancy. The

PP12 content of amnion and chorion laeve was similar to that in decidua.

Synthesis and release of PP12 in tissue culture

The initial PP12 concentration in decidua from early pregnancy was 4928±(555) ng/100 mg tissue. After 28 h incubation the total amount of PP12 released into the medium was 16556±(4635) ng/100 mg tissue and the residual PP12 in the tissue was 735(±253) ng/100 mg. PP12 release was reduced in the presence of cycloheximide in the culture medium. ^{35}S methionine was incorporated into immunoprecipitable PP12. In SDS-polyacrylamide gel electrophoresis the immunoprecipitated material had a migration mobility corresponding to an apparent molecular weight of 29-30000 (Fig.2). Trophoblast explants from the same pregnancies showed no incorporation of ^{35}S methionine into PP12.

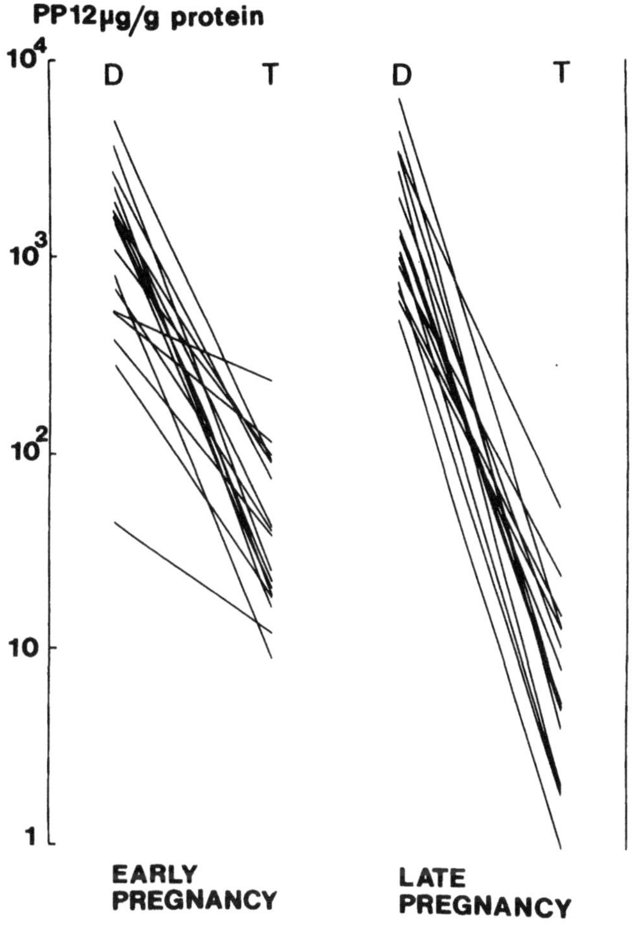

Figure 1. Concentrations of PP12 (µg/g protein) in extracts of matched decidua (D) and trophoblast (T) from early (weeks 7-17) and late (weeks 37-40) pregnancy.

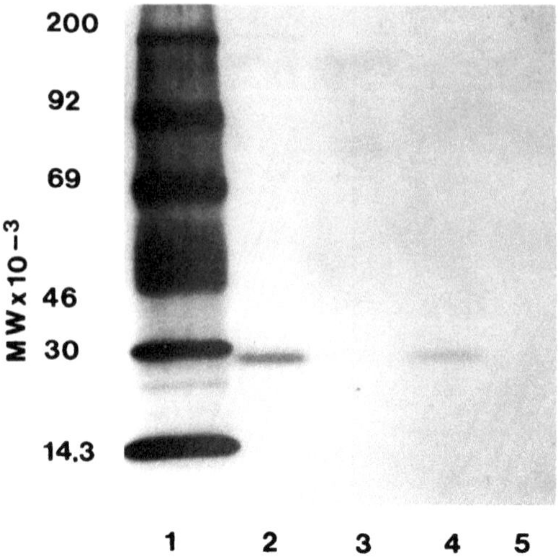

Figure 2. Fluorography of immunoprecipitated [^{35}S]PP12 from two different decidual culture media. Lane 1: ^{14}C-methylated marker proteins. Lanes 2 and 4: Immunoprecipitated [^{35}S]PP12. Lanes 3 and 5: No band was observed when normal rabbit serum was used instead of anti-PP12 antiserum.

Discussion

The present and previous findings (4,6,11-13) suggest that the term 'placental' protein 12 is a misnomer. The confusion has arisen because PP12 was initially isolated from soluble extracts of term human placentae (1). This material contains tissues from two different origins: trophoblast, and membranes from the fetus and decidua overlying the chorion laeve from the mother. The gradient of PP12 from decidua to trophoblast clearly shows that the former is the site of origin of PP12. Furthermore, term pregnancy decidua, but not trophoblast, can synthesize PP12 in tissue culture. In addition, the endometrium contains PP12 during the secretory phase of a non-fertile menstrual cycle; the amount increases towards the end of the secretory phase. This agrees with the results obtained by immunohistochemistry which show appearance of PP12 in the endometrium on the fourth post-ovulatory day (4). The cellular origin of PP12 is still uncertain. In secretory phase endometrium, the positive staining has been observed in the glandular epithelium (4). During pregnancy stromal cells also show positive staining (3).

Acknowledgement

This work was supported by grants from the Sigrid Jusélius Foundation, the Wellcome Trust, and the Academy of Finland.

References

1. Bohn, H., V. Kraus. 1980. Arch. Gynecol. 229, 279

2. Inaba, N., T. Renk, H. Bohn. 1980. Arch. Gynecol. 230, 109.

3. Wahlstrom, T., H. Bohn, M. Seppala. 1982. In: Pregnancy Proteins: Biology, Chemistry and Clinical Application (J.G. Grudzinskas, B. Teisner and M. Seppälä, eds.). Academic Press, Sydney, p.415.

4. Wahlström, T., M. Seppälä. 1984. Fertil. Steril. 41, 781.

5. Seppälä, M., T. Wahlström, A.I. Koskimies, A. Tenhunen, E-M. Rutanen, R. Koistinen, I. Huhtaniemi, H. Bohn, U-H. Stenman. 1984. J. Clin. Endocrinol. Metab. 58, 505.

6. Rutanen, E-M., R. Koistinen, T. Wahlström, H. Bohn, T. Ranta, M. Seppälä. 1985. Endocrinology 116; 1304.

7. Rutanen, E-M., H. Bohn, M. Seppälä. 1982. Am.J.Obstet.Gynecol. 144, 460.

8. Howell, R., L. Perry, N. Choglay, H. Bohn, T. Chard. Placental protein 12 (PP12): A new test for the prediction of the small for gestational age infant. Br. J. Obstet.Gynaecol. In press

9. Laemmli, U.K., 1979. Nature. 227, 680.

10. Bonner, W.M., R.A. Laskey. 1974. Eur. J. Biochem. 46, 83.

11. Rutanen, E-M., R. Koistinen, T. Wahlström, J. Sjöberg. U-H. Stenman, M. Seppälä. 1984. Brit.J. Obstet. Gynaecol. 91,377.

12. Rutanen, E-M., R. Koistinen, T. Wahlström, H. Bohn, M.Seppälä. 1984. Brit. J. Obstet. Gynaecol. 91, 1240.

13. Rutanen, E-M., R. Koistinen, T. Wahlström, U-H. Stenman, M.Seppälä. 1984. Brit. J. Obstet. Gynaecol. 91, 1025.

PROGESTAGEN-ASSOCIATED ENDOMETRIAL PROTEIN: DETECTION AND BINDING IN IMMATURE HUMAN PLACENTA

R.S. Raikar Ph.D., Janet F. Bank M.S. and S.G. Joshi Ph.D.

Departments of Obstetrics & Gynecology, and Biochemistry
Neil Hellman Medical Research Building
Albany Medical College
Albany, New York 12208

Introduction

In pregnant women, the implantation of the fertilized ovum is followed by rapid growth and differentiation of endometrial cells and placental trophoblastic cells. The endometrial activity during early pregnancy is regulated mainly by progesterone secreted by the corpus luteum (CL). The CL activity, in turn, is supported by the human chorionic gonadotropin (hCG) secreted by the trophoblastic cells of placenta. We have previously demonstrated that the uterine decidua of early pregnancy synthesizes and secretes a protein which we have designated Progestagen-associated endometrial protein or "PEP" (1). This protein was detected in the peripheral blood of cycling and pregnant women. Furthermore, we found that the serum concentrations of PEP increased rapidly during early pregnancy (4-8 weeks) at a rate comparable to that of hCG (2). This finding suggested that the placenta also synthesizes PEP or that PEP regulates the synthesis and secretion of placental hormones, including hCG. In this preliminary communication we have studied the possibility that immature placenta synthesizes or binds PEP.

Materials and methods

Immature human placentae, obtained from women undergoing elective termination of early pregnancy (6-10 weeks), were washed several times with Hank's balanced salt solution (Grand Island Biological Company, Grand Island, NY) containing streptomycin sulfate, penicillin G and mycostatin B, and dissected free of decidual and fetal components. Portions of the washed tissues were utilized for immunochemical localization of PEP and hCG, organ culture, and binding studies.

PEP and hCG were localized in tissues by the immunoperoxidase technique described elsewhere (3).

The ability of placental tissues to synthesize PEP was studied by the organ culture method and viability of the cultured tissue implants was assessed by the measurement of amount of hCG or PEP released into the culture medium by the tissue. In each of the five experiments (expt. nos. 1-5) explants of villous tissue from one pregnant women (100-120mg wet weight) were cultured for 5 days by the method of Huot et al (4). The spent media were clarified by centrifugation and stored at $-10^{o}C$ until used for PEP and hCG estimations. The cultured tissues were homogenized in PBS (0.01M phosphate, 0.154M NaCl, pH7.4) and centrifuged at high speed (130,000g for 60 min, $4^{o}C$). The clear supernates and tissue extracts were stored at $-10^{o}C$ to be subsequently assayed for PEP and hCG by specific radioimmunoassays as described elsewhere (2).

Binding of PEP to placental tissue was studied by an indirect method in which crude membranes isolated from the placental tissue were exposed to non-radioactive, partially purified PEP and the PEP which was bound to the membranes was quantitated by further exposing the membranes to radioiodinated, purified anti-PEP antibodies. The reagents required for binding studies and the cell membranes were prepared as follows:

PEP was partially purified from uterine decidua of early pregnancy as described elsewhere (5).

Anti-PEP antibodies were purified from rabbit anti-PEP serum by immunoaffinity chromatography, using partially purified PEP coupled to CNBr-activated Sepharose 4B (Pharmacia Fine chemicals) as the immunoabsorbent. The purified antibodies were radioiodinated to specific activity of approximately 30uCi/ug, using Iodobeads (Pierce Chemical Co. Rockford, Illinois, USA), as described by Markwell (6).

Cell membranes from placental tissue were prepared by the method essentially similar to that described by Arogona and Friesen (7). In brief, the placental tissue was washed thoroughly and hand homogenized in a teflon-glass homogenizer in five volumes of ice-cold 25mM Tris HCl, pH7.4, containing 0.25M Sucrose, 10mM $MgCl_2$ and 1mM p-hydroxymercury benzoate (homogenizing buffer). After filtering through four layers of cheesecloth the homogenate was centrifuged at $4^{o}C$ for 15 min at 700xg. The pellet was discarded and the supernatant was further centrifuged at 1000xg for 30 min at $4^{o}C$. The sediment was designated as "membrane fraction I". A portion of 1000xg supernatant was centrifuged further at 7000xg for 30 min at 4°C. The supernatant was discarded and the sediment was designated as "membrane fraction II". Both membrane fractions were suspended in homogenizing buffer supplemented with 0.01% (w/v) NaN_3. Aliquots of the suspensions of the membrane fractions were stored for up to 10 days at -20°C until used.

Portions of membrane fractions I and II were treated with 5M $MgCl_2$ for 2 hours at 4°C as described by Katikineni et al (8) and the treated membranes were centrifuged at appropriate speeds and the supernates were saved for PEP analysis. The sediments containing PEP-stripped membranes were washed twice and suspended in binding assay buffer (25mM Tris HCl, pH7.4 containing 10mM $MgCl_2$ and 0.1% BSA). In all binding experiments Fc receptors on membranes were blocked by incubating aliquots (1ml) of membrane suspensions (1mg protein) with diluted normal rabbit serum (1:20 dilution in assay buffer used as a source on non-immune IgG) in a ratio of 1:2 (v/v). After 3 hours incubation at 4°C the membranes were washed, resuspended in binding assay buffer and used for binding studies.

In binding experiments, triplicate aliquots (100ul) of membrane fraction I or II were incubated at 37°C with 40µl of non-radioactive PEP (100µg) or diluted adult male serum (lacking PEP; 1:4 dilution is assay buffer) for 30 minutes in a final volume of 200µl of assay buffer. Following incubation the membranes were washed twice and resuspended in 200µl of assay buffer. The amount of PEP bound to membranes was quantitated by incubating the membranes at 37°C for 60 minutes with excess ^{125}I rabbit anti-PEP (100,000cpm) in a final volume of 300µl of assay buffer. At the end of the incubation period the membranes were washed twice with the binding assay buffer and the radioactivity in the membranes was counted in a gamma counter. The total binding (TB) of PEP to membranes represents the amount of ^{125}I anti-PEP bound to membranes preexposed to PEP. The non-specific binding (NSB) represents the amount of ^{125}I anti-PEP bound to membranes preexposed to diluted male serum. Specific binding equals TB minus NSB.

Membrane protein concentrations were determined by solubilizing membrane proteins with 0.1M NaOH solution containing 0.1%(w/v) sodium docecyl sulfate at 100°C for 30 minutes, and protein was determined by the protein-dye binding method described by Bradford (9).

Results

In immunochemical studies, the sections of the placental villous tissue stained for PEP showed a positive immune reaction in the syncytiotrophoblast as well as in the stromal connective tissue, the reaction being more intense in the syncytiotrophoblast (Fig 1) whereas the sections stained for hCG showed a strong positive reaction only in the cytoplasm of the syncytiotrophoblasts but not in stromal tissue. The tissue sections exposed to nonimmune rabbit serum, which served as controls, showed no reaction (results now shown).

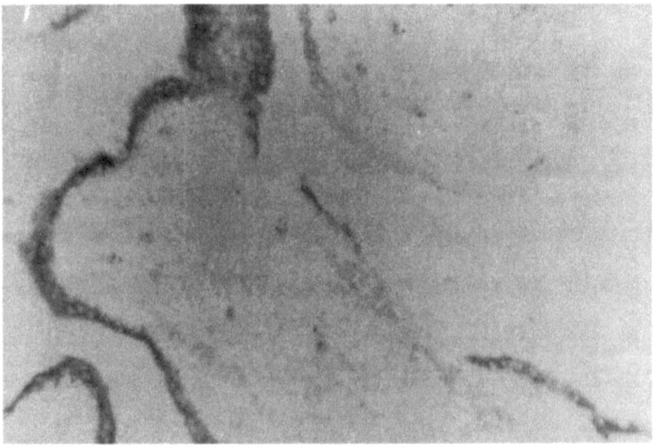

Figure 1. Localization of PEP in syncytiotrophoblasts as demonstrated by immunoperoxidase technique.

Subsequently, the possibility that immature placenta synthesizes PEP was investigated. In each of the experiments (Exp. nos. 1-5) the placental tissue explants were cultured for five days as described earlier and the spent media and the cultured tissue explants were analyzed for hCG and PEP. The tissues cultured in the basic culture medium showed a steady release of hCG into the medium during a 5-day period (Fig 2A). The total amount of hCG released was approximately 6-fold higher than that present initially in the tissue. The cultured tissue did not contain significant amount of hCG (results not shown). Addition of cycloheximide, a known inhibitor of protein synthesis, to the cultured medium caused more than 90% inhibition of hCG release suggesting that there was a de novo synthesis and secretion of hCG by the placental tissues.

The tissue explants cultured in the basic culture medium also released PEP into the medium (Fig 2B). However, addition of cycloheximide to the culture medium did not significantly alter the release of PEP (Fig 2B) nor did it influence the tissue concentrations of PEP (results not shown). This suggests that the placental tissue in culture does not synthesize or

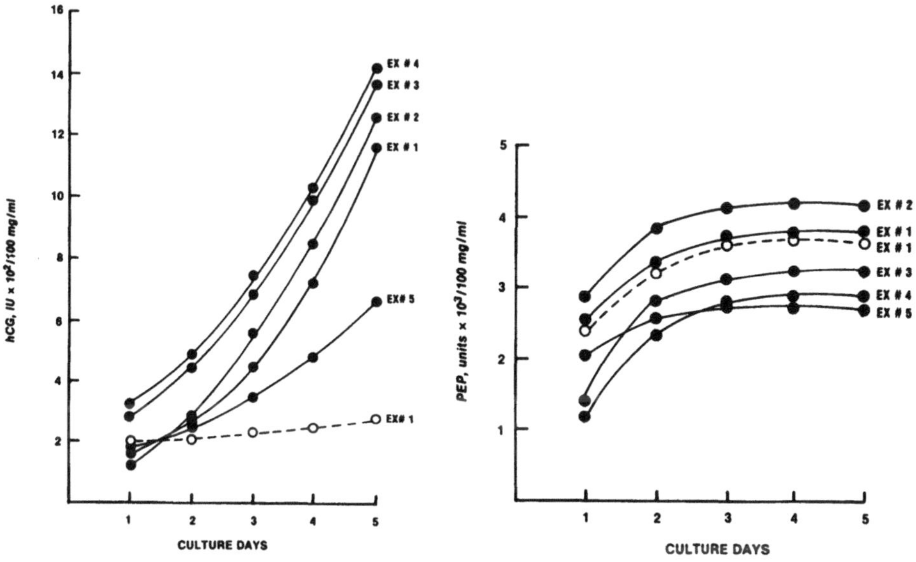

Figures 2. Accumulation in the culture medium of hCG (Fig 2A. Left panel) and PEP (Fig 2B. Right panel) released by placental explants cultured in vitro in experiments 1 through 5. ●—● = Release in absence of cycloheximide. o---o = Release in presence of cycloheximide.

secrete PEP but it releases PEP during the first 3 days of culture which has been accumulated or sequestered in vivo.

Subsequently the ability of the placental tissue to bind PEP was investigated using the indirect method as described earlier. Both membrane fractions, I and II, bound about the same amount of PEP per unit membrane protein. However, most of the binding studies which are reported in this communication were carried out using the membrane fraction I which could be recovered in higher yield from smaller amounts of placental tissue available to us.

The data illustrated in Figure 3 indicate the specificity of interaction of rabbit anti-PEP with the PEP-bound membrane complexes. The amount of ^{125}I-anti-PEP bound to PEP membrane complexes decreased in the presence of increasing concentrations of non-radioactive antibody to PEP. No significant inhibition in the binding of radiolabeled antibody was however observed in the presence of comparable concentrations of non-immune IgG.

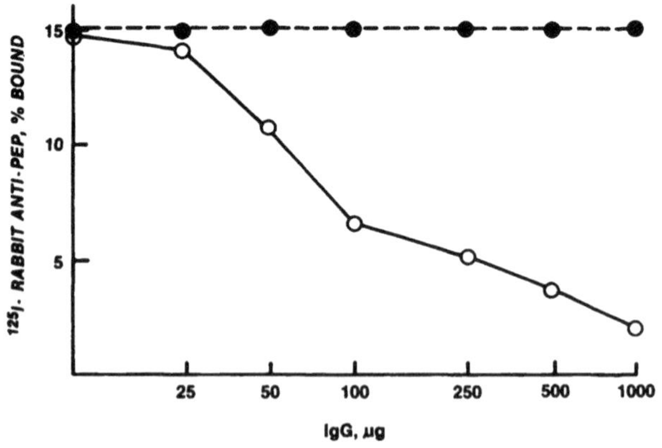

Figure 3. Specificity of interaction of ^{125}I-rabbit anti-PEP with PEP-bound to membranes. Aliquots of membrane fraction I (10µg) were first incubated with PEP (100µg) and then with ^{125}I-anti-PEP (100,000cpm) in presence of increasing amounts of non-radioactive rabbit anti-PEP (o——o) or normal rabbit serum IgG (●---●).

Figure 4 shows the relationship between ^{125}I-anti-PEP binding and the concentration of membranes in the reaction mixture. The non-specific binding ranged from 2-17% of the total counts added, depending upon the membrane concentration. The specific binding of PEP to placental membrane proteins increased progressively (8-30%) with the increased membrane concentration and reached maximum at 250µg membrane protein. No further increase in PEP binding was observed when the membrane concentration in the reaction mixture was raised beyond 250µg per 200ul.

Figure 5 incorporates the data on temperature dependency of PEP binding. Both non-specific as well as specific binding were higher and more rapid at 37°C than at 4°C and it occurred in a time dependent manner. The specific binding at 37°C reached a plateau all in about 30

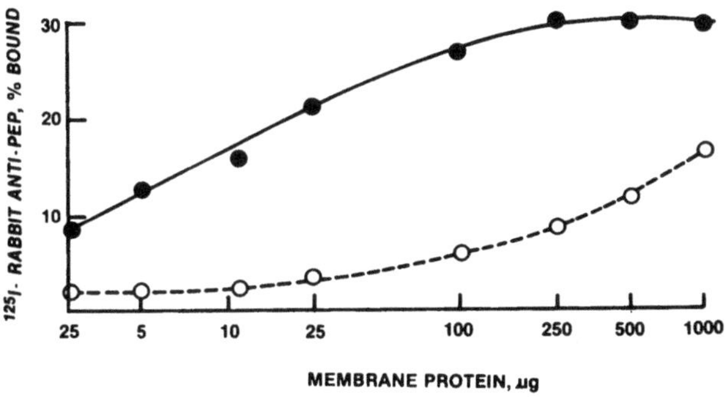

Figure 4. Binding of ^{125}I-rabbit anti-PEP to PEP-membrane complexes as a function of membrane concentration. Aliquots of membrane fraction I (2.5-1000 µg) were sequentially incubated with PEP (100µg) (●——●) or diluted male serum (o---o) and ^{125}I-rabbit anti-PEP (100,000cpm)

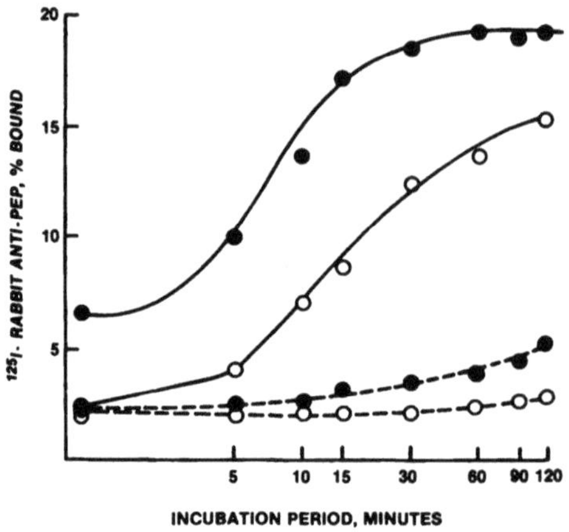

Figure 5. Effect of temperature on binding of PEP to membranes. Aliquots of membrane fraction I (10μg) were first incubated with PEP (100μg) (●——● and o——o) or diluted male serum (●---● and o---o) at 4°C or at 37°C for 5-120 min and then with ^{125}I-rabbit anti-PEP (100,000cpm) at 37°C for 60 min. ●——● and ●---● = binding at 37°C; o——o and o---o = binding at 4°C.

minutes of incubation and remained at about the same level up to 120 minute incubation. In contrast, the binding at 4°C did not reach a plateau during incubation for 120 minutes. Preheating the membranes at 100°C for 10 minutes before exposure to PEP completely abolished PEP binding (results not shown).

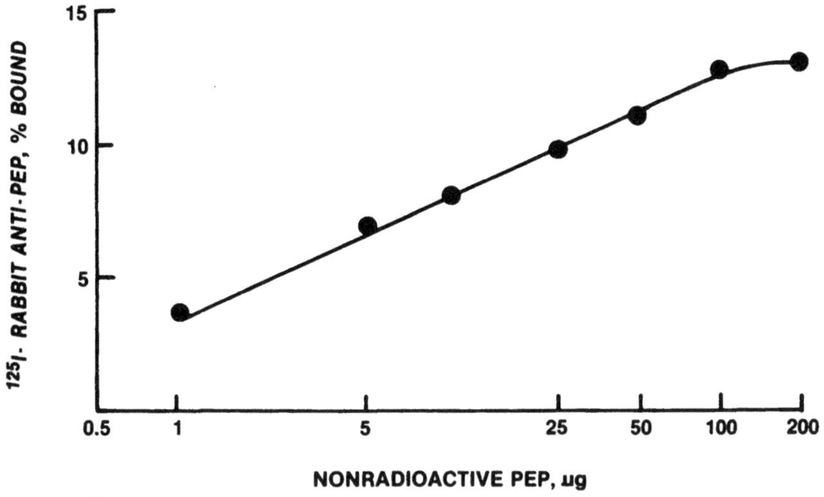

Figure 6. Specific binding of PEP to membrane as a function of PEP concentration. Aliquots of membrane fraction I (10µg) were first incubated with increasing amounts of PEP (0.5 - 200µg) and then with ^{125}I-rabbit anti-PEP. Non-specific binding was determined as described in the text.

Finally binding of PEP to the placental membrane protein increased consistently with the increasing concentration of PEP in the reaction mixture (Fig 6). The specific binding was maximum at 100µg PEP per reaction volume and it remained unchanged even when the concentration of up to 200µg PEP was used, indicating a saturation of binding sites for PEP in membrane fraction.

Discussion

The present immunocytochemical study clearly demonstrates that both PEP and hCG are associated with syncytiotrophoblastic cells. Furthermore, in vitro culture experiments show that the explants of immature placenta synthesize and secrete hCG but they do not synthesize PEP. However, such cultured explants release into the culture medium PEP which has been accumulated in vivo. Therefore we conclude that the immature placenta does not contribute significantly to the rising levels of PEP in the maternal serum during early pregnancy. Apparently, the detection of PEP in trophoblastic cells is due to non-specific adsorption or specific binding of endogenous PEP onto trophoblastic cells. The present in vitro binding studies demonstrate that the crude membrane fractions prepared from the immature placenta are capable of binding PEP in a specific dose-and-temperature-dependent manner. In addition to PEP at least two other proteins, namely placental protein-12 (PP-12) (10) and a pregnancy-associated plasma protein-A (PAPP-A) (11,12) have been found to be associated with both the placenta and endometrium (10,11,12). It is likely that both these proteins, like PEP, are synthesized and secreted by the endometrium and subsequently sequestered by the placenta. The relationship of PEP to PP12 and PP14 and the significance of binding of PEP to placental membranes is being investigated.

Acknowledgement

These studies were supported by Grants Nos. HD 09622 and CA 3224 from NIH, Bethesda, Maryland.

References

1. Joshi S.G. 1983. Sem. Reprod. Endocrinol. 1, 221.

2. Joshi, S.G., J.F. Bank, E.S. Henriques, A. Makarachi, G. Matties. 1982. J. Clin. Endocrinol. Metab. 55, 642.

3. Mazurkiewicz, J.E., F.F. Bank, S.G. Joshi. 1981. J. Clin. Endocrinol. Metab. 52, 1006.

4. Huot, R.I., J.M. Foidart, K. Stromberg. 1979. In Vitro 15, 497.

5. Joshi, S.G., F.F. Bank, D.H. Szarowski. 1981. J. Clin. Endocrinol. Metab. 52, 1185.

6. Markwell, M.K. 1982. Anal. Biochem. 125, 427.

7. Aragona, C., H.G. Friesen. 1975. Endocrinol. 97, 677.

8. Katidineni, M., T.F. Davis, I.T. Hohtaniemi, K.J. Calt. 1980. Endocrinol. 107, 1980.

9. Bradford, M.M. 1976. Anal. Biochem. 72, 248.

10. Seppala, M., E.M. Rutanen, J.I. Siiteri, T. Wahlstrom, R. Koistinen, R. Pietila, H. Bohn. 1983. Ann. N.Y. Acad. Sci. 47, 368.

11. Lin, T., S.P. Halbert. 1976. Science 173, 1249.

12. Bischoff, P., S. DuBerg, M.T. Sizonenko, A.M. Schiudler, W.L. Hermann, P.C. Sizonenko. 1984. Am. J. Obstet. Gynecol. 148, 13.

SECRETORY PROTEINS OF THE HUMAN ENDOMETRIUM AND DECIDUA DURING MENSTRUAL CYCLE AND PREGNANCY: CHARACTERIZATION OF PREGNANCY-ASSOCIATED ENDOMETRIAL α_1- AND α_2-GLOBULINS (α_1- AND α_2-PEG)

S.C. Bell[‡*], J.O. Drife [*]

Department of Biochemistry[‡] and Department of Obstetrics and Gynaecology[*], University of Leicester, Clinical Sciences Building, Leicester Royal Infirmary, PO Box 65, Leicester, LE2 7LX, U.K.

Introduction

Much research effort has been expended during the past two decades in identifying and assessing the clinical relevance of secretory products of the placenta, however only scanty information is available concerning the secretory products of the uterine endometrium during the menstrual cycle and pregnancy (1). Progesterone-directed differentiation of the endometrium results in the formation of the secretory endometrium (zona spongiosa) and decidualized endometrium (zona compacta), tissues which probably function primarily in implantation and pregnancy establishment in the first trimester.

Evidence from laboratory rodents have suggested that functions of these tissues, especially the latter, may be mediated in part by secreted polypeptides and proteins (2-6). However, most studies on protein synthesis by the human endometrium have concentrated upon the synthesis of cytosolic proteins and the effect of steroid hormones in vitro and in vivo. Many investigations have reported the synthesis of a protein of M_r under reducing conditions of 50-55 K by the secretory endometrium and induced by exposure to progesterone (7-9). Joshi and colleagues (10) have also reported the synthesis of a progestagen-dependent endometrial protein (PEP) by the secretory and decidualized endometrium, and although direct evidence of secretion was not reported, its detection in amniotic fluid and

Pregnancy Proteins in Animals

blood suggested that it is secreted (11,12). Decidualized endometrium has been demonstrated to synthesize and secrete prolactin (13) and it is suggested to be the origin of amniotic fluid and peripheral serum diamine oxidase activity (14). Only one study has been directed toward the specific identification of proteins synthesized and secreted in vitro by the endometrium during the menstrual cycle (15). In this report, a number of polypeptides, whose rate of synthesis and secretion by the endometrium of the menstrual cycle is modulated by in vitro exposure to progesterone, were described.

Assuming that the roles of the secretory endometrium and decidua in human ovo-implantation and pregnancy as in rodents may also in part be mediated by secretory protein products, in this study, an analysis and characterization of the proteins synthesized de novo, by employing radiolabelled amino acids, and secreted by these tissues in vitro, has been undertaken. Employment of these techniques ensured that proteins detected would represent unequivocally endometrially-derived proteins and examination of secretory proteins imply that they may ultimately be detectable in peripheral serum. This study represents the first attempt to comprehensively analyse the total secretory protein products of the endometrium during the menstrual cycle and pregnancy. These secretory proteins may provide markers to determine whether defects in endometrial differentiation and function may be involved in clinical disorders such as infertility and first trimester abortion, and pathological conditions only apparent later in pregnancy such as pre-eclampsia. They will also provide molecular markers for studies on hormone-dependent differentiation and gene transcription in the human endometrium.

Material and Methods

Specimens

Endometrium of pregnancy (decidua parietalis) was isolated from specimens obtained from women undergoing therapeutic termination of pregnancy (6-12 weeks) and women undergoing total hysterectomy during pregnancy (2 specimens). Endometrial tissue during the menstrual cycle was obtained by curettage prior to sterilisation.

Amniotic fluid was obtained during therapeutic termination of early pregnancy (6-14 weeks), during amniocentesis for diagnosis of neural tube defect and genetic disease (15-20 weeks) and during Caesarean section and normal labour. Amniotic fluids were centrifuged at 1000 g for 20 min and stored at -20°C.

Incubation. 100-150 mg of tissues were incubated with 2 ml Trowell's T8 (Gibco Europe) supplemented with penicillin (100 /ml), streptomycin (100 mcg/ml) and fungizone (0.25 mcg/ml). Vials were pre-incubated for 2 hours at 37°C and medium replaced with medium containing 40 uCi [^{35}S]-methionine (Amersham Int., spec. act. 800 Ci/mmol) and incubation carried out at 37°C in 95% O_2 - 5% CO_2 atmosphere. Progesterone was added at 0.1 ug/ml in ethanol. Aliquots of medium were removed, centrifuged at 100,000 g for one hour and stored or dialysed where necessary. For estimating total incorporation into protein, aliquots were precipitated with 10% (w/v) TCA and washed three times with 5% TCA before counting.

Protein separation. For one dimensional SDS-PAGE, samples were heated at 100°C for 3 min with β-mercaptoethanol/glycerol/Trisma sample buffer and electrophoresed on a 5-15% gradient PAGE slab gel containing 0.1%SDS. Two dimensional isoelectric focusing-SDS-gradient PAGE was performed as previously described (16). 1D gels were stained in Coomassie blue, 2D gels in silver stain reagent (Biorad) and both were processed for fluorography (17).

Biochemical separation techniques. Ion exchange chromatography: Concentrated ^{35}S-methionine labelled incubation medium was dialysed and applied to a DE52 column in 0.1 M-Tris-HCL, pH 8.0 and eluted with 0-0.2 m Nacl gradient. Gel filtration: A Sephadex GI50-SF (Pharmacia) column in 0.1 Tris-HCL, 0.5 m Nacl, pH 8.0 was employed. Concanavalin A sepharose chromatography : A con A-sepharose column (Pharmacia) was employed equilibrated in 0.5 M-Tris-HCL, pH 7.4, containing 1 m Nacl, and eluted with a gradient of 0-0.2 M - α -D-methylmannoside in the same buffer. Chromatofocusing: A column of polybuffer exchanger PBE 94

(Pharmacia) equilibrated in 0.025 m-imadazole, pH 7.4 was eluted with 12.5 column volumes of polybuffer 74-HCL, pH 4.0. Further removal of bound material was achieved by elution with 1 m-Nacl.

Immunochemical techniques. Antisera were raised in New Zealand White rabbits against cytosol preparations of first trimester pregnancy endometrium, and purified preparations of α_1- and α_2-PEG. Rocket immunoelectrophoresis was performed with these antisera in 1% (w/v) agarose in 0.06 m-barbital buffer, pH 8.6, with a voltage gradient of 100 v for 18 hours on 10 x 10 cm plates.

Results

Identification and classification of polypeptides synthesized and secreted in vitro by endometrium - menstrual cycle and pregnancy

Incubation of endometrium with ^{35}S-methionine resulted in the secretion of radiolabelled protein after a delay of 2-4 hours and was linear for up to 24 hours. Aliquots of incubation medium were analysed on 1D gradient/SDS PAGE under reducing conditions. For a typical experiment using pregnancy endometrium (decidua parietalis) the majority of proteins on stained gels were of serum origin, ie, predominant albumin band. However, a fluorograph (Fig 1) of the same gel revealed a dramatically different pattern with 17 labelled polypeptide bands present in the medium of a 9-10 week pregnancy endometrium (termed endometrial polypeptides EP). The very different fluorographic pattern of cytosol preparations demonstrated that the radiolabelled medium polypeptides did not arise by cytoplasmic loss during the incubation (last track, Fig 1).

Two dimensional isoelectric-focusing/gradient SDS PAGE revealed that a number of endometrial polypeptides exhibited polymorphism. For example, EP 14 was comprised of two forms of differing pI of identical molecular weight (M_r = 32,000), whereas EP 15 exhibited both charge and size polymorphism being comprised of 3 major and 7 minor spots on 2D gels (average M_r = 28,000).

Fig 1. SDS-gradient polyacrylamide slab gel electrophoresis of pregnancy endometrium secreted medium proteins.

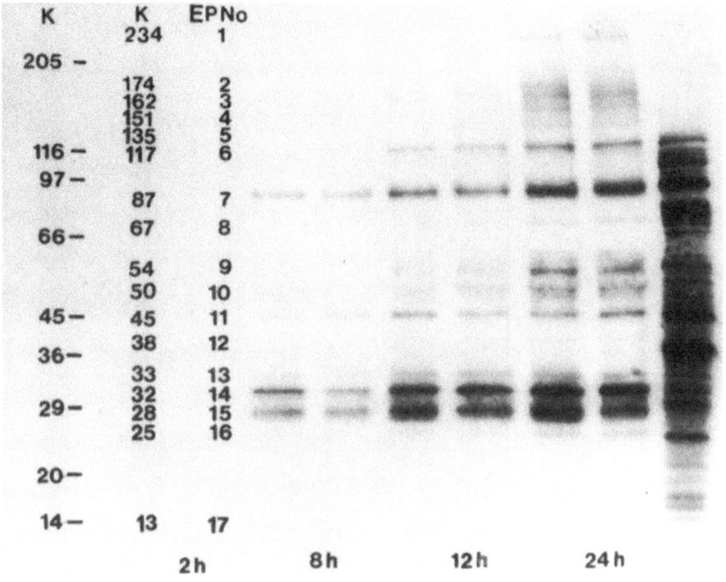

Aliquots of medium and cytosolic fractions obtained at timed intervals (2, 8, 12 & 24h) during incubation of pregnancy endometrium (week 9-10) with and without progesterone were analysed by electrophoresis and subjected to fluorography. For each time interval, the right hand track contains fractions from incubations where progesterone was included, the left where progesterone was excluded. Positions of molecular weight markers were recorded from stained gel (myosin, β-galactosidase, phosphorylase b, bovine serum albumin, ovalbumin, glyceraldehyde-3-phosphate, carbonic anhydrase, trypsin inhibitor and α-lactalbumin). Right hand track represented aliquot of cytosol for internal comparison. Seventeen radiolabelled secreted endometrial proteins are shown together with molecular weight estimates.

Fluorographic analysis of media from cultures of mid-proliferative and mid-secretory endometrium from the menstrual cycle revealed similar patterns but major quantitative differences were apparent in the synthesis and secretion of certain polypeptides especially in their response to exposure to progesterone (Fig 2).

Fig 2. One-dimensional polyacrylamide gel analysis of polypeptides synthesised and secreted by human endometrium incubated in vitro and effect of progesterone.

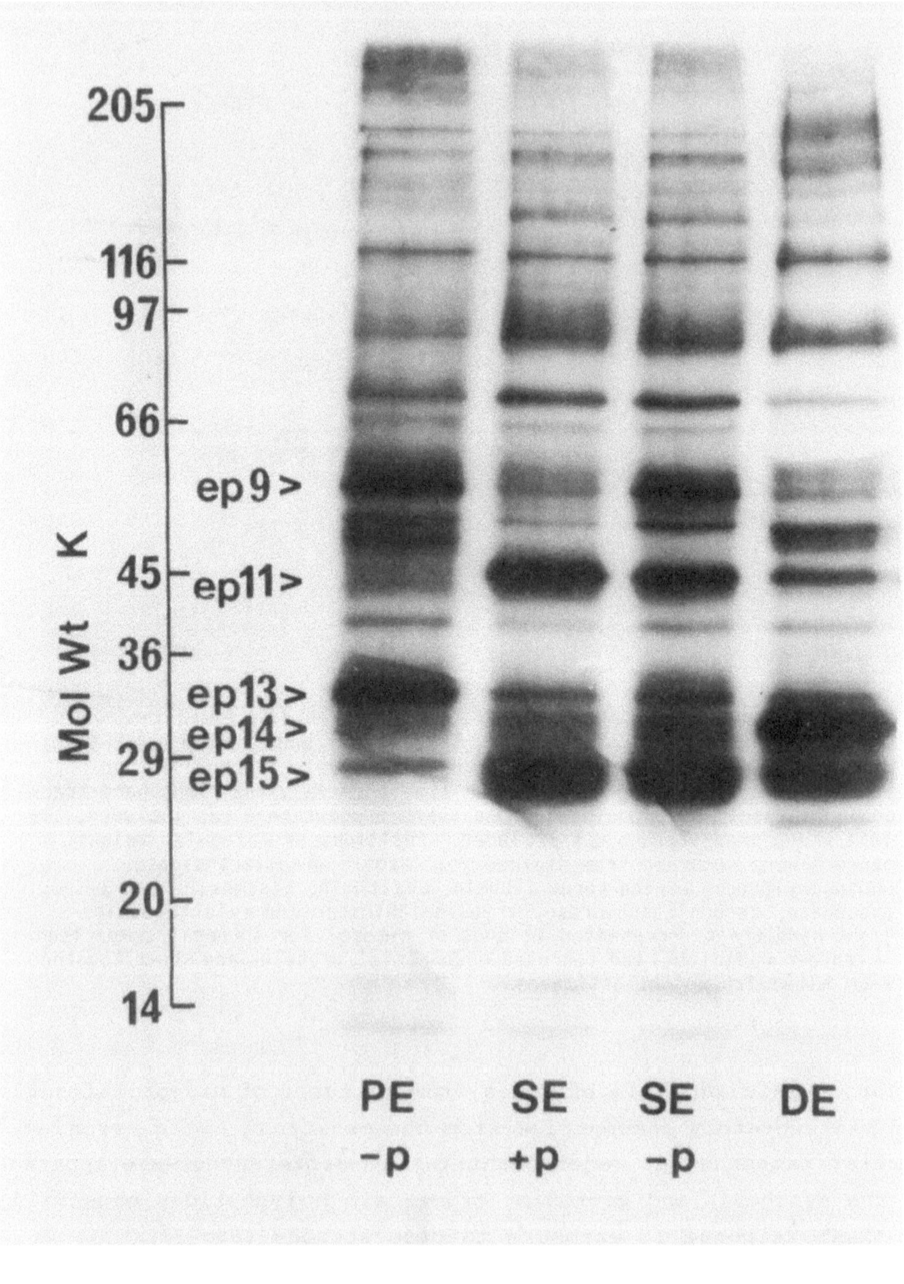

Human endometrium was obtained during the non-pregnant menstrual cycle (PE-proliferative endometrium day 11, SE secretory endometrium day 21) and early pregnancy (DE-decidualized pregnancy endometrium week 8) and incubated in vitro with [^{35}S]-methionine with (+P) and without (-P) progesterone for 18 hours. Incubation medium was subjected to gradient SDS-slab PAGE and fluorography. Endometrial polypeptides referred to in text are arrowed and molecular weight axis is calculated from migration of marker proteins.

We thus classified the EP synthesized and secreted by the endometrium during the menstrual cycle and early pregnancy according to the following groups (Table 1 for examples):

Table I Classification of endometrial polypeptides synthesized and secreted in vitro by endometrium during menstrual cycle and pregnancy

Group	Endometrial polypeptide No.	Mr	Linked to histology of endometrium	Regulated by short-term progesterone in vitro
I	EP 6	117 K		
	EP 12	38 K		
II	EP 13	33 K	+ (proliferative)	-
	EP 14	32 K	+ (decidualized)	-
	EP 15	28 K	+ (secretory)	-
III	EP 9	54 K		+ (↓)
	EP 11	45 K		+ (↑)

Group I: Polypeptides synthesized and secreted constitutively by endometrium irrespective of histology and source and unaffected by short-term progesterone exposure, eg, EP 6 and 12.

Group II: Polypeptides whose synthesis and secretion were linked to the histological classification but unaffected by short-term progesterone exposure, eg, EP 15 principally associated with secretory endometrium of menstrual cycle and maintained in pregnancy; EP 14 principally associated with decidualized endometrium of pregnancy. Although these polypeptides were unaffected by short-term progesterone exposure, ie, 24 hour, presumably their synthesis and secretion is ultimately progesterone-dependent since endometrial differentiation itself is principally caused by long-term progesterone action.

Group III: Polypeptides synthesized and secreted by endometrium irrespective of histology and source but affected dramatically by short-term progesterone exposure, eg, EP11 whose synthesis and secretion was induced by progesterone whereas EP 9 synthesis and secretion was repressed by progesterone exposure.

Synthesis and Secretion of the Major First-Trimester Pregnancy Endometrial Polypeptides EP 14 and 15

Examination of a number of first trimester pregnancy endometrial specimens revealed that EP 14 and 15 accounted for about 60% of the ^{35}S-methionine incorporated into secretory polypeptides and thus represented the major secretory polypeptides of the endometrium during early pregnancy. However, dramatic differences in the rates of synthesis and secretion of EP 14 and 15 were observed during the first trimester (Fig 3).

Fig 3. One dimensional polyacrylamide gel analysis of polypeptides synthesized and secreted by human pregnancy endometrium incubated in vitro

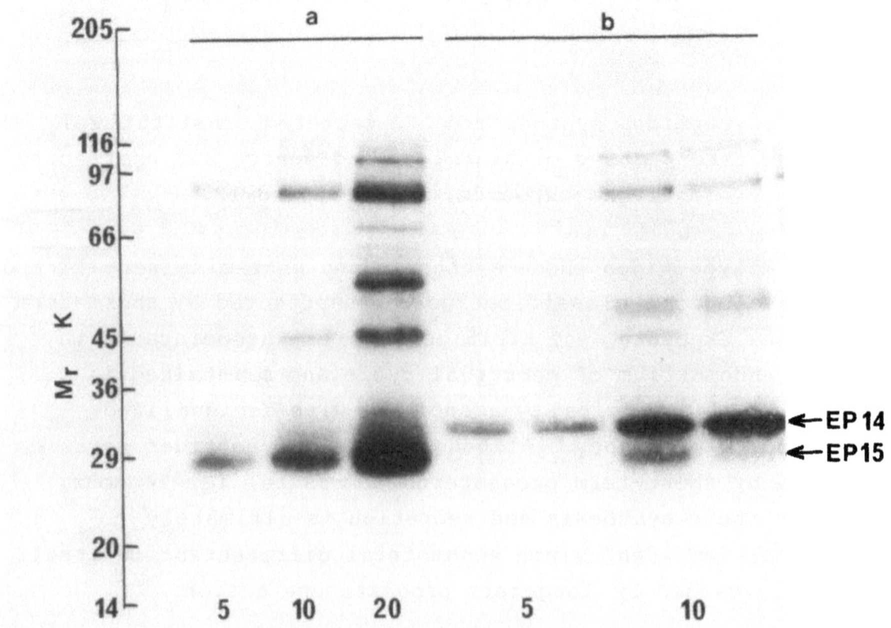

Human endometrium was obtained during early pregnancy at a) week 7 and b) week 10, incubated in vitro with [^{35}S]-methionine and the incubation medium analysed by gradient SDS-slab PAGE and fluorography. Left-hand axis refers to molecular weight (K) and lower numbers to the number of hours of incubation. Note predominant production of EP15 in a) and of EP14 in b).

Over the period examined (weeks 6-15), the rate of synthesis and secretion of EP 15 decreased to a minimum at weeks 12-15 whereas for EP 14, the rate increased to a maximum at weeks 12-15 (see Fig 4).

Fig 4 Relationship between rate of de novo synthesis and/or secretion of EP14 and EP15 (α_1- and α_2-PEG) by endometrium in vitro and amniotic fluid levels of α_1- and α_2-PEG

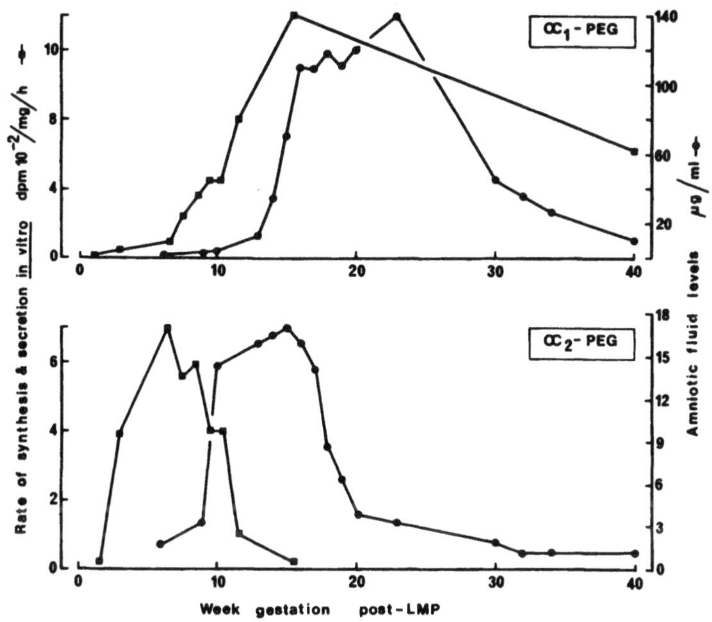

The rates of in vitro synthesis and secretion of α_1- and α_2-PEG were calculated from densitometer scanning of fluorographs of medium analysed on 1D SDS-PAGE gels to determine the % incorporation of [^{35}S]-methionine into EP14 (α_1-PEG) and EP15 (α_2-PEG) and multiplying by the total incorporation of radioactivity into TCA precipitable medium proteins. The rate was then determined by performing this analysis at timed intervals during the incubation. Levels of α_1- and α_2-PEG in amniotic fluids were determined by rocket immuno-electrophoresis with monospecific antisera. Estimations were performed on 146 amniotic fluids and points represented mean values.

Histological examination of specimens suggested that this could be due to an alteration in the contribution of the two layers of decidua parietalis, the decidua spongiosa and decidua compacta during this period of pregnancy, with the compacta dominating towards the end of this period. This interpretation is compatible with the synthesis of EP 15 by mid-secretory phase endometrium of the menstrual cycle whose secretory zona spongiosa is similar to the decidua spongiosa of pregnancy. This would imply that EP 14 is associated with decidua compacta, and EP 15 with the decidua spongiosa regions of the decidua parietalis. Equivalent incubations of first trimester trophoblast with ^{35}S-methionine demonstrated that neither EP 14 or 15 was synthesized or secreted by this tissue.

Identification and Characterization of Endometrial Proteins From Which EP 14 and 15 Are Derived - Pregnancy-Associated Endometrial α_1- and α_2-Globulins (α_1- and α_2-PEG)

EP 14 and 15 were identified on SDS-PAGE gels under reducing conditions and therefore they may not represent the native form of the proteins. To determine the properties of the native form, bulks of medium from in vitro incubation of pregnancy endometrium with ^{35}S-methionine were subjected to biochemical separation techniques. The purification of the proteins from which EP 14 and 15 were derived were monitored by radioactivity estimations and fluorographic examination of gradient/SDS PAGE gels of column fractions. Initial anion exchange chromatography (DE52, Whatman) completely separated the proteins from which EP 14 and 15 were derived, ie, eluted at 0.14 m and 0.11 m-Nacl respectively. Subsequent purification of the protein from which EP 14 was derived, demonstrated a M_r of 36,000 for this protein, it failed to bind to concanavalin A-sepharose and eluted from a chromatofocusing column in two pI forms. EP 15 however was derived from a dimeric protein with a M_r of 56,000 (subunit M_r 28,000) which bound to concanavalin A-sepharose and eluted from a chromatofocusing column over a wide pI range.

To confirm that the proteins from which EP 14 and 15 were derived were produced in vivo, rabbit antisera were prepared against a cytosol preparation of first trimester pregnancy endometrium. After absorption of these antisera with male serum, this antisera reacted against two antigenic determinants in pregnancy endometrium of α_1 and α_2 electrophoretic mobility and were termed antigens A and B. Subsequently by monitoring the fractions obtained during the purification of the proteins from which EP 14 and 15 were derived, it was unequivocally demonstrated that these two major antigens A and B were identical to the proteins from which EP 14 and 15 were derived, ie, EP 14 (M_r 32,000) from a α_1-globulin of native M_r 36,000, and EP 15 (M_r 28,000) from a dimeric glycoprotein of native M_r 56,000. In view of their demonstrated de novo synthesis and secretion by the endometrium, these have been termed pregnancy-associated endometrial α_1- and α_2-globulins (α_1- and α_2-PEG) and represent the major true secretory protein products of the endometrium during pregnancy. These terms refer to the observed dramatic increase in their synthesis and secretion associated with pregnancy, to their endometrial origin defined by strict biochemical criterion and their electrophoretic mobilities.

Levels of α_1-PEG and α_2-PEG in Amniotic Fluid During Gestation

Employing rabbit antisera raised against purified α_1- and α_2-PEG and rocket immunoelectrophoretic techniques, it was determined that amniotic fluid possessed high concentrations of these endometrial proteins. The levels of the proteins were then examined in this compartment during gestation and a dramatically different pattern of their levels was observed, as shown in Fig 4.

Peak levels of α_2-PEG were detected between weeks 10 and 15, and thereafter fell steeply to week 20. Levels were undetectable by this assay from week 32 (limit of sensitivity 2 µg/ml). In direct contrast α_1-PEG levels did not rise significantly until week 13 and peak levels were observed between weeks 20 and 24. Levels of this protein were detectable throughout the third trimester and peak levels of α_1-PEG were approximately 8-fold higher than peak $\alpha 2$-PEG levels.

Relationship Between α_1- and α_2-PEG and Other Amniotic Fluid Proteins

Given the high concentrations of α_2-PEG in amniotic fluid, consideration has been given to their possible relationship with other proteins exhibiting similar profiles in amniotic fluid whose ultimate tissue origin has not been demonstrated by strict biochemical criterion. α_1-PEG appears immunochemically related to PP 12 and α_2-PEG related to PP 14 (Fig 5).

Fig 5 Immunochemical relationship between endometrial and placental proteins, α_1-PEG, α_2-PEG, PP12 and PP14

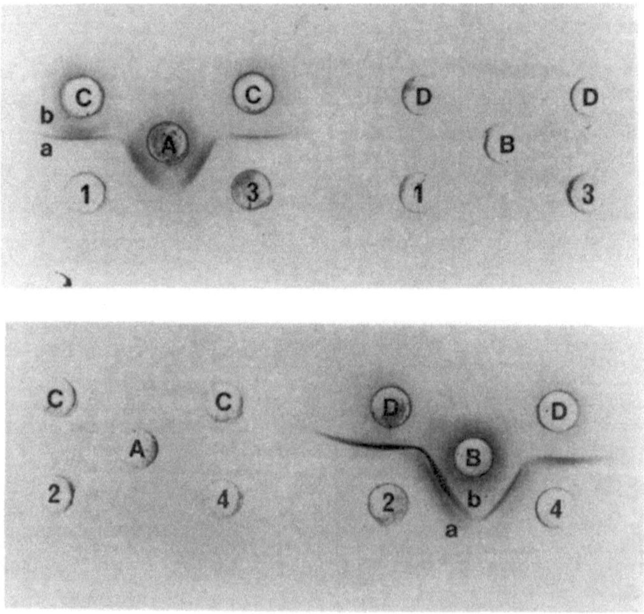

Antisera: A. anti-α_1-PEG, B. anti-α_2-PEG, C. anti-PP12A&B, D. anti-PP14;
Antigens: 1. α_1-PEG, 2. α_2-PEG, 3. PP12, 4. PP14.

PP 12 and PP 14 were isolated from term placenta (18) and since no synthesis and secretion of α_1- and α_2-PEG has been demonstrated by trophoblastic tissue, their isolation from term placenta may reflect either contamination of placenta by endometrial tissue or trophoblast-bound forms of α_1 and α_2-PEG. However, it must be noted that the physiochemical properties of PP 12 and PP 14 are not similar to those described for α_1- and α_2-PEG.

Discussion

This study has demonstrated that endometrial differentiation initiated during the luteal phase of the non-fertile menstrual cycle and extended during the first trimester of pregnancy is associated with dramatic alterations in the de novo synthesis and secretion in vitro of specific polypeptide populations. This differentiation is principally dependent upon progesterone and two groups of secretory polypeptides appear dependent upon this steroid. Group III polypeptides as exemplified by EP 9 and 11 are affected by short-term in vitro exposure to progesterone and this appears to be irrespective of the histological stage of endometrial differentiation, ie, synthesis and/or secretion of EP 11 is induced and EP 9 is repressed by progesterone in either the proliferative or secretory endometrium. The synthesis and/or secretion of Group II polypeptides are linked to the histological stage of endometrial differentation and unaffected by short-term exposure to progesterone. However, since endometrial differentiation is progesterone dependent, then presumably the expression of these genes are ultimately progesterone dependent, eg, EP 15 with zona spongiosa of secretory endometrium and decidua spongiosa of decidua parietalis (see Fig 6). Studies of these secretory polypeptide groups may be of differential clinical value since Group II polypeptides would indicate presence of a certain stage of endometrial differentiation, ie, EP 15 with secretory endometrium, whereas Group III polypeptides would reflect the action of progesterone upon any endometrium (Table I).

The Group II polypeptides EP 14 and 15 represent the major secretory products of the endometrium during the first trimester and initial studies have concentrated upon their characterization. These polypeptides are not in vitro artefacts since using antisera raised against cytosols of pregnancy endometrium, the most immunogenic components of this tissue have been demonstrated to represent the proteins from which EP 14 and 15 are derived (since

Fig 6 Postulated hierachy of genes coding for secreted endometrial proteins

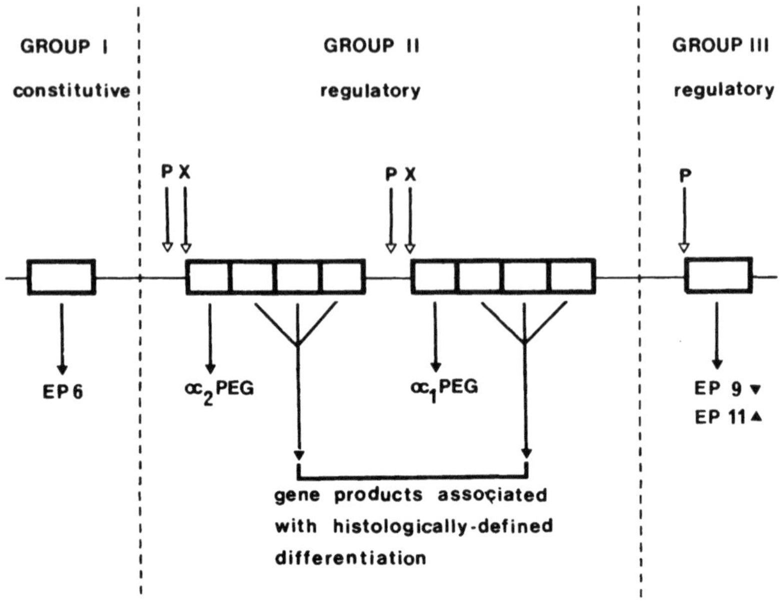

P - progesterone, X - undefined pregnancy-associated inducers α_2 and associated gene products expressed during differentiation of glandular epithelium, α_1-PEG and associated gene products expressed during differentiation of stromal cells to decidual cells.

polypeptides were detected employing SDS-PAGE under reducing conditions, we did not initially detect the proteins in their native form). EP 14 is derived from a M_r 36,000 protein of α_1-mobility existing in two pI forms and EP 15 from a dimeric glycoprotein M_r 56,000 (subunit M_r 28,000) of α_2-mobility and exhibiting multiple forms, and have been termed α_1- and α_2-PEG. The in vitro pattern of synthesis and/or secretion of those proteins alters dramatically during the first trimester, with synthesis and/or secretion switching from principally α_2-PEG (assayed as EP 15) at week 6/7 to principally α_1-PEG (assayed as EP 14) at week 11/12. What regulates this switch is unknown but it may be related to the association of α_1- and α_2-PEG synthesis and secretion by particular regions of the endometrium. α_2-PEG is principally synthesized by the spongiosa layers of the endometrium and immunohistological studies have localized this protein to the glandular epithelium of this layer, epithelial tissue in the decidualized compacta layer is inactive (S C Bell, A Kubicki & R Walker, unpublished observations). This tissue is most prominent during the first trimester, so involution of the secretory glands in the spongiosa layer may result in decreased $_2$-PEG synthesis and secretion. α_1-PEG is associated with the decidualized compacta layer of the decidua parietalis and this tissue persists throughout pregnancy. However, whether the increase in synthesis from weeks 6-15 is due to progressive decidualization during this period or presence of an inducer is unknown.

The principal direction of secretion of these major secretory endometrial proteins appears to be the amniotic fluid with the low levels in peripheral serum detectable only by radioimmunoassay (S C Bell, unpublished observations). Three other proteins demonstrated to be synthesized or secreted by the endometrium during pregnancy are also primarily secreted in the amniotic fluid, ie, decidual prolactin (13), diamine oxidase (19) and PEP (11). It is therefore apparent that a general principle may be involved, which is that endometrially-derived proteins are directed into the amniotic fluid. This would imply that pregnancy-associated plasma protein A, PAPP-A, recently suggested to originate in the decidualized endometrium (20) is not synthesized de novo and secreted by this tissue since peripheral serum levels of this protein are far greater than amniotic fluid levels (21). Also there is no evidence for de novo synthesis and secretion of PAPP-A from the decidualized endometrium. The profiles of α_1- and α_2-PEG in amniotic fluid lag behind the profile of in vitro synthesis and secretion by 5-6 weeks. This could be due in part to the half-life of the proteins in amniotic fluid and alterations in the mechanism of transport of the proteins from the endometrium. If amniotic fluid levels are principally derived from the decidua parietalis, then a limiting factor would be the rate of transport across the amniochorion. The mechanism of transport of proteins across the amniochorion is unknown. However, the fusion of the decidua parietalis with the amniochorion during the second trimester could alter the efficiency of transport and account for the increasing levels of α_1-PEG in the amniotic fluid from week 15.

Apart from prolactin, which is anyway not tissue-specific, these proteins represent the first proteins to be unequivocally identified as endometrial proteins and demonstrated to be synthesized de novo and secreted in vitro by pregnancy-endometrium. However, consideration must be given to their possible relationship to proteins isolated from the first trimester and term placenta which are also found at high concentrations in amniotic fluid and whose ultimate tissue source has not been equivocally demonstrated by strict biochemical criteria. It is apparent that α_1-PEG and α_2-PEG are immunochemically related to the term placental proteins PP 12 and PP 14 respectively. PP 12 and PP 14 have previously been demonstrated to exhibit immunochemical identity with chorionic α_1-microglobulin (CAG-1)/placental specific α_1-microglobulin (PAMG-1) and chorionic α_2-microglobulin (CAG-2)/placental specific α_2-microglobulin (PAMG-2) respectively, isolated from first trimester placental tissue (22,23). These observations are supported by the detection of both CAG-2 (24) and PP 12 (25) in the endometrium. If α_1-PEG and α_2-PEG are identical to CAG-1/PAMG-1/PP 12 and CAG-2/PAMG-2/PP 14, it would imply that these "placental proteins" which are immunohistological localized to placental trophoblast (23) are synthesized and secreted by the endometrium and are bound to the trophoblast since we have demonstrated that trophoblast preparations do not synthesize α_1- and α_2-PEG. However, the relationship between these proteins at present is based solely on immunochemical criterion and in fact, their reported properties are dramatically different (Table 2), therefore, the possibility must be considered that these proteins represent the trophoblast-bound forms of α_1- and α_2-PEG and whose properties have been modified during transport and binding to the trophoblast. α_1- and α_2-PEG however represent the true endometrially synthesized and secreted forms since they have been identified by biochemical techniques and not relied upon immunochemical techniques. α_2-PEG may be identical to PEP since the molecular properties are similar and although in vitro

Table 2 Relationship of pregnancy-associated endometrial α_1- and α_2-globulins to placental and endometrial proteins - properties and tissue origin

References	Protein nomenaclature (with molecular weights)		Tissue source
28,29	Pregnancy-associated endometrial α_1-globulin (α_1-PEG) M_R 36,000 (32,000 in SDS/PAGE)	Pregnancy-associated endometrial α_2-globulin (α_2-PEG) M_R 56,000 (2 subunits M_R 28,000)	Endometrium-de novo synthesis & secretion Amniotic fluid
18	Placental protein 12 (PP12) M_R 25,200 or [51,000 in SDS/PAGE (dimer?)]	Placental protein 14 (PP14) M_R 42-43,000	Term placenta-extraction and localization. Amniotic fluid Maternal serum
	18	18	
22,23	Chorionic α_1-microglobulin (CAG-1) or placental-specific α_1-microglobulin (PAMG-1) M_R 20,000	Chorionic α_2-microglobulin (CAG-2) or placental-specific α_2-microglobulin (PAMG-2) M_R 25,000	First trimester placenta-extraction asnd localization. Amniotic fluid
10,11		Progestagen-associated endometrial protein (PEP) M_R 48,000 (2 sub-units M_R 27,000)	Endometrium-de novo synthesis Amniotic fluid Maternal serum
		27	
26		Alpha uterine protein (AUP) M_R 50,000 (2 subunits M_R 25,000)	Endometrium-extraction Amniotic fluid

secretion has not been reported, it is synthesized de novo by the pregnancy endometrium. These unequivocally endometrially-derived proteins may provide not only protein markers to study hormone-dependent differentiation in the human endometrium but also valuable clinical markers for the assessment of endometrial differentiation and function in pregnancy and the menstrual cycle.

As pregnancy-associated proteins, α_1- and α_2-PEG should provide very different clinical information than the range of trophoblast-derived pregnancy-associated proteins already available to the clinician to assess fetal placental function. We suggest that it may be very useful to classify pregnancy-associated proteins according to their biochemically demonstrated site of synthesis since dramatic diferences in function and clinical applications may reflect this classification. If strict biochemical criterion are applied in assignment of proteins to such classes, the very tenuous nature of the evidence that is available at present for certain proteins is apparent. These classes could include Class I: trophoblast-derived, eg, hcg, hpl, Class II: endometrially-derived, eg, α_1-PEG, α_2-PEG, Class III: maternal immune system, eg, α_2-PAG, Class IV: maternal liver, eg, ceruloplasmin.

These endometrially-derived secretory proteins and their patterns of synthesis and secretion during pregnancy and their directional secretion to the amniotic fluid and trophoblast would suggest that although they presumably fulfil different functions during pregnancy, they may be intimately involved in the interactions between the trophoblast and endometrium and fetus and maternal host (see Fig 7). These proteins now offer an opportunity to examine the nature of these interactions.

Acknowledgements

The authors are grateful to the medical and theatre staff at the Leicester General and Leicester Royal Infirmary Hospitals for help in the collection of specimens, and clinical colleagues in the Department of Obstetrics and Gynaecology, especially Dr P H Kirwan

Fig 7 Feto-placental-ovarian-endometrial interactions during human pregnancy

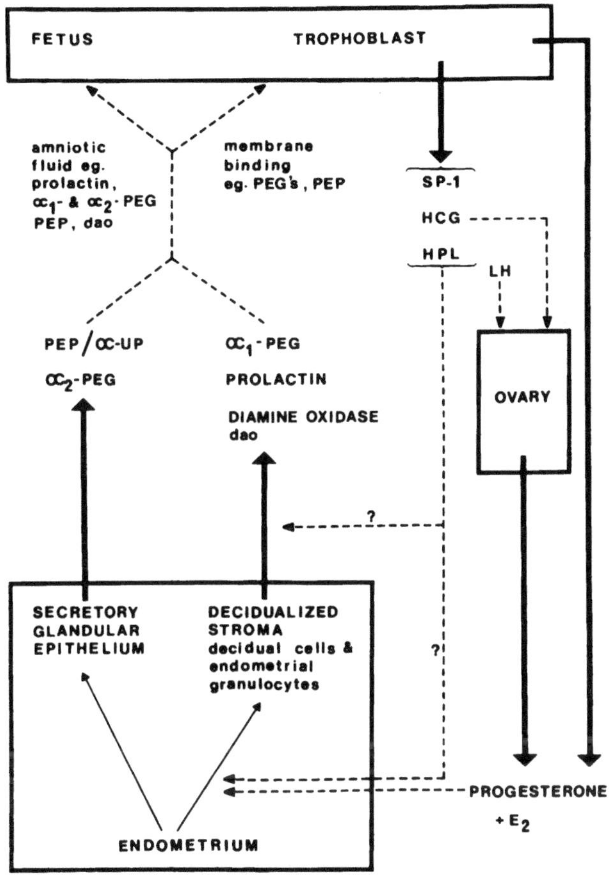

Endometrially derived proteins directed into amniotic fluid (from decidua parietalis) or bound to trophoblast (from decidua basalis). Binding endometrial protein to trophoblast could modulate trophoblastic function ie, secretion of trophoblast derived products. In turn these products may affect endometrial function in synthesis and secretion of endometrial proteins.

and Mr C R Stewart, to Miss M Hales and Miss S Patel for expert technical assistance, to Miss K Mann and Mrs D Flude for secretarial assistance, to Professor J MacVicar for valuable advice and support and the University of Leicester and the Trent Regional Health Authority for financial support. Part of this study has been accepted for publication by Brit J Obs Gynaec (28) and J Reprod Fert (29) and is reproduced with their permission.

References

1. Bell, S.C.. 1985. Comparative aspects of decidualization in rodents and human: cell types, secreted products and associated function. In: Human ovo-implantation. (R.G. Edwards, J. Purdy and P.C. Steptoe, eds). Academic Press, London. (In press).

2. Bell, S.C.. 1979. Protein synthesis during deciduoma morphogenesis in the rat. Biol. Reprod. 20, 811-821.

3. Bell, S.C.. 1979. Synthesis of "decidualization-associated protein" in tissues of the rat uterus and placenta during pregnancy. J. Reprod. Fert. 56, 255-262.

4. Bell, S.C.. 1979. Immunochemical identity of "decidualization-associated protein" and α_2-acute-phase macroglobulin in the pregnant rat. J. Reprod. Immunol. 1, 193-206.

5. Bell, S.C.. 1983. Decidualization: Regional differentiation and associated function. In: Oxford Reviews of Reproductive Biology, Vol. 5 (C.A. Finn, ed). Oxford University Press, Oxford. pp. 220-271.

6. Bell, S.C.. 1983. Decidualization and associated cell types: Implications for the role of the placental bed in the materno- fetal immunological relationship. J. Reprod. Immunol. 5, 185-194.

7. Hirsch, P.J., I.L.C. Fergusson, and R.J.B. King. 1977. Protein composition of human endometrium and its secretion at different stages of the menstrual cycle. Ann. N. Y. Acad. Sci. 286, 233-245.

8. O'Grady, J.E., S.C. Bell, A.D.T. Govan and W. Black. 1978. Protein synthesis in human endometrium during the menstrual cycle. J. Endocrinol. 77, 21-22p.

9. Shapiro, S.S. and S.H. Forbes. 1978. Alterations in human endometrial protein synthesis during the menstrual cycle and in progesterone-stimulated organ culture. Fertil. Steril. 30, 175-180.

10. Joshi, S.G., K.M. Ebert and D.P. Swartz. 1980. Detection and synthesis of a progestagen-dependent protein in human endometrium. J. Reprod. Fert. 59, 273-285.

11. Joshi, S.G., R.A. Smith and D.K. Stokes. 1980. A progestagen-dependent endometrial protein in human amniotic fluid. J. Reprod. Fert. 60, 317-321.

12. Joshi, S.G.. 1983. A progestagen-associated protein of the human endometrium: basic studies and potential clinical applications. J. Steroid. Biochem. 19, 751-157.

13. Riddick, D.H., D.C. Daly and C.A. Walters. 1983. The uterus as an endocrine compartment. Clinics in Perinatology 10, 627-639.

14. Weisburger, W.R., G. Mendelsohn, J.C. Eggleston, and S.B. Balin. 1978. Immunohistological localization of histaminase (diamine oxidase) in decidual cells of human placenta. Lab. Invest. 38, 703-706.

15. Strinden, S.T. and S.S. Shapiro. 1983. Progesterone-altered secretory proteins from cultured endometrium. Endocrinology, 112, 862-870.

16. O'Farrell, P.H.. 1975. High resolution two-dimensional electrophoresis of proteins. J. Biol. Chem. 250, 4007-4021.

17. Chamberlain, J.P. 1979. Fluorographic detection of radioactivity in polyacrylamide gels with the water-soluble fluor, sodium salicylate. Anal. Biochem. 98, 132-135.

18. Bohn, H., W. Kraus and W. Winckler. 1982. New soluble placental tissue proteins: their isolation, characterization, localization and quantification. Placenta Suppl. 4, 67-81.

19. Tujvesson, G. 1978. Purification and properties of human amniotic fluid diamine oxidase. Scand. J. clin. Lab. Invest. 38, 463-472.

20. Schindler, A.M., P. Bordignon and P. Bischof. 1984. Immunohistological localization of pregnancy-associated plasma protein-A in decidua and trophoblast. Comparison with human chorionic gonadotrophin and fibrin. Placenta 5, 227-236.

21. Bischof, P., S. Duberg, W. Hermann and P.C. Sizonenko. 1982. Amniotic fluid and plasma concentrations of pregnancy-associated plasma protein-A PAPP-A throughout pregnancy: comparison with other fetoplacental products. Brit.J. Obstet. Gynaecol. 89, 358-363.

22. Petrunin, D.D., I.M. Gryaznova, Yu. A. Petrunina and Yu. S. Tatarinov. 1978. Comparative immunochemical and physiochemical characteristics of human chorionic α_1 and α_2-microglobulins. Bull. exp. Biol. Med. USSR 5, 658-661.

23. Tartarinov, Yu. S., G.A. Kozljaeva, D.D. Petrunin and Yu. A.

Petrunina. 1980. Two new human placenta specific - globulins: identification, purification characteristics cellular localization and clinical investigation. In: The Human Placenta: Proteins and Hormones (A. Klopper, A. Genazzani and P.G. Crosignani, eds) Academic Press, London. pp 35-46.

24. Petrunin, D.D, G.A. Kozljaeva, N.V. Mesrjankina and O.P. Shevchenko. 1980. Detection of chorionic α_2-microglobulin in the endometrium in the secretory phase of the menstrual cycle and in the male sperm. Akusherstvoi Gynekologija USSR, 3, 22-23.

25. Rutanen, E.-M., R. Koistinen, T. Wahlstrom, J. Sjoberg, U.-H. Stenman, and M. Sappala. 1984. Placental protein 12 PP12 in the human endometrium: tissue concentration in relation to histology and serum levels of PP12, progesterone and oestradiol. Brit. J. Obstet. Gynaec. 91, 377-381.

26. Sutcliffe, R.G., A.E. Bolton, F. Sharp, L.V.B. Nicholson and R, Mackinon. 1980. Purification of human alpha uterine protein. J. Reprod. Fert. 58, 435-442.

27. Sutcliffe, R.G., S.G. Joshi, W.F. Paterson and J.F. Bank. 1982. Serological identity between human alpha uterine protein and human progestagen-dependent endometrial protein. J. Reprod. Fert. 65, 207-209.

28. Bell, S.C., M.W. Hales, S. Patel, P.H. Kirwan and J.O. Drife. 1985. Protein synthesis and secretion by the human endometrium and decidua during early pregnancy. Brit. J. Obstet. Gynaec. (In press).

29. Bell, S.C., S. Patel, M.W. Hales, P.H. Kirwan and J.O. Drife. 1985. Immunochemical detection and characterization of pregnancy-associated endometrial α_1- and α_2-globulins secreted by human endometrium and decidua. J. Reprod. Fert. 74. In press.

PLACENTAL PROTEIN 14 (PP14) IN THE HUMAN REPRODUCTIVE TRACT

A.E.Bolton*, K.J.Clough

Department of Biology & Biochemistry, North East London Polytechnic, Romford Road, London E15 4LZ, U.K.

H.Bohn

Research Laboratories, Behringwerke AG, D-3550 Marburg/Lahn, FRG.

M.G. Chapman

Department of Obstetrics, Guy's Hospital Medical College, London Bridge, London SE1, U.K.

Introduction

Placental Protein 14 (PP14) is one of a series of proteins isolated and partically characterised from the human term placenta, and has a proposed glycoprotein structure with a Mr of 42-43000 (1). Following the development of a radioimmunoassay for PP14 (2), this molecule has been detected in the peripheral circulation of non-pregnant women, indicating a non-placental origin. However, a rapid increase in circulating levels of PP14 has been demonstrated during the first few weeks of pregnancy rising to a maximum at about week 9 of gestation (3), suggesting a physiological association with early pregnancy. We have investigated levels of PP14 in extracts of tissues taken from the female reproductive tract, and fluids associated with reproduction, in an attempt to determine the tissue distribution of this protein. We have also investigated the secretion of PP14 by female reproductive tissue maintained in explant culture to determine the site of production. Preliminary experiments exploring the potential biological function of PP14 indicate a possible immunosuppressive role for this protein.

---

*Present address: Department of Biological Sciences, Sheffield City Polytechnic, Pond Street, Sheffield, S1 1WB, U.K.

Pregnancy Proteins in Animals

Materials and Methods

Tissue samples were taken either at legal pregnancy termination in the first trimester or from non-pregnant women at curretage. Tissue samples were rinsed rapidly in cold saline containing 0.1 mmol/l phenylmethylsulphonyl fluoride (PMSF) as protease inhibitor and stored frozen at -20 C. Extracts were prepared by homogenisation of the tissue in phosphate buffered saline containing PMSF followed by removal of cell debris by centrifugation at 9,000 x g for 2 mins. The supernatants were collected and immediately assayed for PP14 by radioimmunoassay as previously described (2).

Explants of decidual tissue were obtained by an aseptic technique and cultured in vitro for 25 hrs as described in detail elsewhere (4). The amount of PP14 released into the medium between 5 and 25 hrs of culture was measured. To distinguish between de novo synthesis of PP14 and the release of pre-existing protein from the tissue, control explants were cultured in the presence of cyclohexamide to inhibit protein synthesis (4).

Two way mixed lymphocyte reactions were set up using mononuclear cells from two healthy unrelated volunteers. Mononuclear cells were prepared by the method of Boyum (5) using density gradient centrifugation on Lymphoprep (Nyegaard). The cells were washed in Dulbecco's medium (Gibco) containing bicarbonate and supplemented with HEPES (DMEM). Cells from the two sources were each diluted to a concentration of 1 x 10^6 and mixed in equal volumes. 0.2 ml aliquots of the mixed cell suspension were pipetted into wells of microtitre plates and 50 µl aliquots of PP14 solution in DMEM at concentrations indicated in the results section were added. Controls were incubated with 50 µl of DMEM, and further controls were provided by using aged (freeze/thawed) preparations of PP14 in DMEM. Plates were incubated for 5 days at 37 C in an atmosphere of 5% carbon dioxide in air. Aliquots (25 µl) of a solution of tritiated thymidine (^{3}H-TdR) in DMEM were added to each well. Incubation was continued for a further 24 hours, after which the cells were harvested, the DNA precipitated with two volumes of 5% trichloroacetic acid, the residue collected by filtration on Whatman GF/C filters and, after washing with ice cold ethanol, the filters were counted by liquid scintillation.

Results

The concentration of PP14 in reproductive tissues from non-pregnant women are shown in Table 1, from which it can be seen that levels of this protein are some 10 times higher during the luteal than the follicular phase of the menstrual cycle. By the tenth to the twelfth week of gestation, the level of PP14 in decidual tissue increases a further 100 fold, and is some 300 times higher in decidua than trophoblast (Table 2). This difference could be explained by deciduum being either the site of biosynthesis of PP14, or being a site of accumulation and storage of protein synthesised in another tissue.

Table 1
The Concentration of PP14 in Non-Pregnant Female Endometrium

Follicular phase	6.5 ± 7.4, n=5
Luteal phase	73.9 ± 55.7, n=4

Values are given as the mass of PP14 in μg per gram wet weight of tissue, ± standard deviation.

Table 2
The Concentration of PP14 in Female Reproductive Tissues Taken at 10-12 Weeks Gestation

Trophoblast	30.9 ± 20.3, n=6
Decidua	8840 ± 6855, n=6

Values are given as mass of PP14 in μg per gram wet weight of tissue, ± standard deviation.

Preliminary observations on the short term culture of decidual explants revealed that in the absence of cyclohexamide significantly more PP14 (about 10 fold) was released into culture media that in the presence of this inhibitor of protein synthesis. This strongly suggests de novo synthesis of PP14 by decidual tissue.

In addition to reproductive tissues, the presence of PP14 was investigated in fluids associated with the reproductive process. From the results (Table 3) it can be seen that this protein is present in high concentration in seminal plasma and also in cervical mucus, but is not detectable in the fluid aspirated from ovarian follicles. Concentrations in amniotic fluid (51.5 ± 30.2 µg/ml at 16 weeks, mean ± s.d., n=8) are some 50 times higher than in the maternal circulation at the same gestational age.

Table 3
The Concentration of PP14 in Fluids Associated with Human Reproduction

Seminal plasma	133.8 ± 10.7 mg/l, n=36
Cervical mucus	1.6 mg/l n=2
Follicular fluid	n.d. (less than 2.5 µg/l, n=8)

PP14 purified by the method of Bohn et al (1) exhibited a dose dependent inhibition of ^{3}H-thymidine incorporation in the mixed lymphocyte reaction (Table 4). Furthermore, unfractionated decidual extracts also showed an inhibitory action in this system, and this was quantitatively related to the PP14 content of the extract as measured by radioimmunoassay. PP14 in both pure and crude preparations was found to be unstable in dilute solution, particularly to freezing and thawing. Using preparations in which the concentration of PP14 as measured by radioimmunoassay had decayed to less than 5% of the original value it was found that their inhibitory action in the mixed lymphocyte reaction had also been lost.

Table 4
Immunosuppressive Activity of PP14 in the Two-way Mixed Lymphocyte Reaction

	Conc. PP14 (μg/ml)	% incorporation of ^{3}H-TdR
Pure	0	100 ± 12.7
	1.0	79.4 ± 11.8*
	5.0	46.2 ± 13.4**
	10.0	32.9 ± 7.0**
Crude decidual extract	5.0	53.5 ± 21.3*
	10.0	37.1 ± 18.7**

Crude decidual extracts were diluted to give the above concentrations of PP14 as measured by radioimmunoassay.
Significances of differences from zero level of PP14 were calculated using Student's t-test. *, $p < 0.01$, **, $p < 0.005$

Discussion

PP14 is a uterine protein present at low levels in the endometrium and which shows a significant variation throughout the menstrual cycle being present at increased concentration in the luteal phase of the cycle (Table 1). There is a marked increase in PP14 in decidual tissue and it would seem that this is a major secretory protein of the deciduum. From studies on the peripheral levels of PP14, it would appear that this increase in synthetic activity occurs very close to the time of implantation (3). It would be surprising if a protein secreted in such large quantities by a specific tissue had no biological function. Furthermore, PP14 appears to be present in substantial amounts in seminal plasma and cervical mucus and would thus appear to be present in the environment of the sperm in the female tract as well as in the uterine tissue at the time of implantation of the fertilized embryo. It can be argued that this temporal distribution of PP14 is compatible with a general immunosuppressive function.

Sperm are highly immunogenic (6), the intra-vaginal route can result in immunization to antigens (7) and hence there would appear to be a need for the prevention of immunological attack on inseminated sperm. This has been attributed to the presence of a variety of potent if ill-defined immunosuppressive agents present in seminal plasma (8). This immunosuppressive activity would need to be present during passage of sperm through the cervix and into the uterine cavity and Fallopian tubes. The data presented here suggest that PP14 may have such a distribution. The fertilized embryo is also a potential target for immunological attack as, unlike the syncytiotrophoblast tissue of the placenta, MHC antigens are expressed during the peri-implantation period (9). A local immunosuppressive activity could thus be important for the successful implantation of the fertilized embryo. Later in pregnancy, survival of the developing foetal allograft appears to depend at least in part on the presence of blocking antibodies secreted by the mother in response to paternal antigens (10). Consequently, immunosuppressive factors may be less important later in pregnancy. The presence of PP14 in seminal plasma, cervical mucus and in decidual tissue reaching a maximum by weeks 8-10 of pregnancy could fit the role of such an immunosuppressive agent.

Many of the pregnancy-specific and pregnancy-associated proteins have been ascribed immunosuppressive properties, for example human chorionic gonadotrophin, pregnancy-associated plasma protein-A, pregnancy-associated α_2-glycoprotein and early pregnancy factor, although in some cases, such as human chronic gonadotrophin, this activity may be associated with minor impurities of the preparation investigated. To this list may now be added PP14 in the light of the data presented here (Table 4), although, again, the activity by impurities in the PP14 samples cannot yet be excluded. Of the proteins claimed to have immunosuppressive activities associated with pregnancy, only PP14 would appear to have the distribution of an immunosuppressant involved in a variety of aspects of reproduction. If these preliminary findings are confirmed, the investigation of PP14 may be central to studies on implantation, infertility and contraception, and PP14 may itself provide the basis of a useful pharmaceutical immunosuppressive agent.

References

1. Bohn, H., W. Kraus, W. & W. Winckler (1982) Placenta Supple 4, 67.

2. Bolton, A. E., M. G. Chapman, R. J. Stoker, C. E. Andrew, D. Wass & H. Bohn (1983) Clin. Chim. Acta 135 283.

3. Chapman, M. G., A. E. Bolton, J. G. Grudzinskas, J. L. Yovich & H. Bohn. This volume.

4. Rutanen, E-M., R. Koistinen, T. Wahlstrom, H. Bohn, T. Ranta, M. Seppala (1985) Endocrinology 116 1304-1309.

5. Boyum, A. (1968) Scand. J. Clin. Lab. Invest. 21 97.

6. Anderson, D. J. & T. M. Tarter (1982) J. Immunol. 128 535.

7. Bell, E. B. & Wolf, B. (1967) Nature 214 423.

8. James, K. & T. B. Hargreaves (1984) Immunology Today 5 357

9. Faulk, W. P., A. Sanderson & A. Temple (1977) Transplantation Proceedings 9 1379.

10. Rocklin, R. E., J. L. Kitzmiller & M. D. Kaye (1979) Am. Rev. Med. 30 375.

HUMAN ENDOMETRIAL PROTEINS: NEWS AND VIEWS

J.G. Grudzinskas

Joint Academic Units of Obstetrics and Gynaecology and Reproductive Physiology, The London Hospital Medical College and St. Bartholomew's Hospital Medical College, London, UK

The human endometrium no longer needs to be considered to be a passive organ responsive only to the commands of the ovary to prepare for implantation. In common with some other mammals, the endometrium before and after decidualisation in humans has been shown to produce a large number of proteins, some of which can be detected readily in the peripheral circulation. If these phenomena herald the development of a biochemical index of endometrial/decidual function accessible in blood or urine, the physiologist and clinician alike can look forward to the ability to study the endometrium non-invasively, previously not possible without resorting to surgical biopsy. Furthermore, the definition of the biological activity of the proteins would lead to a better understanding of the role of the endometrium and decidua in the control of fertility, the maintenance of pregnancy and parturition, as well as in disorders of the endometrium including neoplasia. This brief review deals with some recent developments in the field discussed at the International Meeting on Pregnancy Proteins in Animals, Copenhagen in April 1985.

It is now possible to list some features of molecules which are, or may be, of endometrial origin (Table 1). It is noteworthy that definitive evidence of endometrial synthesis of a particulr protein is provided only by the identification of the relevant messenger ribonucleic acid in that tissue, and that all other criteria merely provide indirect on presumptive evidence with varying degrees of conviction. By applying such arbitrary criteria it is possible to identify substances which may be synthesised by endometrial tissues from those which have been considered as candidates on the basis of rather flimsy evidence (Table 2). Definitive evidence for the synthesis of prolactin by decidualised endometrium is well-established by contrast to the other molecules listed in Table 2.

Pregnancy Proteins in Animals

Table 1. Characteristics of endometrial and decidual secretory proteins

1. Identification of the relevant messenger RNA in the endometrium
2. Demonstration of the incorporation of radiolabelled amino acids in newly synthesised molecules identified immunologically by size and charge
3. Detection in tissue culture media
4. High concentrations in extracts of presumed tissue or origin, i.e. endometrium
5. Histochemical localisation
6. Higher concentrations in amniotic fluid than in peripheral blood.

Table 2. Proteins of endometrial/decidual origin

1. Prolactin (1)

2. Alpha-2 mobile proteins:

   Progestagen associated endometrial protein (PEP) (2)

   Alpha-uterine protein (AUP) (3)

   Placental protein 14 (PP14) (4)

   Alpha-2 chorionic microglobulin (α_2 CMG) (5)

   Endometrial polypeptide 15 (EP15) (6)

   Alpha$_2$ pregnancy-associated endometrial globulin (α_2 PEG) (6)

3. Alpha-1 mobile protein

   Placental protein 12 (PP12) (4)

   Alpha-1 chorionic microglobulin (α_1 CMG) (5)

   Endometrial polypeptide 14 (EP14) (6)

   Alpha-1 pregnancy associated endometrial globulin (α_1 PEG) (6)

A comprehensive review of the physicochemical characteristics of the molecules listed as α_2 mobile proteins reveals some striking similarities with respect, at least, to molecular size (approx. 50 KD). Immunological identity has been demonstrated between PEP and AUP, PEP and PP14, α_2 PEG and PP14, and PP14 and α_2 CMG. It is tempting to speculate that this group of proteins does represent the same substance, the minor differences in structure being a function of analytical or preparatory purification techniques. Nevertheless, it is likely that these substances are at least a closely related group. The biological and physiological aspects of PEP are discussed in detail elsewhere in this volume, but it is noteworthy that there is an apparent influence of progesterone on synthesis, and trends in blood levels such as the first trimester peak bear a strong similarity with the blood level trends of human chorionic gonadotrophin. The information on the α_1 mobile glycoproteins is not as extensive as for PEP, but the common features of molecular size, charge and in some cases antigenic identity (PP12 and α_1 CMG; PP12 and α_1 PEG) once again lead to similar conclusions that this group of proteins is at least related. In contrast with PEP, PP12 secretion may reflect the functioning mass of the decidualised endometrium, peak blood levels being seen in the second trimester of pregnancy.

These preliminary findings suggest that not only will it be possible to further investigate the important role of the endometrium but to do so by techniques which are not invasive. These developments await further detailed characterisation of these molecules with respect to structure and function, and a resolution of the impending problem of nomenclature. There is no shortage of questions in human reproduction waiting to be addressed.

Acknowledgement

Some of these views came from a panel discussion which included S.C. Bell (UK), H. Bohn (W. Germany), A.E. Bolton (UK,) M.G. Chapman (UK), G.S. Joshi (US), E.M. Rutanen (Finland), and others.

References

1. Healy, D.L. 1984. Aust.N.Z.J.Obstet.Gynaec. 24, 111.
2. Joshi, S.G., Ebert, K.M., Swartz, D.P. 1980. J.Reprod.Fertil. 59, 273.

3. Sutcliffe, R.G., Joshi, S.G., Paterson, W.R., Bank, J.F. 1982. J. Reprod.Fertil. 65, 207.

4. Bohn, H., Winckler, W. 1982. Placental Suppl. 4, 67.

5. Tatarinov, Y.S., Kozljaeva, G., Petrunin, D., Petrunina, Y. 1982. In: Pregnancy Proteins: Biological and Clinical Application (Grudzinskas, J.G. et al, eds). Academic Press, London, p.179.

6. Bell, S.C., Patel, S., Hales, M.W., Kirsan, P.L., Drife, J.O. 1985. J.Reprod.Fertil. 74, 261.

PREGNANCY-RELATED CHANGES IN MURINE AND HUMAN ENDOMETRIUM REVEALED BY DIFFERENTIAL BINDING OF FLUORESCEINATED LECTINS.

Ivan Damjanov and Mao-Chi Lee

Department of Pathology and Laboratory Medicine, Hahnemann University School of Medicine, Philadelphia, PA 19102 USA

Introduction

Plant and animal lectins labeled with fluorescein isothiocyanate (FITC) or horseradish peroxidase have been used for histochemical studies of many normal tissues of both animal and human origin (1,8,9). We have applied derivatized lectins to early mammalian embryology (10,11) and have subsequently extended those investigations to murine (2,12) and human (3,5) reproductive organs. In this report we summarize our lectin histochemical data on mouse and human endometrium (4,7) and show that lectins could be used as markers for pregnancy associated changes in the uterus.

Materials and Methods

Mouse endometria were obtained from virgin or timed-pregnant Swiss Webster mice. Normal human endometrial tissue in proliferative phase was obtained by curettage performed for various reasons. Endometrial tissue of pregnancy was obtained during the voluntary interruption of pregnancy in the first and early second trimester of pregnancy. All tissues were snap frozen in 2-methylbutane precooled in liquid nitrogen, sectioned in a cryostat, fixed in cold absolute acetone for 10 minutes and incubated with FITC lectins purchased from E.Y. Laboratory, San Mateo, California. Control sections were incubated in a solution that combined the FITC labeled lectin and 0.2 M of the specific inhibitory sugar (1) as listed in Table 1. The slides were washed extensively in phosphate buffered saline, mounted in glycerin and examined with a microscope outfitted with UV epi-illumination.

Pregnancy Proteins in Animals

TABLE 1: LECTINS USED AND CORRESPONDING INHIBITING SUGARS

LECTIN	SOURCE	INHIBITING SUGAR
Suc.ConA	Canavalia ensiformis	Methyl -Mannopyranoside
LCA	Lens culinaris	Methyl -Mannopyranoside
WGA	Triticum vulgare	D-GlcNAc
PWM	Phytolacca americana	D-GlcNAc
LAA	Laburnum alpinum	D-GlcNAc
STA	Solanum tuberosum	D-GlcNAc
DBA	Dolichos biflorus	D-GalNAc
SBA	Glycine max	D-GalNAc
HPA	Helix pomatia	D-GalNAc
LBA	Phaseolus limensis	D-GalNAc.
VVA	Vicia villosa	D-GalNAc
BPA	Bauhinia purpurea	D-GalNAc
HAA	Helix aspersa	D-GalNAc
SJA	Sophora japonica	D-GalNAc
PNA	Arachis hypogaea	D-Gal
GSA-I	Griffonia simplicifolia	D-Gal
MPA	Maclura pomifera	D-GalNAc
RCA-I	Ricinus communis	D-Gal
RCA-II	Ricinus communis	D-Gal
UEA-I	Ulex europaeus	L-Fucose
PHA-E	Phaseolus vulgaris	D-GalNAc
PHA-L	Phaseolus vulgaris	D-GalNAc

Results

Most lectins reacted in an identical manner with tissues from pregnant and non-pregnant uteri. Only five lectins listed in Table 2 reacted differentially with the epithelium of pregnant and non-pregnant endometrial glands. MPA and BPA did not react with the glands of the non-pregnant uterus but were moderately to intensely reactive with glands of pregnancy (Fig. 1), indicating that these two lectins recognize pregnancy specific and/or associated glycoproteins. UEA-I reacted strongly with endometrial glands in the non-pregnant uteri, but was unreactive with endometrial glands of pregnancy (Fig. 2).

TABLE 2: LECTINS SHOWING DIFFERENTIAL BINDING TO NON-PREGNANT AND PREGNANT MOUSE ENDOMETRIUM

LECTIN		NON-PREGNANT	PREGNANT
WGA	Stroma	++	++
(Triticum vulgare)	Epithelium	+	++
RCA-I	Stroma	++	++
(Ricinus communis)	Epithelium	+/-	++
MPA	Stroma	+	+
(Maclura pomifera)	Epithelium	-	++
BPA	Stroma	-	-
(Bauhinia purpurea)	Epithelium	-	+
UEA-I	Stroma	-	-
(Ulex europaeus)	Epithelium	++	-

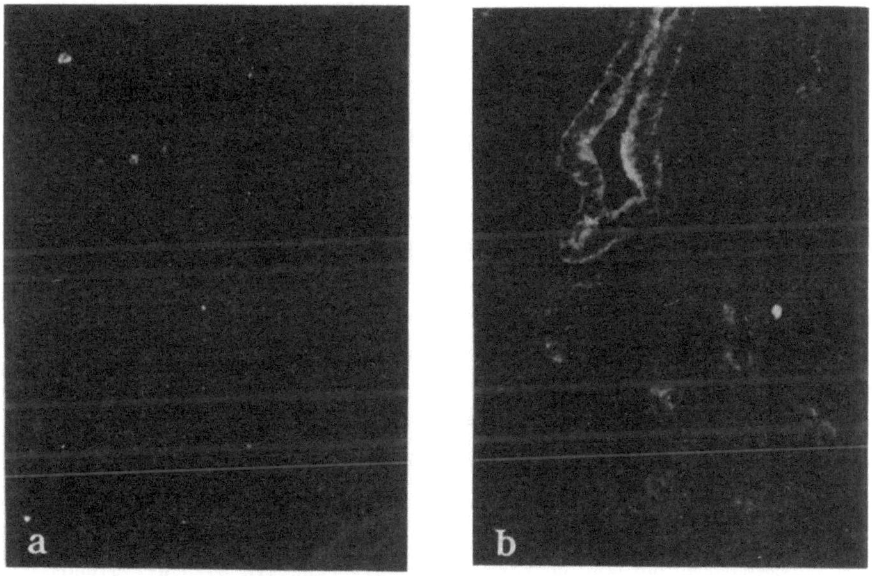

Fig. 1 Mouse non-pregnant (a) and 5 day pregnant endometrium (b) stained with FITC-labeled BPA. There is no reactivity in the non-pregnant endometrium. Glands of the pregnant endometrium are moderately reactive.

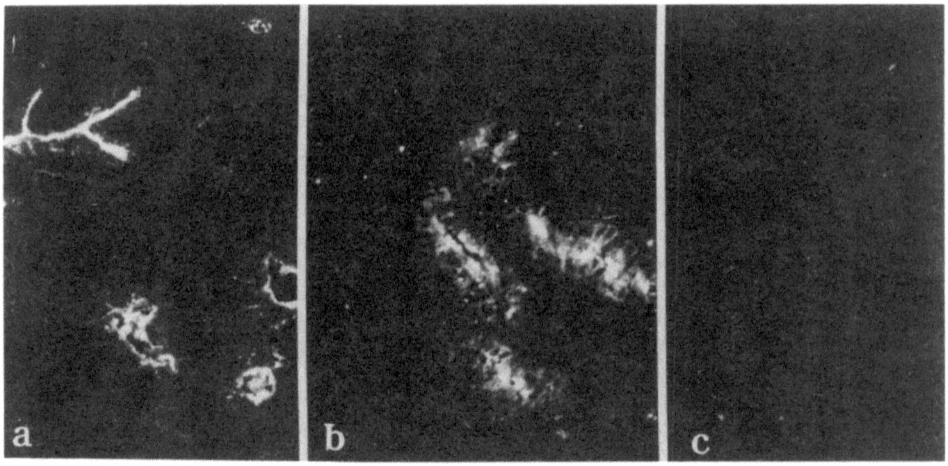

Fig. 2 Mouse non-pregnant (a), day 3 pregnant (b), and day 7 pregnant (c) endometrium stained with FITC-labeled UEA-I. Glandular and surface epithelium of the non-pregnant endometrium show strong reactivity. In day 3 pregnant endometrium, the reactivity of the surface epithelium has disappeared while the glands still show strong reactivity. In day 7 pregnant endometrium there is no visible reactivity.

Most lectins tested on human endometrium of pregnancy showed the same type of reactivity as in the proliferative phase endometrium. However, lectins from soybean (SBA), Vicia villosa (VVA) and peanut (PNA) appeared more reactive with pregnant than proliferative endometrium. SBA did not react with proliferative phase endometrial glands but only the surface epithelium (Fig. 3). This lectin was strongly reactive with all glands in pregnant endometrium. VVA reacted faintly with the basement membrane and the cytoplasm of glands in proliferative phase endometrium, but was strongly reactive with endometrial glands of pregnancy (Fig. 4). PNA reacted only with scattered glands in the proliferative endometrium, but was uniformly strongly reactive with all glands in the endometrium of pregnancy. Lectins from lentils (LCA) and the two lectins from kidney beans (PHA-E and PHA-L) reacted with endometrial glands of proliferative phase but not with the glands of pregnancy.

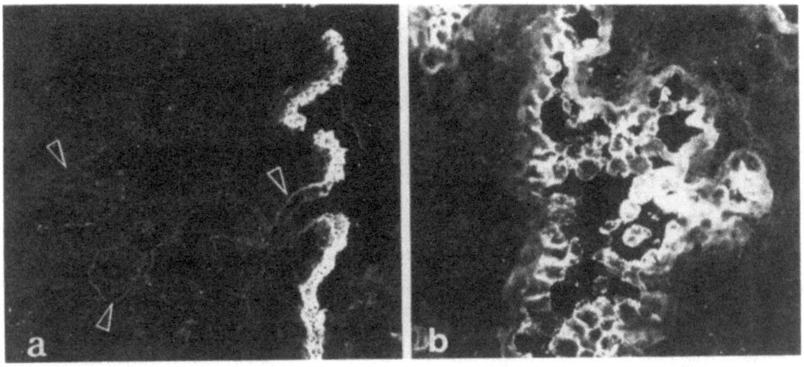

Fig. 3 Human proliferative phase endometrium (a), and endometrium of pregnancy (b) stained with FITC-labeled SBA. In the proliferative endometrium the reactivity is limited to surface epithelium and the basement membranes of endometrial glands (arrowheads). Endometrial glands of pregnancy show strong reactivity most prominently along the periluminal cell surface.

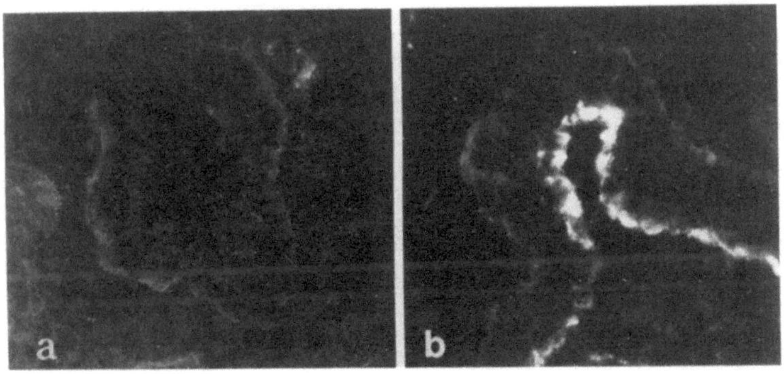

Fig. 4 Human proliferative phase endometrium (a), and endometrium of pregnancy (b) stained with FITC-labeled VVA. Proliferative endometrium shows weak staining of glandular basement membranes and some faint cytoplasmic reactivity of stromal and glandular cells, contrasting with the strong reactivity along the periluminal border of the glands of pregnancy.

Discussion

In the present report we have shown that several derivatized lectins react selectively with pregnant or non-pregnant uterine glands recognizing pregnancy specific and/or associated changes in both mouse and human uteri. Salient features of our study could be summarized as follows: a) Pregnancy associated and/or specific changes recognized by FITC-labeled lectins in human endometrium differ from those noted in mouse endometrium most likely reflecting species differences. Similar species specific differences have been previously recorded for other tissues such as, e.g., sperm (2). b) Some lectins react stronger with pregnant than non-pregnant endometrial glands, whereas others react stronger with non-pregnant than pregnant endometria. This indicates that certain glycoconjugates, recognized by these lectins appear on the endometrium of pregnancy, whereas others disappear from the surface of these cells. This is consistent with present day concepts about the cell surface changes in endometrium during pregnancy (4). However, more work is indicated to specifically define the nature of glycoconjugates on the endometrial glands and their exact function. c) Some lectins react differentially with surface epithelium as compared with the glandular epithelium of the non-pregnant uteri, and also show that the pregnancy associated changes in the glandular epithelium differ from those in the surface epithelium of the endometrium. These differences probably reflect distinct functions of various components of the endometrium.

The present report describes our first steps toward identification and characterization of pregnancy associated cell surface changes on endometrial cells. Following this preliminary screening phase, we are using the lectin histochemistry data as leads for chromatographic isolation of functionally important glycoconjugates from murine and human endometria.

References

1. Goldstein, I.J. and C.C. Hayes. 1978. The lectins: carbohydrate-binding proteins. Adv. Carbohydr. Chem. Biochem. 35, 127-340.

2. Lee, M.-C. and I. Damjanov. 1983. Anatomic distribution of lectin binding sites in mouse testis and epididymis. Differentiation 27, 74-81.

3. Lee, M.-C. and I. Damjanov. 1984. Lectin histochemistry of human placenta. Differentiation 28, 123-128.

4. Lee, M.-C. and I. Damjanov. 1985. Pregnancy-related changes in the human endometrium revealed by lectin histochemistry. Histochemistry 82, 275-280.

5. Lee, M.-C. and I. Damjanov. 1985. Lectin binding sites on human sperm and spermatogenic cells. Anat.Rec. (in print).

6. Lee, M.-C., T.-C. Wu,Y.-J. Wan and I. Damjanov. 1983. Pregnancy-related changes in the mouse oviduct and uterus revealed by differential binding of fluoresceinated lectins. Histochemistry 79, 365-375.

7. Spicer, S.S. and B.A. Schulte 1982. Identification of cell surface constituents. Lab. Invest. 47, 2-4.

8. Watanabe, M., T. Muramatsu, J., Shirane and K. Ugai. 1981. Discrete distribution of binding sites for Dolichos buflorus agglutinin (DBA) and for peanut agglutinin (PNA) in mouse organ tissues. J. Histochem. Cytochem. 29, 779-790.

9. Wu, T.-C., M.-C. Lee, Y.OJ. Wan and I. Damjanov. 1984. Lectin binding sites of the mouse ovary, intraovarian and ovulated ova. Histochemistry 80, 527-533.

10. Wu, T.-C., Y.-J. Wan and I. Damjanov. 1983. Fluorescein conjugated Banderiraea simplicifolia lectin as a marker of endodermal, yolk sac and trophoblastic differentiation in the mouse. Differentiation 24, 55-59.

11. Wu, T.-C.,Y.-J. Wan and I. Damjanov. 1983. Distribution of Bandeiraea simplicifolia lectin binding sites in genital organs of female and male mice. Histochemistry 77, 233-241.

THE ACUTE PHASE RESPONSE IN HEALTHY PREGNANT WOMEN AND IN PREGNANT PATIENTS WITH RHEUMATIC DISEASE.

M. Østensen, G. Husby, G. Marhaug

Department of Rheumatology, University of Tromsø, Tromsø, Norway

B. von Schultz

Department of Obstetrics and Gynaecology, University of Umeå, Sweden

Introduction

In inflammatory rheumatic diseases, several acute phase proteins are employed as indicators of disease activity. Their measurement has been found useful in monitoring longitudinally the disease activity in a given patient where they may provide guidelines for successful treatment (1). Many studies suggest that particularly the serum concentration of C-reactive protein (CRP) reflects quite precisely the extent of inflammation and parallels clinical parameters of disease activity both in rheumatoid arthritis (RA) and ankylosing spondylitis (AS) (1).

The present work refers to a prospective study of pregnant patients with rheumatic diseases followed with regular intervals during pregnancy and after parturition. Previous own studies (2) and studies done by others (3) had revealed that RA and AS behave quite different during gestation. While about 80% of RA patients improve strikingly from disease symptoms, the majority of AS patients has unaltered course or even aggravation of their disease during pregnancy (2). However, both groups of patients experience a flare of disease symptoms, often associated with an acute exacerbation 6-12 weeks after delivery.

The purpose of the present study was to evaluate the interaction

Pregnancy Proteins in Animals

between pregnancy and rheumatic disease both clinically and biochemically. Special attention was paid to the 2 major acute phase reactants CRP and serum amyloid related protein A (SAA) and their behaviour during pregnancy in healthy women as well as in rheumatic patients. Thirty-seven pregnant patients were compared to 40 healthy pregnant women studied longitudinally during and after pregnancy.

Patients and methods

The study comprised 12 women with definite RA, 18 with AS and 7 women with other seronegative arthropathies (SNA) including psoriatic arthritis. Their age ranged 18 to 38 years. Fourty randomly selected healthy pregnant women aged from 18 to 38 years, were studied for control. There were no differences between patients and controls with respect to parity.

When possible, all pregnant women were seen at 4 to 6 weeks intervals. At each control, venous blood samples were drawn and immediately analysed for various laboratory parameters including ESR, haptoglobin, orosomucoid, ceruloplasmin, complement factors C3 and C4, and CRP. Aliquots of serum were stored at -20° C until assayed for SAA.
Orosomucoid was assayed by an immunoprecipitation technique while ceruloplasmin, CRP and complement factors were measured by radial immunodiffusion using commercial plates and standards. The range of CRP assay was 6-100 mg/l, values below 6 mg/l were within the normal range. SAA was determined in 26 patients and in 28 controls by a radioimmunoassay (4). Serum levels below 3 mg/l were within the normal range. Alpha-2-pregnancy associated globulin (α_2-PAG) was measured by a radioimmunoassay. Disease activity of RA, AS and SNA was assessed by various clinical parameters using a four-graded scale (6). Active rheumatic disease was defined as a total score of > 8 while remission was defined as a total score of < 6.

After delivery, the patients continued their regular 6 week interval visits. Blood samples were obtained from the healthy controls either immediately after parturition (referred to as week 1) or 6 weeks after delivery. Fifteen of the healthy controls had an additional visit 12 months after delivery to obtain non-pregnant baseline levels.

Results

10 of the RA patients, 6 of the AS and 4 SNA patients improved markedly from disease symptoms during the 1st and 2nd trimester of pregnancy and had their best period in the 3rd trimester. Two RA, 12 AS and 3 SNA patients experienced no change or even aggravation of disease activity during gestation. A disease flare-up occurred in most patients, regardless of diagnosis, 6 to 12 weeks after delivery.

Biochemical changes. During pregnancy, most laboratory measurements including ESR, haptoglobin, orosomucoid, ceruloplasmin and complement factors in pregnant rheumatic patients paralleled those found in healthy controls and were typical for normal gestation. The 3 groups of women showed a gradual increase in ESR, ceruloplasmin and complement factors (results not shown) and a decrease of haptoglobin and orosomucoid (Table 1). The median pregnancy values of haptoglobin and orosomucoid never exceeded the normal non-pregnant range, even in the patient group with active disease. However, marked elevations of haptoglobin and orosomucoid were observed in individual patients with an acute flare of their disease during pregnancy. In contrast, ESR and ceruloplasmin clearly exceeded the normal range, regardless of disease activity, in the 2nd and 3rd trimester both in patients and controls.

Table 1. Median values of several acute phase proteins during pregnancy in patients with rheumatoid arthritis (RA) and ankylosing spondylitis (AS).

Parameter	Patient	1. trimester	2. trimester	3. trimester	Range	Range in pregnant controls
ESR	RA	23	36	46	7-67	4-50
MM/HR	AS	22	34	55	4-120	
Hapto-globin	RA	0.94	0.74	0.75	0.38-1.41	0.26-0.97
G/L	AS	0.74	0.65	0.75	0.39-1.75	
Orosom-ucoid	RA	0.74	0.52	0.47	0.38-1.36	0.25-0.86
G/L	AS	0.64	0.56	0.60	0.33-1.41	
Cerulo-plasmin	RA	0.56	0.69	0.62	0.40-0.82	0.38-0.92
G/L	AS	0.55	0.75	0.80	0.34-1.07	
CRP	RA	8.7	11.1	9.1	6-25	< 6-8
MG/L	AS	8.2	9.5	10.3	6-40	
SAA	RA	10.9	8.5	3.5	1.2-105	0.2-3.3
MG/L	AS	2.5	4.5	2.0	2-22	

In healthy pregnant women, serum levels of CRP and SAA were within the normal range throughout gestation with the exception of 3 women who had an increase of CRP and SAA at episodes of urinary tract infection. Three further women showed a slight rise of CRP without any sign of infection 2 weeks before delivery. In 5 women, SAA was measured the day before delivery and was found to be normal. In rheumatic patients, CRP and SAA measurements correlated significantly to the disease activity score (Fig. 1), but RA patients had generally higher SAA and CRP values than AS and SNA patients. Rheumatic patients with inactive disease during pregnancy showed a clear decrease or even normalization of both CRP and SAA while patients with acute flares or continously active disease had elevated CRP and SAA

levels. Great individual variations of CRP and SAA levels were noted.
Interestingly, 5 of the patients who were followed during 2 of their subsequent pregnancies, displayed the same individual pattern both for the acute phase proteins and immunoglobulins. The trauma of parturition caused a marked increase in ESR and all of the acute phase proteins both in patients and controls. While levels of haptoglobin, orosomucoid and complement factors increased two- or threefold and ceruloplasmin by about 50%, CRP showed a tenfold increase. The most impressive rise was seen for SAA which increased several 100 fold and up to 1,000 fold the non-pregnant values both in healthy women and in pregnant patients (Fig. 1). The highest SAA levels were measured in women with prolonged labour or who delivered by Cesarean section.

When comparing acute phase proteins in pregnant healthy controls and pregnant patients, significant differences between the groups were found only for CRP and SAA. All other acute phase reactants were similar in patients and controls during gestation, although somewhat higher median values were generally found in RA patients. None of the proteins measured showed any negative correlation to parameters of disease activity, neither alone or in combination with other serum factors. In healthy women, the acute phase proteins had returned to normal at 6 weeks after delivery.
In rheumatic patients, a shift from the pregnant to the non-pregnant pattern occurred during the first two weeks after delivery and then followed the clinical parameters of disease activity.

In search for a remission inducing factor, α_2-PAG was measured during and after pregnancy in 10 RA, 15 AS and 7 SNA patients. The pregnancy levels of α_2-PAG in patients who remitted and those who did not are shown in table 2. No significant differences between the groups could be detected. Interestingly, patients with active disease had higher α_2-PAG levels than patients in remission. This association between active joint inflammation and elevated α_2-PAG was clearly demonstrated in a SNA patient who could be followed in 2 of her subsequent pregnancies. In the first pregnancy, she improved from arthritis and had an α_2 -PAG mean value of 533 µg/ml while she got an acute flare in the following pregnancy with α_2-PAG rising to 1,700 µg/ml. Thus, α_2-PAG appeared to behave like an acute phase reactant in these patients.

Table 2. α_2-PAG levels (µg/ml) in patients with rheumatoid arthritis (RA) ankylosing spondylitis (AS), seronegative oligoarthritis (SNA) and controls during pregnancy

Diagnosis	In remission	Not in remission
RA	Mean: 376.3 Range: 68-1450 (N=7)	Mean: 489.0 Range: 190-1040 (N=3)
AS	Mean: 302.9 Range: 145-670 (N=5)	Mean: 424.3 Range: 76-1720 (N=10)
SNA	Mean: 391.3 Range: 56-550 (N=3)	Mean: 629.2 Range: 200-1050 (N=4)
Controls	Mean: 434 Range: 70-1000 (N=14)	

Fig. 1. Disease activity scores and serum concentrations of SAA during and after pregnancy in 26 women with rheumatic diseases. A, disease activity score in 12 patients with rheumatic arthritis (RA) and 14 patients with ankylosing spondylitis (AS). In RA patients, the disease activity score decreased during pregnancy as a sign of improvements. In AS patients, the disease activity score increased due to active disease during pregnancy. B, serum concentrations (median values) of SAA in 12 RA patients, 14 AS patients and 28 healthy women during and after pregnancy.

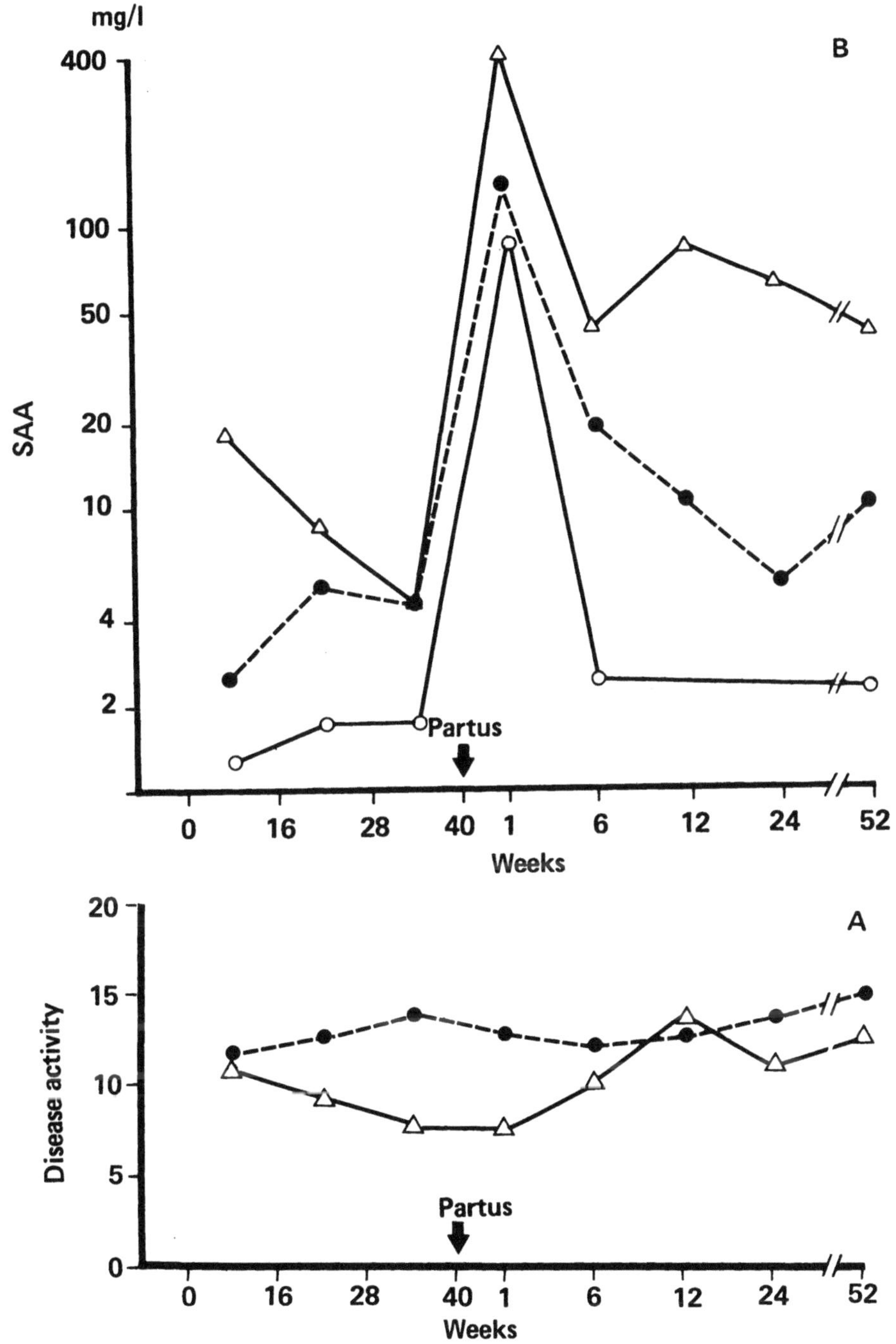
mg/l
B
SAA
400
100
50
20
10
4
2
Partus
0
16
28
40
1
6
12
24
52
Weeks
A
Disease activity
20
15
10
5
0
Partus
0
16
28
40
1
6
12
24
52
Weeks

Conclusion

From the present study, it can be concluded that normal pregnancy alters serum concentrations of most acute phase proteins but not of CRP and SAA. In contrast to previous cross-sectional studies of selected pregnant women near term (for reference, see 7), we found elevation of these two sensitive acute phase reactants in uncomplicated pregnancy. Intercurrent infections in pregnant women caused a transient rise in both CRP and SAA. Also, these proteins normalized rapidly after parturition. This means that CRP and SAA can be employed as useful monitors of inflammatory states occurring during normal gestation.

With regard to pregnant rheumatic patients, CRP and SAA proved to be the most reliable indicators of disease activity during gestation as they were not influenced by pregnancy itself. The clear decrease of these 2 acute phase proteins in patients who improved during pregnancy can be regarded as an objective sign of remission. It has been suggested that some of the acute phase reactants could be involved in the suppression of inflammation which is regularly seen in pregnant RA patients. Yet, the present data provide no evidence for any of the acute phase proteins measured to be a remission inducing factor in inflammatory rheumatic disease.

Finally, α_2-PAG could not be confirmed as an remission-inducing factor in the RA or AS patients studies, in contrast, it appeared to behave like an acute phase reactant.

References

1. Pepys, M.D. 1981. Clin.Immunol.Allergy. 1, 77.

2. Østensen, M., G. Husby. 1984. Klin.Wschr. 62, 891.

3. Cecere, F.A., R.H. Persellin. 1981. Clin.Rheum.Dis. 7, 747.

4. Marhaug, G. 1983. Scand.J.Immunol. 18, 329.

5. Damber, M.G., B. von Schultz, T. Stigbrand, K. Carlstrøm. 1976. Clin.Chim.Acta. 66, 85.

6. Østensen, M., G. Husby. 1983. Arthr.Rheum. 26,1155.

7. Østensen, M., G. Marhaug, G. Husby. 1985. Acta.Path.Microbiol. Immunol. Scand. Sect. C. 93, 1.

PREGNANCY PROTEINS AND THE ACUTE-PHASE RESPONSE

*Gillian T. Waites and S.C. Bell

Department of Obstetrics & Gynaecology and Biochemistry, University of Leicester, University Road, Leicester LE1 7RH, U.K.

*Present address: AFRC Institute of Animal Physiology, Babraham, Cambridge CB2 4AT, U.K.

Introduction

Threats to the integrity of the organism such as tissue injury or infection are met by a local acute inflammatory response and a number of systemic and metabolic changes which together constitute the acute-phase response. The primary aim of this response is to permit immediate post-trauma survival, prior to the development of specific protection afforded by the immunological response, by the containment or destruction of infectious agents, removal of damaged tissue, and repair of the affected organ. The liver is central to the acute-phase response as reflected by the increase in hepatic weight following infection and by the ultrastructural and metabolic alterations observed (see 1). Amongst these changes are dramatic increases or decreases in the concentration of certain serum proteins largely of hepatic origin (2, 3) although other sites of synthesis for some of these proteins, termed acute-phase reactants, have been reported (4, 5). Although the precise function of many of these proteins in the acute-phase response is uncertain, the consistency of their response and the deleterious effects ensuing from failure of this indicate their importance.

During pregnancy implantation of the blastocyst, development of the haemochorial placenta, involving as it does extensive invasion of the uterine endometrium, or parturition *per se* may be expected to be potential inducers of local acute inflammation and the systemic acute-phase response. Also antigenically foreign trophoblast tissue coming into contact with the maternal blood may induce a specific immunological response. This suggestion of a

Pregnancy Proteins in Animals

relationship between the acute-phase response and pregnancy raises a number of questions: Are elements of the acute-phase response induced by pregnancy and do acute-phase reactants play a role here in protection against tissue damage? Are any pregnancy-associated proteins also acute-phase reactants? Can pregnancy modify an acute-phase response and does this involve pregnancy proteins? It is the aim of this chapter to tackle such questions by investigating the relationship between acute-phase reactants and pregnancy-associated proteins, principally in animals, to show the value of pregnancy-related studies in providing insights into the mechanisms of regulation of inflammation and the acute-phase response.

Acute-Phase Reactants

Many proteins have been described whose serum concentration alters non-specifically in response to most forms of trauma and these can be classified according to the magnitude and rapidity of their response. Table 1 shows a classification scheme for human and rodent acute-phase reactants based on the system defined by Kushner (1) for human proteins. The majority of proteins mentioned are well documented (2) and will not be discussed here; rather attention will be focused upon a group of proteins termed "pentraxins" (6) or more correctly "pentaxins" (3) due to their cyclic pentameric symmetry. This group of proteins, to which C-reactive protein (CRP) and serum amyloid P-component (SAP) belong, is the subject of an excellent recent review to which the reader is referred (3).

The pentaxins exhibit extensive homologies in their amino acid sequence and analogues have been detected in all species of vertebrates and invertebrates examined, although they are not always acute-phase reactants. In the human CRP is the major archetypal acute-phase reactant increasing up to 1000-fold after injury, whilst in the mouse SAP is the major reactant increasing up to 50-fold. In the rat neither CRP nor SAP behave as acute-phase reactants and instead α_2-acute-phase macroglobulin (α_2MG; 7) seems to uniquely represent the major acute-phase reactant. However, unusually high levels of CRP are found in both normal and

Table 1. Classification of some of the Major Acute-Phase Reactants in Humans and Rodents

	Positive			Negative
	Class I (50% increase)	Class II (up to 10-fold increase)	Class III (up to several hundred fold increase)	
<u>Human</u>	Caeruloplasmin	Haptoglobin	CRP	Albumin
	C3	α_1-antitrypsin	SAA	Transferrin
		fibrinogen		α_2-HS glycoprotein
		α_1-antichymotrypsin		
		α_1-acid glycoprotein (orosomucoid)		
		α_2-PAG		
<u>Mouse</u>		Haptoglobin	SAP	
			SAA	
			α_1-PAP	
<u>Rat</u>	CRP	α_1-acid glycoprotein	Acute-phase	Transcortin (CBG)
	caeruloplasmin	fibrinogen	α_2-macroglobulin	α_1-PAP
	α_1-antitrypsin	haptoglobin		FP
	fibrinogen	α_1-glycoprotein		
<u>Hamster</u>		FP (♂)		FP (♀)

injured rats. In the hamster a single protein (hamster female protein, FP) has been described which seems to be structurally and functionally analogous to CRP/SAP in other species, although exhibits a female-specificity and a divergent acute-phase response not observed for the other pentaxins (8).

Both CRP and SAP have been shown to be of hepatic origin and it has been suggested, to explain stimulation by inflammation occurring at peripheral sites, that this involves the activation of local mononuclear phagocytes to produce a monokine, possibly interleukin-1 (IL-1), which is then transported to the liver where

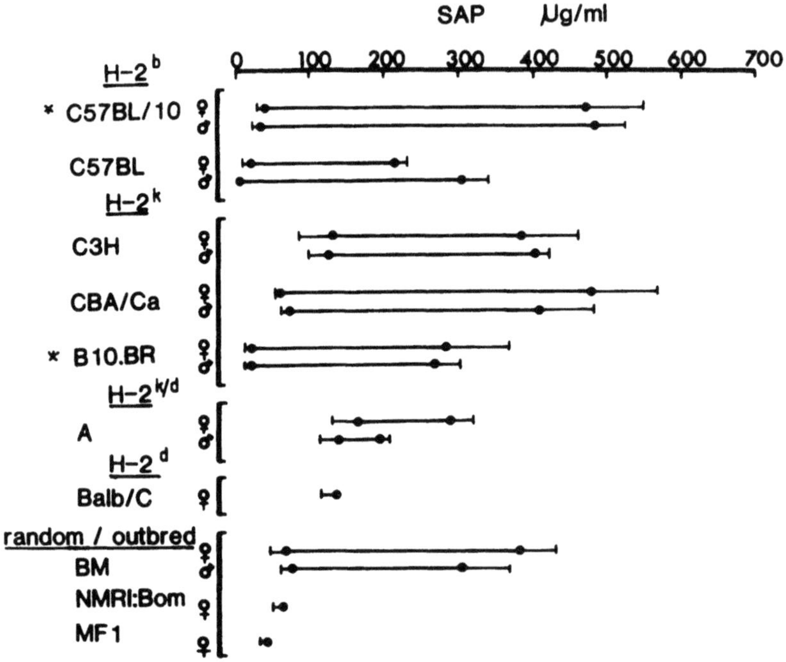

Figure 1. Endogenous serum levels of SAP in various inbred and random bred strains of mice and the increase following s.c. injection of 0.1 ml turpentine/arachis oil (1:1; v/v) 72 and 24 h prior to sampling. Differences in the endogenous levels can be seen to be strain-dependent. Animals with a BL10 non-H-2 background (C57BL/10 and B10.BR) have significantly lower levels of SAP than most other strains and a secondary influence is exerted by H-2 genes. The increase following injury can be seen to be dependent on the endogenous levels (see also Figure 2), not evident during pregnancy (Figure 4). *indicates congenic mice.

it triggers synthesis of acute-phase proteins (see 3) in a manner similar to, but distinct from that regulating another human acute-phase protein, serum amyloid-A (SAA; 9). Recently it has been suggested that prostaglandins may play a crucial role in macrophage stimulation (10).

While there are no major sex-dependent differences in the CRP or SAP response, strain-dependent variations in the endogenous levels of murine SAP have been reported (11) which, by studying a large

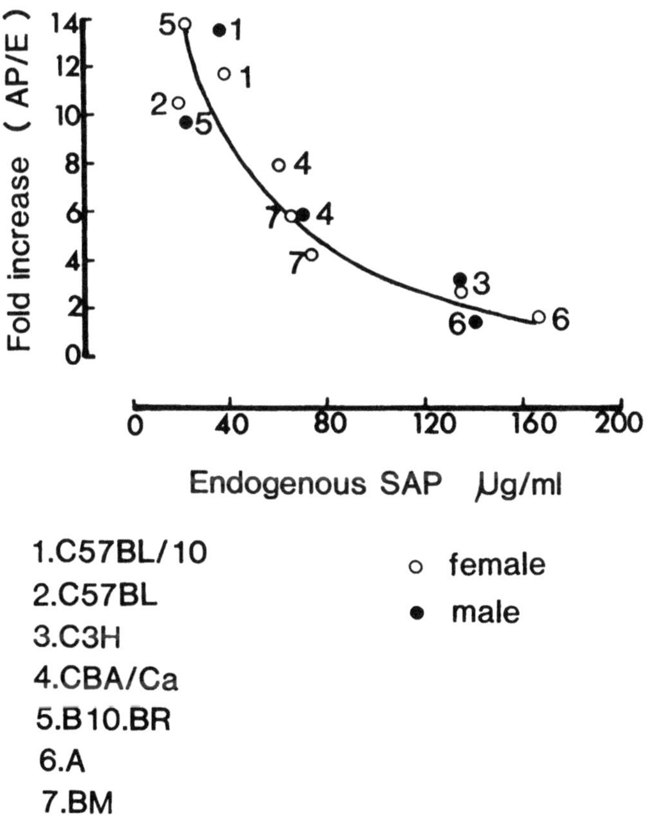

Figure 2. Relationship between endogenous SAP levels and the increase on acute-phase induction (AP/E) in the strains shown in Figure 1. No such relationship could be demonstrated during pregnancy (see Figure 4).

number of inbred strains, were found to be most strongly influenced by the non-H-2 background and more weakly by the H-2 complex (12). Furthermore, this latter study also demonstrated an inverse relationship between the endogenous levels and the magnitude of acute-phase increase attained and these differences have been suggested to reflect strain-dependent variations in IL-1 synthesis by macrophages. A similar dependence on endogenous levels has been found with hamster FP, although in this case the high normal female levels actually decrease on acute-phase stimulation while the low male levels increase (8). As part of our investigations on acute-phase reactants in pregnancy, levels of SAP were determined in a number of murine strains before and after injury and as Figure 1 shows the strain-dependent variations observed by other workers were confirmed. The relationship between endogenous levels and the acute-phase increase can be seen expressed graphically in Figure 2.

Pregnancy Proteins in the Acute-Phase Response

In addition to these classic acute-phase reactants there are a number of proteins first identified as pregnancy-associated proteins, which have subsequently been found to also respond to acute-phase stimulation.

a) Human

Elevated serum levels of human pregnancy-associated α_2-glycoprotein (α_2-PAG) may occur after surgery and also in various malignancies and disease states (see 13) and the protein was put forward as an acute-phase reactant by Bohn (14) although not classical in its response (15). Nevertheless, the protein has been shown to stabilize lysosomes (16), to effect polymorphonuclear leucocyte (PMN) and monocyte functions (17, 18) which are central components in the acute-phase response, and would therefore seem to exert some influence on elements of the acute-phase response. Another protein, pregnancy-associated plasma protein A (PAPP-A), although not classified as an acute-phase reactant, is structurally and functionally very similar to human α_2-MG (19), itself similar to the acute-phase macroglobulin of the rat. The finding of a pregnancy-associated protein with prostaglandin synthetase inhibitory activity (PAPSI) by Stimson (20) adds further support to the suggestion of an important association between pregnancy and the acute-phase response.

b) Animals

Several animal proteins have recently been described which appear to represent rodent analogues of human α_2-PAG (Waites and Bell, this volume). Murine α_1-pregnancy-associated protein (α_1-PAP), originally identified in mid-pregnancy sera, was found to be elevated in females of the inbred C57BL/10 and random bred BM strains in the acute-phase of an inflammatory response (21), where it exhibited similar physico- and immunochemical properties to the pregnancy protein, although was present at much lower levels. In contrast to human α_2-PAG, the profile of serum concentrations of α_1-PAP following injury in females paralleled that of the classic acute-phase reactant SAP, but the proteins were shown to be distinct (21). The female specificity of α_1-PAP in the acute-phase response was found to be due to the regulatory effects of steroids with oestradiol in the male and progesterone in the female, in the absence of an overriding effect of testosterone, being required to enable injury to induce synthesis of α_1-PAP (22). Although SAP does not seem to be influenced by steroid hormones α_1-PAP is not unique in this type of regulation of synthesis; as already seen hamster FP is similarly inhibited by testosterone and rat α_2-MG is oestradiol-suppressed in the female (23). However, recent studies have shown that administration of exogenous oestradiol or progesterone or both to normal female mice failed to induce α_1-PAP synthesis and although a small increase in protein levels was detected in the sera of tubally-ligated pseudo-pregnant females, this was greatly enhanced by the presence of artificially-induced deciduoma suggesting that hormones alone do not provide the primary stimulus for α_1-PAP synthesis during gestation, not negating an indirect requirement for the maintenance of decidua. This may be suggestive of a different type of synthesis in the acute-phase response to that in pregnancy. Some hepatic synthesis has been demonstrated during pregnancy (Udagawa, Waites and Armstrong et al., this volume) and it is likely that acute-phase stimulation operates via this route, as with other acute-phase reactants, but the low levels induced imply that the other sites of synthesis are not triggered by injury. Although the stimuli may differ, synthesis by the liver may be hormone-dependent in both the acute-phase response and pregnancy, which would account for the low levels of α_1-PAP seen in non-decidualized pseudopregnant females. If a mechanism of regulation of α_1-PAP synthesis in the acute-phase response similar to that of SAP is assumed then hormones would either exert their effects on macrophages or directly on the liver.

The acute-phase response of α_1-PAP determined in a number of other inbred strains is shown in Figure 3. There was no evidence for genetic regulation of endogenous α_1-PAP levels similar to that

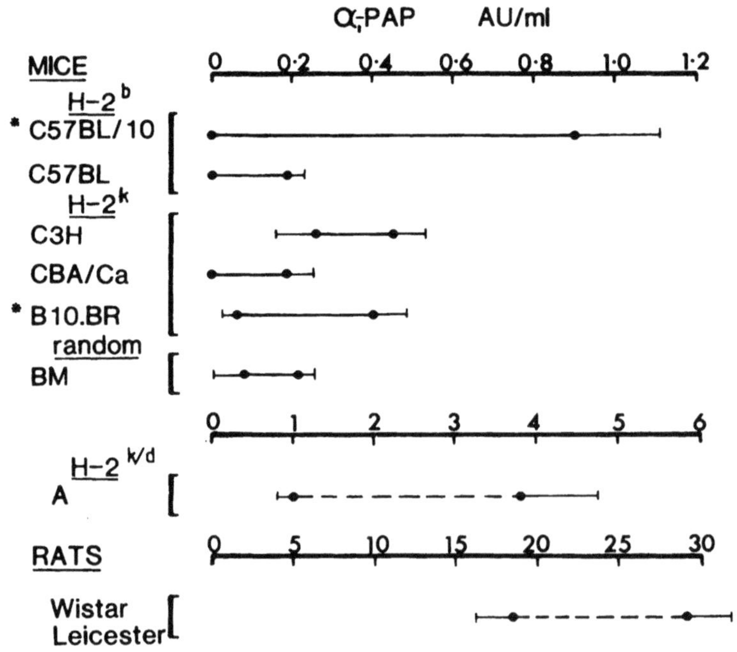

Figure 3. Alterations in the serum concentration of α_1-PAP following s.c. injection of 0.1 ml turpentine/arachis oil (1:1; v/v) 72 and 24 h prior to sampling. Values are mean ± sem (5 < N < 15) where 1 AU is the peak rocket height obtained with a 1:30 dilutuion of a standard of pooled murine pregnancy serum. Hatched lines represent a decrease. *indicates congenic mice.

described for SAP, and the magnitude of the acute-phase response did not seem to be dependent on endogenous levels. However, in females possessing substantially higher endogenous α_1-PAP levels, the inbred A strain and the outbred MF1 strain (not shown) α_1-PAP levels actually decreased on acute-phase induction which did show some hint of individual correlation with high endogenous levels reminiscent of the response of hamster FP. Rat α_1-PAP behaved similarly, being high in all non-pregnant females and exhibiting a mean decrease of 24% following injury. Of the other suggested rodent analogues of human α_2-PAG described, rat PAG, also high in non-pregnant females, has been found to be a negative acute-phase reactant (24) but rat α_2-PAG described by Sarcione and Delluomo (25) is absent from both non-pregnant and acute-phase females.

The major rat acute-phase reactant previously mentioned, α_2-MG, was in fact originally described as a pregnancy protein (26) where it is synthesized by the fetal liver (27), and possibly decidual tissue and the metrial gland (28, 29) at levels in excess even of the high acute-phase levels. This protein is exemplary of an association between pregnancy the the acute-phase response.

Acute-phase Reactants and the Acute-Phase Response in Pregnancy

It is well-known that pregnancy is capable of modulating the inflammatory response, as exemplified by the amelioration or exacerbation of a number of inflammatory disorders during gestation (30). It is now generally accepted that a majority of rheumatoid arthritis (RA) patients experience a remission of disease activity during pregnancy, and while patients with disorders such as systemic lupus erythematosus (SLE) and ankylosing spondylitis (AS) often also experience some modification of their disease this is not consistently a remission. There is considerable evidence to suggest that such modifications are the result of an effect of pregnancy-associated serum components. High levels of human α_2-PAG have been correlated with remission of RA during pregnancy (31, 32) and the protein has been shown to block not only PMN and monocyte functions as discussed, but also the *in vitro* sensitization against synovial components in RA (33).

However, such a correlation was not found by all workers (34, 35), this latter study showing instead a positive correlation between α_2-PAG levels and total disease activity during pregnancy which is more in agreement with its function as an acute-phase reactant, similar to CRP and SAP which have been shown to reflect disease activity in RA (36). Whether α_2-PAG is primarily responsible for modulation of inflammatory disorders during pregnancy remains controversial and other suggestions such as the release of soluble factors from fetal suppressor cells (37) or indeed from the maternal suppressor cells found in uterine draining lymph nodes and decidua of mice (38) and humans (39) could be implicated. In any event there is a definite modulation of the acute-phase response by pregnancy and, being a good measure of this response, the changes in serum levels of acute-phase reactants during normal pregnancy is of relevance.

a) Humans

The few studies undertaken have on the whole failed to show any dramatic changes in the serum concentrations of most acute-phase reactants, although albumin and α_1-acid glycoprotein have been reported to decrease while transferrin, caeruloplasmin and CRP increase towards parturition (40, 41, 42). There is some evidence to suggest that the CRP increase is indicative of infection or premature rupture of the membranes (43) and is not therefore a feature of normal pregnancy. This and the finding of elevated α_1-acid glycoprotein in pregnancies complicated by acute inflammation (44) suggests that pregnancy does not obscure certain elements of a normal acute-phase response.

b) Animals

Recent studies in the mouse showed an interesting triphasic increase in the serum levels of SAP and haptoglobin associated with primiparous allogeneic pregnancy in C57BL/10 females (45). That these increases corresponded to period of expression of paternal alloantigens gave rise to the hypothesis that foreign antigens could serve as an alternative acute-phase stimulus triggering maternal hepatic synthesis, although intraperitoneal administration of allogeneic spleen cells could not substitute for the pregnancy stimulus. This was thought to be due to the route of stimulation, however more recent studies to determine the strain-dependency of this pregnancy-induced increase, for comparison with that in the acute-phase response, revealed a more complex situation. Figure 4 shows the increase in levels of SAP on day 18 (previously found to represent maximal synthesis) in a variety of allogeneic and syngeneic pregnancies. The original observation of elevated SAP levels in C57BL/10 allogeneic pregnancies was confirmed using another male partner and levels also increased during pregnancy in a number of other inbred females examined. However, in no other females were the allogeneic/ syngeneic differences as dramatic as in the C57BL/10 strain, primarily due to the elevated SAP levels in syngeneic as well as allogeneic pregnancies. The pregnancy-induced increase did not reflect the acute-phase elevation of SAP levels either in its magnitude or strain-dependency (Figure 1) nor was there any

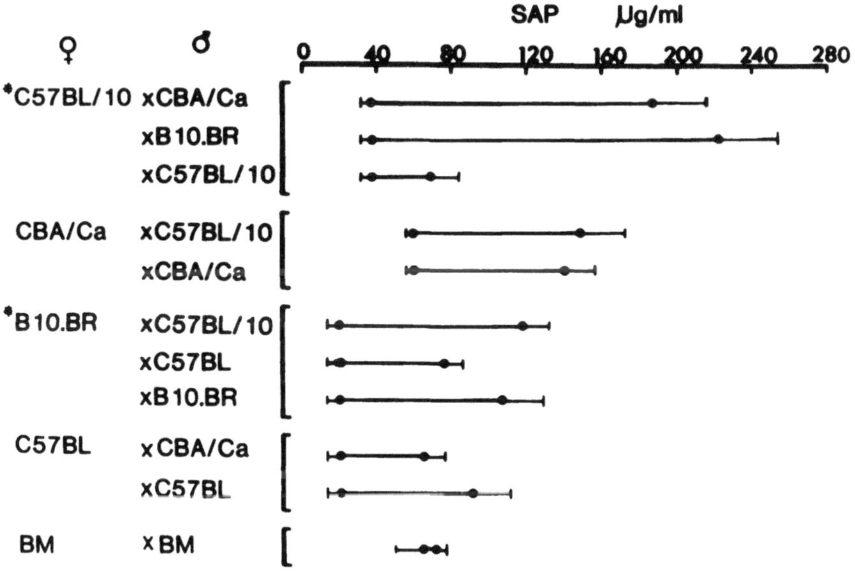

Figure 4. The increase in serum levels of SAP in various inbred and random bred strains on day 18 p.c. of pregnancy. With the exception of the C57BL/10 female no differences in serum SAP levels are observed between allogeneic and syngeneic pregnancies. Neither H-2 nor non-H-2 genes seem to have any influence on the pregnancy-induced SAP levels. Values are mean ± sem (4 < N < 15). *indicates congenic mice.

correlation between endogenous and pregnancy levels, as demonstrated for SAP in the acute-phase response (shown in Figure 2). This suggests different regulation of SAP synthesis during pregnancy and the acute-phase response, as proposed for α_1-PAP. The nature of this remains unknown, although the failure of the random bred BM strain to produce SAP in pregnancy may provide a clue to the induction of synthesis in pregnancy.

Administration of an acute-phase stimulus during the first half of pregnancy resulted in an increase in the SAP levels similar to that normally seen in an acute-phase response. In contrast levels of α_1-PAP were significantly reduced compared to normal pregnancy levels (46). This differed from the positive acute-phase response normally seen in the C57BL/10 female employed and could be interpreted as a reflection of the dependence on the pre-acute-phase α_1-PAP levels, elevated due to the pregnancy causing a negative acute-phase response similar to that noted in high endogenous females. Whether the SAP response would be similarly negative if assessed later in pregnancy is not yet known. It is an interesting corollary that hamster FP increases during pregnancy only in older females where endogenous levels are lower (Coe, personal communication).

The nature of the acute-phase response of rat α_1-PAP was apparently unaffected during the first half of pregnancy, being negative as usual although levels were somewhat lower than in non-pregnant injured females (46) and pregnancy levels of α_2-MG did not alter in response to acute-phase stimulation, although in this instance altered maternal synthesis could not be excluded (23). These findings suggest that while a normal acute-phase response can occur during pregnancy, there can be a transient interference as seen with α_1-PAP.

Concluding Remarks

There is no doubt that pregnancy can modulate the acute inflammatory response. It has been seen that a number of acute-phase reactants, but not all, alter their serum concentration throughout gestation, and conversely certain pregnancy-associated proteins seem to have a role in the acute-phase response. Studies on murine α_1-PAP imply that although in its response to injury the protein can be considered as a classic acute-phase reactant, the route of stimulation and control of synthesis during pregnancy differs from that in the acute-phase response. Furthermore, α_1-PAP differed from the classical SAP both in its apparent absence of any genetic regulation of endogenous levels and in the inability of these to control the acute-phase response, although this latter may occur. This and the fact and that the SAP response in pregnancy differed from that following acute-phase stimulation, implied the existence of two control mechanisms. It would not be unusual to have proteins synthesized by different routes in different physiological situations as exemplified by rat acute-phase α_2-MG, although this does not exclude the possibility of one site of synthesis, such as the liver, being triggered by different stimuli. The nature of the pregnancy stimulus is yet to be elucidated but it would not seem to be exposure to alloantigens as originally proposed for SAP. Pregnancy certainly does not substitute for an acute-phase stimulus but the fact that acute-phase stimulation during pregnancy is capable of modifying the serum α_1-PAP levels, strongly suggests some overlap between the two mechanisms of control.

The important question to be answered regards the function of these proteins and whether this is the same in both the acute-phase response and pregnancy. Some of the acute-phase reactants have defined roles such as transport of haemoglobin and copper by haptoglobin and caeruloplasmin respectively, a greater demand for which can be envisaged during pregnancy. The proteinase-binding properties of α_2-MG implicate it in the regulation of proteolytic activity in inflammation and tissue re-organisation (47) and it is possible that in pregnancy α_2-MG regulates tissue invasion by the trophoblast. The role of the pentaxins has not yet been determined although recently SAP has been shown to bind fibronectin (48) which could imply an involvement in the formation and maintenance of cellular contacts, again of importance both in the acute-phase response and pregnancy. The observed diminution of the SAP response in T-cell deprived mice (49) may imply some intervention by immunological factors in this response, which would then be of relevance during pregnancy. It is probable that the inflammatory and immune responses operate in unison and immunosuppressants such as human α_2-PAG would therefore be of relevance in both situations. Of additional importance is the effect of human α_2-PAG and its rodent analogues on inflammatory cell functions. It is believed that by studying proteins such as these described in rodent pregnancy much could be learnt about their function and mechanisms of control both during pregnancy and the acute-phase response.

Acknowledgements

G.T.W. was in receipt of a Postgraduate Studentship from the Medical Research Council.

References

1. Kushner, I. 1982. The phenomenon of the acute-phase response. Ann. N.Y. Acad. Sci. 389, 39-48.

2. Koj, A. 1974. Acute-phase reactants: their synthesis, turnover and biological significance. In: Structure and Function of Plasma Proteins, Vol. 2 (A.C. Allison, ed.). Plenum, New York. pp. 73-132.

3. Pepys, M.B. and M.L. Baltz. 1983. Acute-phase proteins with special reference to C-reactive protein and related proteins (pentaxins) and serum amyloid A protein. Adv. Immunol. 34, 141-212.

4. Hovi, C.H.W., D. Mosher and A. Vaheri. 1977. Cultured human monocytes synthesize and secrete α_2-macroglobulin. J. Exp. Med. 145, 1580-1589.

5. Isaacson, P., D.B. Jones and M.A. Judd. 1979. α_1-antitrypsin in human macrophages. Lancet ii, 964-965.

6. Osmand, A.P., B. Friedenson, H. Gewurz, R.H. Painter, T. Hoffman and E. Shelton. 1977. Characterization of C-reactive protein and the complement subcomponent Clt as homologous proteins displaying cyclic pentameric symmetry (pentraxins). Proc. Natl. Acad. Sci. U.S.A. 74, 739-743.

7. Weimer, H.E. and D.C. Benjamin. 1965. Immunochemical detection of an acute-phase protein in rat serum. Am. J. Physiol. 209, 736.

8. Coe, J.E. 1983. Homologs of CRP: A diverse family of proteins with similar structure. In: Contemporary Topics in Molecular Immunology, Vol. 9 (F.P. Inman and T.J. Kindt, eds.). Oxford University Press, New York · Oxford. pp. 341-361.

9. Tatsuta, E., J.E. Sipe, T. Shirahama, M. Skinner and A.S. Cohen. 1983. Different regulatory mechanisms for serum amyloid A and serum amyloid P synthesis by cultured mouse hepatocytes. J. Biol. Chem. 258, 5414-5418.

10. Whicher, J.T., A.M. Bell, M.F.R. Martin, L.A. Marshall and P.A. Dieppe. 1984. Prostaglandins cause an increase in serum acute-phase proteins in man, which is diminished in systemic sclerosis. Clin. Science 66, 165-171.

11. Pepys, M.B., M. Baltz, K. Gomer, A.J.S. Davies and M. Doenhoff. 1979. Serum amyloid P-component: an acute-phase reactant in the mouse. Nature 278, 259-261.

12. Mortensen, R.F., K. Beisel, N.J. Zeleznik and T. Le. 1983. Acute-phase reactants of mice II strain-dependence of serum amyloid P-component (SAP) levels and response to inflammation. J. Immunol. 130, 885-889.

13. Thomson, A.W. and C.H.W. Horne. 1980. Biological and clinical significance of pregnancy-associated α_2-glycoprotein (α_2-PAG) - a review. Invest. Cell Path. 3, 295-309.

14. Bohn, H. 1972. Characterization of the pregnancy-associated glycoproteins as acute-phase reactants. Archiv. für Gynäkol. 213, 54-72.

15. Horne, C.H.W., H. Bohn, A.L.C. McLay, E.H. Wood and W.D. Thompson. 1975. Serum pregnancy-associated α_2-glycoprotein (α_2-PAG) - an acute-phase reactant? Behring Inst. Mitt. 58, 50-53.

16. Hempel, K.H., L.A. Fernandez and R.H. Persellin. 1970. Effect of pregnancy sera on isolated lysosomes. Nature 225, 955-956.

17. Persellin, R.H. and J.K. Leibfarth. 1978. Studies on the effects of pregnancy serum on polymorphonuclear leukocyte functions. Arth. Rheum. 21, 316-325.

18. Persellin, R.H. and J. Rhodes. 1981. Inhibition of human monocyte Fc receptor and HLA-DR antigen expression by pregnancy alpha-2 glycoprotein. Clin. Exp. Immunol. 46, 350-354.

19. Sutcliffe, R.G., B.M. Kukulska-Longlands, J.R. Coggins, J.B. Hunter and C.H. Gore. 1980. Studies on human pregnancy-associated plasma protein-A. Purification by affinity chromatography and structural comparison with α_2-macroglubulin. Biochem. J. 191, 799-809.

20. Stimson, W.H. 1983. The influence of pregnancy-associated serum proteins and steroids on the maternal immune response.

In: Immunology of Reproduction, (T.G. Wedgmann and T.J. Gill III, eds.). Oxford University Press. pp. 281-302.

21. Waites, G.T. and S.C. Bell. 1984. Identification of a murine pregnancy-associated serum protein (α_1-PAP) as a female-specific acute-phase reactant. J. Reprod. Fert. 70, 581-589.

22. Waites, G.T. and S.C. Bell. 1984. Regulation of murine α_1-pregnancy-associated protein by gonadal steroids during the acute-phase response. J. Endocr. 101, 315-318.

23. Bell, S.C. 1980. Effect of oestradiol and progesterone on the concentration of α_2-macroglobulin in the sera of injured male and female rats. J. Endocr. 86, 189-191.

24. Schade, R., F. Gotz, B. Porstmann, A. Friedrich and E. Nugel. 1980. The acute-phase proteins of the rat, their regulation by hormones and biological significance. Agents and Actions Suppl. 10, 213-231.

25. Sarcione, E.J. and D.S. Delluomo. 1981. Demonstration of pregnancy-associated α_2-glycoprotein (α_2-PAG) in pregnant oestrogen-treated and breast cancer bearing rats. Onco. Dev. Biol. Med. 2, 345-354.

26. Beaton, G.H. and A. Selby. 1960. A new serum protein in pregnant rats. Fedn. Proc. 19, 74.

27. Sarcione, E.J. and M. Bohne. 1969. Synthesis of alpha$_2$ (acute-phase) globulin by foetal and neonatal rat liver in vitro. Proc. Soc. Exp. Biol. Med. 131, 1454-1456.

28. Bell, S.C. 1979. Synthesis of 'decidualization-associated protein' in tissues of the rat uterus and placenta during pregnancy. J. Reprod. Fert. 56, 255-262.

29. Bell, S.C. 1979. Immunochemical identity of 'decidualization-associated protein' and α_2-acute-phase macroglobulin in the pregnant rat. J. Reprod. Immunol. 1, 193-206.

30. Cercere, F.A. and R.H. Persellin. 1981. The interaction of pregnancy and the rheumatic diseases. Clinics in Rheumatic Diseases 7, 747-769.

31. Persellin, R.H., D.A.F. Wiginton, J.E. Rutstein, M.J. Cohen, R. Homer, R. Goehrs, A.D. Steele, V.K. Philips and R.P. Kradel. 1982. Pregnancy alpha-glycoprotein (PAG) and rheumatoid arthritis (RA) activity - A prospective analysis during gestation. Arthritis Rheum. Suppl. 25, S6.

32. Unger, A., A. Kay, A.J. Griffin and G.S. Panayi. 1983. Disease activity and pregnancy-associated α_2-glycoprotein in rheumatoid arthritis during pregnancy. Br. Med. J. 286, 750-752.

33. Appelboom, T. and R.H. Persellin. 1984. Leukocyte sensitization against synovial components in rheumatoid arthritis:

blocking by pregnancy serum. Int. Archs. Allergy appl. Immunol. 74, 200-205.

34. Horne, C.H.W., A.W. Thomson, C.B.J. Hunter, A.M. Tunstall, C.M. Towler and M.E.J. Billingham. 1979. Pregnancy-associated α_2-glycoprotein (α_2-PAG) and various acute-phase reactants in rheumatoid arthritis and osteoarthritis. Biomedicine 30, 90-94.

35. Østensen, M., B. von Schoultz and G. Husby. 1983. Comparison between serum α_2-pregnancy associated globulin and activity of rheumatoid arthritis and ankylosing spondylitis during pregnancy. Scand. J. Rheumatol. 12, 315-318.

36. Levo, Y., M. Ehrenfeld and S. Wollner. 1982. Serum amyloid P-component levels in rheumatoid arthritis. Isr. J. Med. Sci. 18, 715-717.

37. Froelich, C.J., J.S. Goodwin, A.D. Bankhurst and R.C. Williams. 1980. Pregnancy, a temporary fetal graft of suppressor cells in autoimmune disease? Am. J. Med. 69, 329-331.

38. Clark, D.A., R. Slapsys, A. Croy, J. Rossant and M. McDermott. 1983. Regulation of cytotoxic T cells in pregnant mice. In: Immunology of Reproduction (T.G. Wegmann and T.J. Gill, eds.). Oxford University Press, New York · Oxford. pp. 341-361.

39. Daya, S., D.A. Clark, C. Devlin, J. Jarrell and A. Chaput. 1985. Suppressor cells in human decidua. Am J. Obstet. Gynecol. 151, 267-270.

40. Mendenhall, H.W. 1970. Serum protein concentrations in pregnancy. I. Concentrations in maternal serum, pp. 388-399. II. Concentrations in cord serum and amniotic fluid, pp. 581-585. Am J. Obstet. Gynecol. 106, 388-399.

41. Fuchs, F., L.L. Cederqvist and H.E. Spiegel. 1980. Maternal plasma proteins. In: The Human Placenta: Proteins and Hormones. Proceedings of the Serono Symposium, Vol. 35 (A. Klopper, A. Genazzani and P.G. Crosignani, eds.). Academic Press, London · New York. pp. 389-399.

42. Haram, K., K. Augensen and S. Elsayed. 1983. Serum protein pattern in normal pregnancy with special reference to acute-phase reactants. Br. J. Obstet. Gynaecol. 90, 139-145.

43. Romen, Y. and R. Artal. 1984. C-reactive protein as a predictor for chorioamnionitis in cases of premature rupture of the membranes. Am. J. Obstet. Gynecol. 150, 546-550.

44. Chu, C.Y-T, V.P. Singla, H-P. Wang, B. Sweet and L.T-Y. Lai. 1981. Plasma $alpha_1$-acid glycoprotein levels in pregnancy. Clin. Chim. Acta 112, 235-240.

45. Waites, G.T., A.M. Bell and S.C. Bell. 1983. Acute phase serum proteins in syngeneic and allogeneic mouse pregnancy.

Clin. exp. Immunol. 53, 225-232.

46. Waites, G.T. 1984. Studies on murine pregnancy-associated serum proteins. PhD Thesis. University of Bristol.

47. Barrett, A.J. and P.M. Starkey. 1973. The interaction of α_2-macroglobuin with proteinases. Characteristics and specificity of the reaction, and a hypothesis concerning its molecular mechanism. Biochem. J. 133, 709-724.

48. De Beer, F.C., M. Baltz, S. Holford, A. Feinstein and M.B. Pepys. 1981. Fibronectin and C4-binding protein are selectively bound by aggregated amyloid P component. J. Exp. Med. 154, 1134-1149.

49. Pepys, M.B., M.L. Baltz, R. Musallam and M.J. Doenhoff. 1980. Serum protein concentrations during Schistosoma mansoni infection in intact and T-cell deprived mice. I. The acute phase proteins C3 and serum amyloid P-component (SAP). Immunology 39, 249-254.

CANINE PREGNANCY ZONE PROTEIN, AN ACUTE PHASE REACTANT IN DEGENERATIVE JOINT DISEASE?

H. Strøm[1], S. Alexandersen[2], O.M. Poulsen[3] and J. Hau[2,4]

1) Small Animal Clinic

2) Institute of Pathology

3) Institute of Dairy Science

4) Laboratory Animal Unit

Royal Veterinary and Agricultural University, Copenhagen,
DK-1870 Frederiksberg C, Denmark

Introduction

Human pregnancy zone protein (PZP, α_2PAG, SP3) was first detected in serum of pregnant women by Olivelli & Rugieri (1) and Smithies (2).

Elevated levels of human PZP have been recorded in serum of patients suffering from a variety of malignancies (3) including chronic inflammatory diseases (4). These observations led Bohn (3, 5) to suggest that PZP should be classified as an acute phase reactant, although not classical in its response (6). PZP has been shown to stabilize lysosomes (7) and to effect polymorphonuclear leucocyte and monocyte functions (8, 9), which are central components in the acute phase response (10). Moreover, an association between active joint inflammation and elevated PZP has been shown (11).

Pregnancy-associated murine protein-1 (PAMP-1) (12) was demonstrated to cross-react immunologically with human PZP (13), and this protein has been demonstrated to be elevated in serum of female mice in the acute phase of an inflammatory response (14).

The existence of a PZP-analogue in the dog was independantly shown by Bohn & Ronneberger (15) and von Schoultz et al. (16), and the present report is the first attempt to use this molecule as an acute phase marker in the dog.

Materials and Methods

Animals

Blood samples were obtained from the brachiocephalic vein from 10 dogs with degenerate joint disease (DJD) and from 12 normal healthy dogs. The diagnosis of DJD was based on the findings of arthrotomi, and no clinical signs of systemic involvement. Studies revealed that the DJD dogs had increased intrasynovial levels of immunoglobulin M (17).

Quantification of dog PZP

Quantitative rocket immunoelectrophoresis (18) was performed using a thin layer agarose gel technique previously described by Alexandersen et al. (19). Rabbit anti dog pregnancy zone protein (PZP) antiserum was kindly donated by dr. H. Bohn, Marburg, FRG.

A pool of serum from 10 hospitalized dogs was arbitrarily assigned a value of 100 arbitrary units/ml and used as the master standard.

Statistics

Student's t-test was used to evaluate statistical differences.

Results

The mean serum level of PZP in healthy dogs was 75 AU/ml (s.d. 18, n = 12), and the mean PZP serum level of dogs suffering from DJD was 136 AU/ml (s.d. 43, n = 10). There was a significantly increased PZP serum level ($p < 0.001$) in the dogs suffering from DJD.

Discussion

Human and murine pregnancy zone protein have recently been shown to be of potential value as markers in inflammatory diseases (11, 14).

This is the first demonstration of increased serum levels of dog PZP in animals of this species suffering from a local joint inflammation. These results indicate that quantitative measurements of dog PZP may be of veterinary clinical value. Moreover, the dog might be useful as a general model for studying the course of joint inflammation using PZP as a marker.

References

1. Olivelli, F. and P. Ruggieri. 1958. Minerva Ginecologica 10, 953-958.

2. Smithies, O. 1959. Advances in Protein Chemistry 14, 65-113.

3. Bohn, H. 1972. Archiv für Gynäkologie 213, 54-72.

4. Horne, C.H.W. 1972. M.D. Thesis, Glasgow University.

5. Bohn, H. 1973. Blut 26, 205-209.

6. Horne, C.H.W., H. Bohn, A.L.C. McLay, E.H. Wood and W.D. Thompson. 1975. Behring Inst.Mitt. 58, 50-53.

7. Hempel, K.H., L.A. Fernandez and R.H. Persellin. 1970. Nature 225, 955-956.

8. Persellin, R.H. and J.K. Leibfarth. 1978. Arth.Rheum. 21, 316-325.

9. Persellin, R.H. and J. Rhodes. 1981. Clin.Exp.Immunol. 46, 350-354.

10. Waites, G.T. and S.C. Bell. 1986. These Proceedings, p.

11. Østensen, M., G. Husby and B. von Schoultz. 1986. These Proceedings.

12. Hau, J., P.E. Svendsen, B. Teisner and S.-E. Svehag. 1978. J.Reprod.Fert. 54, 239-243.

13. Hau, J., J.G. Westergaard, P. Svendsen, Annelise Bach and B. Teisner. 1981. J.Repdrod.Immunol. 3, 341-349.

14. Waites, G.T. and S.C. Bell. 1984. J.Reprod.Fert. 70, 581-589.

15. Bohn, H. and H. Ronneberger. 1973. Archiv für Gynäkologie 215, 277-284.

16. Von Schoultz, B., T. Stigbrand, K. Martinsson and N. Holmgren. 1976. Acta Endocrinol. 81, 379-384.

17. Strøm, H., S. Alexandersen, O.M. Poulsen and J. Hau. 1986. In preparation.

18. Laurell, C.-B. 1972. Scand.J.Clin.Lab.Invest. 29, Suppl. 124, 21-37.

19. Alexandersen, S., J. Hau, O.M. Poulsen and B. Aasted. 1985. Electrophoresis 6, 535-538.

FEMALE PROTEIN OF SYRIAN HAMSTER: A SERUM PENTRAXIN NOT PREGNANCY SPECIFIC, BUT UNDER SEX STEROID AND ACUTE PHASE CONTROL

J. E. Coe
NIH, NIAID, Laboratory of Persistent Viral Diseases, Rocky Mountain Laboratories, Hamilton, Montana 59840

Sex hormones control the synthesis of proteins in a number of species and usually these sex-limited proteins are thought to be important in reproduction. For example in insects, female specific proteins have been found during egg production which are necessary for successful maturation of the egg (1,2). A well characterized serum phosphoprotein in the laying hen (serum vitellin) (3) was found to be an important serum carrier for constituents of the egg yolk (4). This protein was also found in male chickens after treatment with diethylstilbestrol (DES) (5). In mammals a variety of serum proteins which are under sex hormone control have been defined and many of these proteins are specifically associated with reproduction. These pregnancy associated proteins theoretically have evolved to maintain a viable and successful pregnancy and pregnancy proteins are the primary theme of this conference. However, other mammalian serum proteins also are controlled by sex hormones but these proteins are apparently not involved in pregnancy. For example, some immunoglobulins and complement associated components are under sex steroid control and serum levels in males differ from those in females (6). The reason for an enhanced expression in one gender or the other is largely unknown but does indicate the fact that the physiology of males and females is different also during nonreproductive intervals. This report will review briefly a particularly unusual example of a sex limited protein which is found in the Syrian hamster. In this species, female serum contains a protein (called female protein or FP) at a concentration (1-3 mg/ml) which is more than 100-fold greater than in male serum. This sex difference in serum levels is due to a testosterone suppression of normal hepatic synthesis in males; thus, male serum levels rapidly increase after castration (or diethylstilbestrol administration) and serum levels revert to normal low levels after subsequent injection of exogenous testosterone (7).

The functional significance of this prominent serum protein of female Syrian hamsters is unknown, however, structural analysis has shown a surprising relationship to a well known family of proteins called pentraxins (8). These strange proteins have an unusual structure; the basic molecule is a cyclic pentamer of noncovalently assembled subunits (for review see 9,10,11). Each subunit has a 25-30 K dalton size and the whole pentameric molecule has a characteristic morphology by electron microscopy. Human C-reactive protein (CRP) and serum amyloid P component (SAP) are the two major proteins in this family and they are homologous. That is, their close (50-70% identity) similarity in primary amino acid sequence indicates that they have a common ancestral gene. Most species of animals appear to have CRP and/or SAP homologs. Actually, pentraxins are ancient proteins as CRP homologs have been found in lower teleost fish (11) and even horseshoe crab (12). Structurally and functionally pentraxins have changed very little during evolution and this conservative evolution implies a functional necessity or importance which is still unknown. These proteins have been found to be involved in a number of known biological activities (10). For example, human CRP was found (and named) because of its capacity to bind to C polysaccharide of pneumococci (13,14,15). This Ca^{++} dependent specificity was subsequently shown to be directed toward the phosphorylcholine (PC) ligand (16) and this reaction of CRP with PC will activate the classical complement cascade (17). Hamster FP also specifically binds to PC (8) and can activate complement (classical pathway) during that interaction (18). In contrast, SAP does not have PC specificity but will react in a Ca^{++} dependent fashion with agarose and will bind to some individual complement components (11).

Another phenomenon expressed by human CRP is its capacity to change in serum concentration 24-36 hr after injury. In this so-called acute phase reaction, CRP levels can increase up to 500-fold due to sudden hepatic synthesis (10). In species other than man, pentraxins are also acute phase proteins, although in the mouse for example, SAP rather than CRP shows the acute phase increase (11). However, hamster FP has an especially novel response to injury because serum levels decrease in females and increase in males (19). This sex limited divergent acute phase response can be reversed; females treated with testosterone have low FP levels and then show a male type (increase FP concentration) injury response whereas males castrated or DES treated will manifest a female type (decrease) response to injury. Hamster FP is the only example so far in which an acute phase reaction results in opposite sex limited serum concentration changes, and actually the only serum pentraxin which decreases after injury.

In addition to various cellular interactions, (lymphocytes, macrophages and platelets), pentraxins also bind to lipoproteins with a PC specificity (20). FP has now been shown to be similar to rat CRP in the formation of a lipoprotein complex; this interaction may be important in lipoprotein metabolism and intravascular transport.

Whereas CRP was discovered because of its specificity for pneumococci, amyloid P component (AP), which is identical to SAP in normal serum, was originally isolated from amyloid in which it is a constant minor constituent (for review see 21). The other (major) component is the fibrillar protein which can be constructed from various normal or abnormal serum proteins as AA, L chain, etc. Amyloid is a ubiquitous pathological deposit in many species and is usually composed of these same two constituents, i.e., AP and fibril. The events leading to amyloid deposition are unknown, although a preliminary or framework deposit of SAP has been proposed (22). FP was found to be an AP-like constituent of hamster amyloid (23). This was shown by fluorescent microscopy using specific anti-FP reacted with hamster tissues containing amyloid deposits. In addition, more FP could be extracted from amyloidotic livers than from normal livers and the amount obtained was directly proportional to the severity of the amyloidosis. Furthermore, intravenously administered ^{125}I FP was rapidly sequestered in amyloid deposits, resulting in a characteristic and, in fact, diagnostic alteration of ^{125}I FP plasma metabolism. This dynamic extravascular transfer of i.v. administered ^{125}I FP to amyloid may be unique for hamster as it has not been observed with SAP proteins in other animal models of amyloidosis. Of particular interest is the high incidence of amyloid in female hamsters. It was observed that during normal aging, or during experimental induction of amyloid (by Na caseinate injection) females are much more susceptible to amyloidosis than males. Actually the prevalence of amyloid in the normal aging female hamster results in distinctly shorter life span in females when compared to males, a rare phenomenon in mammals (24). Also, experimental production of male amyloidosis was associated with treatments that elevate their serum FP levels. For example, diethylstilbestrol is an especially effective inducer of amyloid formation in males and this treatment is associated with marked elevation of FP serum levels. Only in the Syrian hamster has DES treatment been associated with amyloid formation.

Therefore, high FP levels are associated with amyloid deposition, although this relationship has not been shown to be etiologic. However, if FP is instrumental in amyloid deposition, it certainly must fulfill an important function in young females and for this reason the gene was preserved during hamster evolution. Although a variety of biological functions have been described for pentraxins since their discovery more than 50 yr ago, their raison d'etre is still unknown. However, it is especially perplexing that this ancient gene, which changed very little during evolution, came under sex steroid control only in the hamster. Furthermore, the uncontrolled FP synthesis in the female resulted in the highest serum pentraxin levels so far described in mammals.

In contrast to the other pregnancy associated (and sex steroid controlled) proteins of mammals, hamster FP does not increase markedly in serum concentration during pregnancy. Daily assay of serum FP during their 16 day gestation showed only a moderate increase (10-20%) during the 4th-8th day in young females, although a greater increase was found in older (1 yr old breeders) females in which serum FP levels usually are lower than in younger females. Recently, this problem has been reexamined and preliminary results do indicate a marked drop in FP concentration at time of delivery. However, this could be simply an expression of an acute phase response which results in fall of serum concentration. The kinetics of the reaction relative to parturition and the plasma metabolism of FP are currently being studied to elucidate the mechanism.

Ongoing experiments are designed to determine the beneficial or detrimental effects of this pentraxin in male and female hamsters. Such studies are relevant to the role of related pentraxins in other species such as man. Actually, the homology of hamster FP and human SAP is sufficiently close that a cDNA probe for human SAP recently was used successfully to identify and isolate the hamster FP gene (Dowton et. al., Science, in press). The unique sex hormone regulation of FP synthesis will permit manipulation of FP levels in male and female Syrian hamsters to determine the role of pentraxins in amyloid induction/maintenance, in lipoprotein metabolism, and in various cellular functions during normal and acute phase interactions.

Table 1. Comparison of Human CRP, SAP and Syrian Hamster FP

	CRP	FP	SAP
Pentraxin structure[*]	+	+	+
Hepatic synthesis	+	+	+
FP homology[**]	50	100	69
Acute phase protein	+	+	0[***]
Calcium dependent binding	+	+	+
PC specificity	+	+	0
Agarose specificity	0	+	+
Complement fixation	+	+	0
Lipoprotein binding	0	+	0
Amyloid component	0	+	+
High serum levels	0	+	0
Sex steroid synthesis control	0	+	0
Sex limited acute phase response	0	+	0

[*]Noncovalent cyclic pentamer of a 20-30 K dalton subunit.

[**]Percent of total amino acid sequence identical to FP (Dowton et al., in press).

[***]In some mammals, as mouse, SAP (rather than CRP) is an acute phase protein.

References

1. de Wilde, J., A. de Loff. 1973. In: The Physiology of Insecta (M. Rockstein, ed.). Academic Press, New York, p. 11.

2. Engelman, F. 1969. Science 165, 407.

3. Laskowski, M. 1935. Biochem. Z. 278, 345.

4. Schjeide, O., M. R. Urist. 1956. Science 124, 1242.

5. Clegg, L. E., P. E. Sanford, R. E. Hein, A. C. Andrews, J. S. Hughes, C. D. Mueller. 1951. Science 114, 437.

6. Shreffler, D. C., C. S. David. 1975. In: Advances in Immunology (F. J. Dixon and H. G. Kunkel, eds.). Academic Press, New York, p. 125.

7. Coe, J. E. 1977. Proc. Natl. Acad. Sci. USA 74: 730.

8. Coe, J. E., S. S. Margossian, H. S. Slayter, J. A. Sogn. 1981. J. Exp. Med. 153, 977.

9. Coe, J. E. 1983. Contemp. Top. Mol. Immunol. 9, 211.

10. Gewurz, H., C. Mold, J. Siegel, B. Fiedel. 1982. Adv. Intern. Med. 27, 345.

11. Pepys, M. B., M. L. Baltz. 1983. Adv. Immunol. 34, 141.

12. Liu, T.-Y., F. A. Robey, C.-M. Wang. 1982. Ann. N. Y. Acad. Sci. 389, 151.

13. Tillett, W. S., T. Francis, Jr. 1930. J. Exp. Med. 52, 561.

14. Abernethy, T. J., O. T. Avery. 1941. J. Exp. Med. 73, 173.

15. MacLeod, C. M., O. T. Avery. 1941. J. Exp. Med. 73, 183.

16. Volanakis, J. E., M. H. Kaplan. 1971. Proc. Soc. Exp. Biol. Med. 136, 612.

17. Kaplan, M. H., J. E. Volanakis. 1974. J. Immunol. 112, 2135.

18. Etlinger, H. M., J. E. Coe. (in press).

19. Coe, J. E., M. J. Ross. 1983. J. Exp. Med. 157, 1421.

20. Saxena, U., A. Nagpurkar, J. Coe, S. Mookerjea. 1985. Fed. Proc. 44, 1456.

21. Glenner, G. G. 1980. N. Engl. J. Med. 302, 1283.

22. Painter, R. H., I. DeScallon, A. Massey, L. Pinteric, S. B. Stern. 1982. Ann. N. Y. Acad. Sci. 389, 199.

23. Coe, J. E., M. J. Ross. J. Clin. Invest. (in press).

24. Kirkman, H., P. K. S. Yau. 1972. Am. J. Anat. 135, 205.

COMPARATIVE EARLY PREGNANCY ENDOCRINOLOGY AND EMBRYO MATERNAL INTERACTIONS IN MAMMALS

A.A. Gidley-Baird, C. O'Neill

Departments of Veterinary Physiology and Obstetrics and Gynaecology, University of Sydney, n.S.W. 2006, Australia

All mammals face the problem of providing an adequate endocrine environment for the maintenance of pregnancy after ovulation and fertilization. In all the species this requires an initial luteal phase which may be part of the oestrous cycle or induced by copulation. During this time the corpus luteum (C.L.) is active and the progesterone produced dominates the uterine environment. This progesterone remains dominant throughout the pregnancy although its course and stimulation varies between species and within a species during the course of pregnancy. The mammalian oestrous cycle can be classified into three main types based on the initiation of ovulation and the luteal phase.

(1) Copulation induced ovulation with a spontaneous luteal phase (e.g. rabbits, cats, ferrets and camels).

(2) Spontaneous ovulation with a copulation induced luteal phase (mice and rats).

(3) Spontaneous ovulation and luteal phase (e.g. primates, marsupials, horses, goats, sheep, pigs and cattle).

Despite these different types of cycles all the mammals share two important common features. Firstly the luteal phase of the cycle is endocrinologically independent of the embryo. This can be artificially induced by stimulation of the cervix in species showing a copulation induced ovulation or luteal phase. Secondly, under normal circumstances, the extension of the progestational phase beyond the normal maternal luteal phase is dependent upon

Pregnancy Proteins in Animals

the embryo. The marsupials have overcome this problem by having, with very few exceptions, a gestation period shorter than the length of their oestrous cycle.

The method by which the luteal phase is extended by the embryo varies between the species and within the pregnancy of a particular species. There are three main ways by which this can occur and a species may use one or more of these methods.

(1) The embryo produces a gonatotrophin which extends the life of the C.L.

(2) The embryo inhibits the release of uterine luteolysins thus extending the life of the C.L.

(3) The feto-placental unit produces its own steroids.

The extension of the luteal phase is critical of successful pregnancy and is the time at which hormonal changes can be used to diagnose pregnancy. The embryo, however, has been affecting the maternal physiology for days before these hormonal changes are apparent. The earliest changes reported are a change in the properties of circulating maternal lymphocytes and a significant thrombocytopenia. These changes can be detected within hours of fertilization and appear to be pregnancy specific. Unlike the different oestrous cycles and methods of extending the luteal phase among the species, these changes appear to be very similar in all the species investigated.

THE ROLE OF BLOOD PLATELETS IN THE ESTABLISHMENT OF PREGNANCY

C. O'Neill

Department of Obstetrics and Gynaecology, University of Sdyney, Royal North Shore Hospital, St Leonards NSW 2065 Australia

Introduction

Blood platelets play a central role in a number of physiological phenomena including maintenance of vascular integrity, modulation of the affector and effector arms of the immune response, and control of cellular growth and development (1). Alterations to platelet count and activity have long been known to occur in pregnancy (2) and particularly in pathological pregnancy states such as eclampsia and pre-eclampsia (3).

It has recently been reported that alterations to platelet physiology occurred immediately following conception in the mouse and human (4,5). The zygote produced a soluble factor which caused platelet activation (6). The platelet aggregates were sequestered by the microvasculature, resulting in a mild thrombocytopenia associated with early pregnancy (4). This factor was partially purified and shown to have similar chemical, biochemical and physiological properties to the potent platelet activating factor 1-O-alkyl-2-acetyl-sn-glyceryl-3-phosphocholine (7).

In a preliminary clincal trial, production of embryo-derived platelet activating factor (PAF) was a necessary pre-requisite for the establishment of pregnancy in humans since only embryos which produce the factor *in-vitro* caused platelet activation following embryo transfer, and only in such cases did pregnancies result (5). Many of the embryos (43%) produced by *in-vitro* fertilization failed to produce the embryo-derived PAF (8)

Pregnancy Proteins in Animals

and such embryos failed to cause pregnancy. Depletion of the maternal platelet count, in the pre- and peri-implantation period of mouse pregnancy, by treatment with anti-platelet anti-sera also caused a marked reduction in the implantation and pregnancy rate (9). These pieces of circumstantial evidence suggest that platelets may play a role in the establishment of pregnancy.

Implantation of the embryo into the uterus is a major hurdle to be overcome in order for pregnancy to ensue. The processes which initiate and control this event are far from clear. It involves modification to the vasculature, maternal immune responses and promotion of cellular and tissue differentiation. The central role that platelets play in all of these processes (1) and the evidence for activation of platelets in the pre- and peri-implantation period (4) make them an attractive candidate for involvement in implantation.

An *in-vitro* model for implantation exists whereby blastocysts, in the presence of a complex culture environment, attach to the substratum of the culture vessel, undergo trophoblast outgrowth and transformation and display primary differentiation of the inner cell mass. This study, using the *in-vitro* model, examines the role of platelet factors on the control of embryo implantation.

Materials and Methods

Embryos

All mice used were of the Quackenbush strain (QO) and were exposed to light from 7.30 to 21.30 h. Blastocysts were collected from 5-6 week old females which had been induced to superovulate by an intraperitoneal injection of 5-10 iu PMSG (Folligon:Intervet, Boxmeer, The Netherlands) given at 15:00 - 17:00 h, followed 46 - 50 hours later by an injection of the same dose of hCG (Chlorulon:Intervet). After the second injection, the females were placed with fertile males and checked the following morning for copulation plugs (= Day 1 of pregnancy). Blastocysts were recovered by flushing the uterine horns with Eagles minimal essential medium (MEM) supplemented with 4 mg BSA/ml between 10:00 and 12:00 h on Day 4 of pregnancy. The blastocysts were washed twice in 1 ml of flush media and then transferred to treatment media.

Serum Collection

Fetal calf serum (FCS) was purchased from Flow Laboratories (Sydney, Australia) and stored at $-20^{o}C$. Following thawing it was heat inactivated at $56^{o}C$ for 20 minutes and then diluted with MEM for culture.

Blood was collected, under light anaesthesia, from the peri-orbital plexus using siliconized (Coatasil, Ajax Chemicals, Australia) glass Pasteur pipettes. Blood was allowed to flow up the pipette by capillarity until it reached the 200 ul calibration mark. The blood was mixed immediately with 20 ul 3.2% (w/v) trisodium citrate as an anti-coagulant, in a sterile siliconized glass centrifuge tube kept at $4^{o}C$. The platelet poor plasma was removed with siliconized Pasteur pipettes and platelet counts performed (Baker Instruments 810 Platelet Analyser). The count was always less than 2,000 platelets/ul (normal range for whole blood 800 - 1200 x 10^{3}/ul). To prepare platelet rich serum (PRS), the buffy coat of the centrifuged blood was then added to the plasma, while for the preparation of platelet poor serum (PPS) an equivalent volume of packed erythrocytes was added to the plasma.

Citrated plasma was then recalcified with 2.5 um $CaCl_2$/ml of plasma and left to clot at $37^{o}C$. The clot was allowed to retract for 45-60 min at $37^{o}C$. The resulting serum was centrifuged at 2000 g for 15 minutes and the supernatant stored at -20°C until use. Such PRS and PPS was prepared from normal cycling females and Day 2 pregnant QS mice.

Platelet Releasate

Blood was collected as described above. The buffy coat was washed thrice with phosphate buffered saline (PBS) containing 0.32% (w/v) trisodium citrate (PBS/citrate). The washed platelets were resuspended in minimal volume of PBS, and 100 ul of 2mM adenosine diphosphate (ADP) in PBS was added. Activation of the platelets was verified microscopically. The mixture was centrifuged at 3000 g for 30 minutes and the supernatant collected. This platelet releasate from 1 ml of blood was added to 1 ml of FCS. A similar volume of PBS-citrate containing ADP, but not platelet releasate material, was added to FCS and acted as controls.

Culture Environment

On the evening prior to embryo culture the sera was thawed and added to freshly prepared MEM to give varying concentrations of serum. Media was set up as 20 ul drops under 10 ml of paraffin oil (BDH, Sydney Australia) in plastic petri dishes (Falcon Plastics) and equilibrated overnight at 37°C in a humidified gaseous environment of 5% CO_2 in air. Up to 6 embryos per drop were cultured for 72 h and then examined by inverted microscope. The number of embryos undergoing trophoblast outgrowth was recorded.

Results

Fetal calf serum is commonly used as the complex protein source for the promotion of blastocyst outgrowth *in-vitro*. In this study there was no significant difference in the number of blastocyst outgrowths when cultured in FCS, PRS from non-pregnant mice and PRS from pregnant mice (Table I). Similar concentrations of serum from platelet depleted plasma of non-pregnant mice was totally incapable of supporting blastocyst outgrowth (Table II).

TABLE I
The effect of FCS, PRS and PPS on mouse blastocyst outgrowth

Serum Concentration (%)	% Outgrowth $\bar{X} \pm$ S.E. (n) FCS	PRS	PPS
10	100 ± 0 (20)	89 ± 7 (30)	0 ± 0 (34)
5	100 ± 0 (20)	78 ± 9 (46)	0 ± 0 (40)
2.5	95 ± 3 (20)	92 ± 2 (50)	0 ± 0 (40)
1.25	92 ± 2 (20)	98 ± 5 (35)	0 ± 0 (25)

PPS from Day 2 pregnant mice was capable of supporting a minor degree of blastocyst outgrowth at concentrations of 2.5 - 10% (v/v) (Table 2). This result is consistent with the observation that platelet activation occurs in the preimplantation phase of pregnancy *in-vivo* (4). Such activation would provide platelet releasate material to PPS. The number of embryos undergoing outgrowth (Table II) and the extent and morphological normality of the outgrowths (Figure I) were considerably less then that promoted by FCS or PRS.

TABLE II

Effect of PRS and PPS From Pregnant and Non-Pregnant Mice on Blastocyst Outgrowth

Serum Concentration (%)	% Outgrowth $\bar{X} \pm$ S.E. (n)		
	Preg PRS	Preg PPS	NP PPS
10	85 ± 8 (25)	30 ± 15 (15)	0 ± 0
5	92 ± 2 (17)	21 ± 9 (18)	0 ± 0
2.5	100 ± 0 (15)	14 ± 8 (15)	0 ± 0
1.25	100 ± 0 (15)	0 ± 0 (15)	0 ± 0

The possibility that the growth promoting effect of PRS was due to the release of material from platelets during the coagulation process was tested by adding platelet releasate material to a minimal concentration of FCS. Such treatment caused an enhancement in the outgrowth rate (Table III) but not to the extent expected.

TABLE III

Effect of Addition of Platelet Releasate and PBS/Citrate (Control) on the Promotion of Outgrowth by 0.1% Fetal Calf Serum

0.1% FCS	FCS (Control)	FCS (Releasate)
22 ± 5	19 ± 6	32 ± 4

Heating PRS for 30 minutes at 56°C had no effect on the ability of this sera to promote outgrowth (10% heated PRS = 89.2 ± 7% outgrowth) while heating at 70°C for as little as

10 minutes rendered the sera incapable of supporting outgrowth.

Discussion

Embryo attachment, outgrowth and differentiation *in-vitro* have been described now for an impressive number of species, although the majority of work has been carried out on mice. This cellular behaviour, which is generally considered to be analogous to implantation *in-utero* (10) occurs on glass and plastic culture surfaces, reconstituted rat collagen, monolayers of mouse uterine cells, live or UV-irridiated HeLa, Chinese hampster and L cells and when embryos are embedded in organ cultures of bovine lens fibre epithelial tissue (11).

Blastocyst outgrowth *in-vitro* was supported by defined media with specific amino acids present and had a requirement for serum macromolecules (12). The provision of serum caused blastocysts to undergo a pleiotypic increase in metabolism (blastocyst activation) (13). It is thought that, provided an appropriate surface is available, outgrowth by blastocysts is an intrinsic consequence of activation.

Outgrowth was reported to be supported by the fetuin fraction of serum, and, on a collagen substrate, by the serum albumin fraction. The molecular homogeneity of the fractions, however, was not known. Hsu (1980) (15) reported that a heat sensitive (100-300 K, M_r) fraction of fetal calf serum was necessary for outgrowth. Little further characterization of the serum factors necessary for embryo growth was carried out.

It has been known for some time that factors in serum are necessary to promote growth of a wide variety of normal diploid cells in culture. It was shown that serum produced from cell free plasma was not capable of supporting growth, while serum produced from platelet rich plasma was. These platelet derived growth factors (PDGF) were defined by their ability to cause a round of cell division or significant ^{3}H-thymidine intake in growth inhibited cells (16).

The preimplantation embryo grows from the 1-cell through to the blastocyst stage in defined culture media without protein supplementation. Once differentiation of the blastomeres in the trophectoderm and inner cell mass of the blastocyst occurs, continued culture in such a medium results in rapid growth inhibition. The results of the present study show clearly that, like other differentiated diploid cells *in-vitro*, blastocysts require serum for continued growth and that molecules from platelets are the principle mitogenic factors involved.

There have been a variety of platelet derived growth factors defined including (1) cationic PDGF: heat stable, PI 9.8-10.2, 31,000 M_r; (2) connective tissue activating factor: PI 8.8, 10,000 M_r; and (3) α and β transforming growth factors characterized by their ability to cause anchorage dependent cells to grow without anchorage in semi-solid medium. The platelet derived growth factors which promote blastocyst growth have not been sufficiently characterized to determine if they are the same as these.

In view of the extensive literature showing that PDGF's are necessary for growth of differentiated cells *in-vitro*, it is not surprising that such factors are necessary for embryo implantation *in-vitro*. It should be noted, however, that serum, as such, is not a physiological solution except at wound sites and the sites of extensive platelet activation. Since embryos produce a platelet activating factor throughout the pre- and peri-implantation period (5), extensive platelet activation occurs at the implantation site. This is further supported by the observation that peripheral blood PBS contains sufficient platelet derived growth factors to cause modest blastocyst outgrowth (Table II). The case for PDGF's having a true physiological role in the initiation and control of implantation *in-vivo*, therefore, warrants consideration.

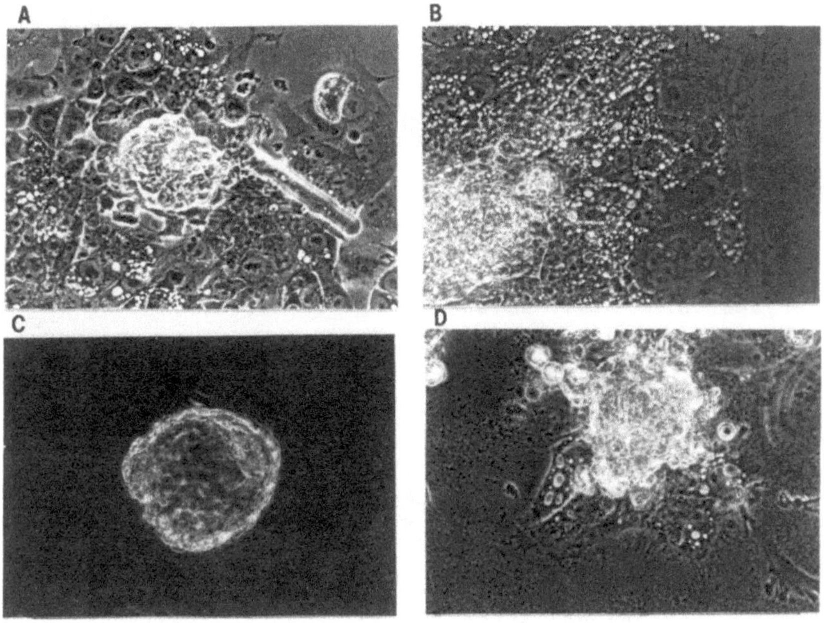

FIGURE I

Blastocyst outgrowth after 72 h culture in (a) 10% FCS, (b) 10% PRS, (c) 10% PPS (nonpregnant), and (d) 10% PPS (Day 2 pregnant).

Acknowledgements

This work was supported by the Australian Research Grants Scheme. I thank Lynne Frizell for typing the manuscript.

References

1. Willis, A.L. 1978. In: Handbook of Experimental Pharmacology, 50/1, Inflammation". (J.R. Vane and S.H. Feineiva, eds). Springer-Verlag:Berlin, p. 138.

2. Fay, R.A., A.O. Hughes, N.T. Farron. 1983. Obstet. and Gynecol. 61, 238.

3. Perkins, R.P. 1979. Obstet. Gynecol. Surv. 34, 101.

4. O'Neill, C. 1985. J Reprod. Fertil. 73, 567-77.

5. O'Neill, C., A.A. Gidley-Baird, I.L. Pike, R.N. Porter, M.J. Sinosich, D.M. Saunders. 1985. J. I.V.F. and E.T.. 2(2), 59.

6. O'Neill, C. 1985. J. Reprod. Fertil. 73, 578.

7. O'Neill, C. 1985. J. Reprod. Fertil. (In Press).

8. O'Neill, C., D.M. Saunders. 1984. Lancet ii, 1035.

9. Gasic, G.J., T.B. Gasic. 1970. Proc. Natl. Acad. Sci. USA 67, 793.

10. Bryson, D.L. 1964. Science N.Y. 144, 1352.

11. Salomon, D.S., M.I. Sherman. 1975. Exp. Cell Res. 90, 261.

12. Jenkinson, E.J., Wilson, I.B. 1973. J. Embryol. Exp. Morph. 30, 21.

13. McLaren, A. 1973. In: "The Regulation of Mammalian Reproduction" (S.J. Segal, R. Crozier, P.A. Corfman, D.F. Condliffe, eds.) C.C. Thomas, Springfield, p 321.

14. Jenkinson, E.J. 1977. In: "Development of Mammals" Vol 2, (M.H. Johnson, ed) North-Holland Pub. Co.:Amsterdam p 151.

15. Hsu, Y.C. 1980. Develop. Biol. 76, 465.

16. Balk, S.D. 1971. Proc. Natl. Acad. Sci. U.S. 68, 271.

THE SIGNIFICANCE OF PLACENTAL ALKALINE PHOSPHATASE AS A PREGNANCY RELATED PROTEIN IN MAN AND ANIMALS

T. Stigbrand, K. Hirano
Department of Physiological Chemistry, University of Umeå, S-901 87 Umeå, Sweden

The abundance of alkaline phosphatases (APs) in nature, including bacteria, plants and animals, indicates that these enzymes are involved in fundamental biochemical processes, although the physiological functions are not yet clear. These enzymes are classified as alkaline phosphatases (EC.3.1.3.1) by their ability to participate in hydrolyze/transferase reactions on a wide variety of phosphate-containing compounds at alkaline pH. The enzymes are mainly present in plasma membranes, in which extensive transport takes place.

Evolutionary aspects

A recent monograph on alkaline phosphatases lists more than 6,000 references (1). In man, alkaline phosphatases have been demonstrated at high concentrations in organs such as liver, bone, kidney, placenta, intestine and in low but significant levels in most other tissues. Two tissue specific alkaline phosphatase isoenzymes have been recognised in man - placental and intestinal. The other organs (liver, kidney, bone, lung) possess an alkaline phosphatase which usually is referred to as "tissue unspecific" phosphatase.

There exists today strong evidence to support the concept of at least three different genes coding for alkaline phosphatase in man. One gene is coding for the tissue-unspecific type of enzyme, and at least two other genes code for the placental and intestinal phosphatases. The location of these three loci in the human genome is so far unknown. A number of observations dealing with catalytic properties of the diverse forms seem to support the three-gene concept (2, 3). L-phenylalanine and L-homoarginine give a major degree of discrimination between tissue-unspecific ALP and the

Pregnancy Proteins in Animals

placental and intestinal enzyme. On the other hand, L-leucyl-glycyl-glycine and L-phenylalanyl-glycyl-glycine discriminate well between placental alkaline phosphatase and the intestinal and tissue-unspecific enzymes. All available structural data seem to indicate the presence of at least three genes. The primary structure as revealed by partial proteolytic peptide mapping of ^{32}P-labeled enzymes in SDS polyacrylamide gels confirms the identity of human liver and kidney enzymes and the nonidentity of the placental and intestinal forms (4, 5). Amino terminal sequences of the liver and placental isozymes are clearly different, Leu-Val-Phe and Ile-Ile-Pro-Val, respectively (6). Finally, the use of monospecific antisera prepared against the different purified enzymes distinguishes three different antigenic groups, in agreement with the results above (7). The placental and intestinal phosphatase, however, have antigenic determinants in common (8, 9) and display partial cross-reactivity in some immunoassays. Anti-placental ALP antisera cross-react with intestinal enzyme, and anti-intestinal ALP antisera are partially active toward the placental form.

The above findings indicate that ALPs in higher animals could be classified as a multilocus enzyme system. It is conceivable that during the course of evolution successive gene duplications from a common ancestral gene have occurred, giving rise to the three loci. This model has been proposed to be applicable when several gene loci code for a set of enzyme proteins, which are rather similar in catalytic or subunit characteristics but structurally distinct. This seems to be the case for ALP in man. After the first duplication from the common ancestral gene, later duplications of one or other of the first pair would account for the occurrence of three loci, and so on. The consequences would be that the enzyme coded for by the genes at the different loci, will show differences in sequence which can be readily accounted for in terms of subsequent mutations in the original duplicated genes. A tentative model of the evolution of alkaline phosphatases is presented in Fig. 1.

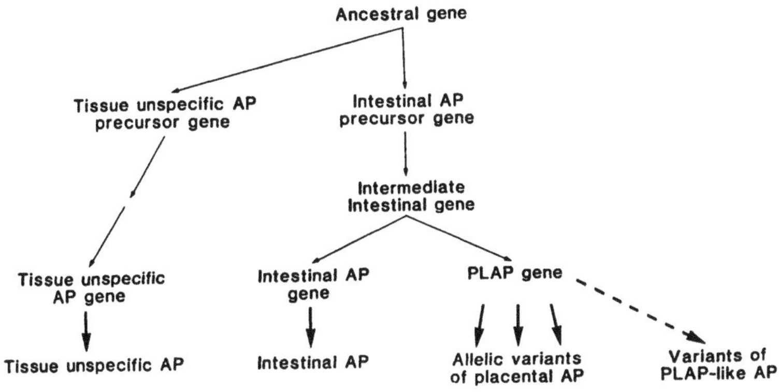

Fig. 1. A tentative model of the evolution of alkaline phosphatases.

Genetic polymorphism of placental alkaline phosphatase

A characteristic property of the placental alkaline phosphatase (PLAP) locus is the extensive genetic polymorphism presenting more electrophoretically defined allelic variants than any other human enzyme studied so far. Three common alleles and at least 18 rare alleles have been identified giving rise to altogether 45 phenotypes of placental alkaline phosphatase. The pronounced polymorphism can at least partially be explained by the fact that this enzyme is a late evolutionary product.

The structural basis for the polymorphism has not yet been elucidated. Both point mutations and intragenic recombinations have been advanced as explanations but structural data are still lacking. Post-translational modifications of the different genetic variants could also occur. These could significantly affect both the surface charge of the molecule and its affinity to certain supporting media used for electrophoresis, such as starch gel. Such post-translational modifications would add a new dimension of complexity and could have a great impact on the classification systems used so far.

Placental alkaline phosphatase in animals

The tissue-unspecific and intestinal type of ALP are found in fish, amphibians, reptiles, birds, and vertebrates, indicating the early appearance of these genes (10).

The placental form on the other hand seems to be a late evolutionary product since it cannot be identified either on structural or catalytic basis in lower animal species (11, 12). In these, placental ALP is closely related to the corresponding liver phosphatase. Rabbit, guinea pig, mouse, and hamster have high levels of ALP in placenta, but the activity is heat-labile and insensitive to inhibition by L-phenylalanine in contrast to human placental ALP (11). In a larger study the conclusions were extended to include also the pig, cat, dog, cow, and sheep (13). The dog placental ALP is thermolabile and has the same inhibition pattern and electrophoretic mobility as the dog tissue-unspecific enzyme after removal of sialic acid by neuraminidase treatment (14). Several other data suggest that lower animals have the tissue-unspecific type of ALP in placenta, supporting the opinion that the human placental type ALP is a newcomer on the evolutionary scene. It is of interest to note that expression of the tissue-unspecific ALP also occurs during the first 12 weeks of development of the human placenta (15). Lower apes seem to lack the term placental enzyme and African green monkey, rhesus monkey, and baboon have heat-sensitive, L-phenylalanine-insensitive placental enzymes (16). Recently Doellgast and Benirshke (17) were able to demonstrate the occurrence of an enzyme immunologically cross-reactive with the placental type ALP that presented similar inhibition profiles in the chimpanzee and orangutan placentas but not in the lowland gorilla or the spider monkey. This observation is of fundamental importance since the occurrence of the unusually high degree of allelic polymorphism at this locus in man might be correlated with its late appearance on the evolutionary scene.

The lack of demonstrable levels of placental alkaline phosphatases has implications also for the interpretation of the appearance of other placenta derived pregnancy related serumproteins in animals, since obviously no analogues to this enzyme exist in lower animals including all common laboratory animals and immunochemical cross-reactivity with placenta derived tissue unspecific phosphatase with phosphatases from other sources should be expected.

An interesting deviation from the evolutionary scheme is provided by the fact that some new world monkeys (marmoset, capuchin, owl monkey and tamarin) present lung alkaline phosphatases which to a significant degree (50 - 80 %) share characteristics with the human placental type phosphatase and not, which would be expected with tissue-unspecific phosphatase in the host animal (25). Several explanations for this could be considered: 1) The enzyme is a product of an earlier gene duplication, occurring in primate evolution since this type of phosphatase is not found in new world monkeys or lower animals; 2) Alternatively the duplication might have occurred much earlier but the placental phosphatase did not become expressed as the major alkaline phosphatase in placenta after the divergence of the lineages man/great apes from that leading to other primates. The second hypothesis includes assumptions of potential mutation(s) in the regulatory mechanism of this specific enzyme system.

Trace expression of placental alkaline phosphatase

Besides the three major forms of alkaline phosphatases, evidence is now accumulating that trace amounts of heat stable APs are also present in normal tissues. Chang et al. (18) observed trace amounts of heat-stable, L-phenylalanine- and L-leucine-sensitive phosphatases in human testis. These enzymes differ from PLAP in their great sensitivity to L-leucine as an uncompetitive inhibitor, a characterization now used to define PLAP-like alkaline phosphatases. The presence of PLAP or PLAP-like enzymes was simultaneously reported in normal endocervix by Goldstein et al. (13), and Nozawa et al. (19). Recently, this search for trace expression has resulted in the demonstration of PLAP or PLAP-like isozymes in the human lung and thymus (20).

Hybridoma technology has allowed the production of monoclonal antibodies which are very specific and sensitive to changes in conformation and/or protein sequence. Several groups are now using different antibodies to elucidate some of the structural aspects of APs (21). The PLAP-like phosphatases demonstrate, in comparison to PLAP, characteristic reactivity patterns with different monoclonal antibodies which make them useful as an alternative approach to evaluate the occurrence of different isozymes (22, 23, 24).

Placental alkaline phosphatase in human pregnancy

During pregnancy placental alkaline phosphatase is synthesized by the syncytiotrophoblast after the 12th week and is shed into the maternal circulation. As shown in Fig. 2, the serum concentrations increase from low basal levels as pregnancy proceeds and the increased production is correlated with the growth of the placenta and the increased synthesis of estrogen. It was also early recognized that deviations from the "prediction curves" could indicate fetal disorders and death in utero. These first observations made by catalytical assays were later confirmed by use of

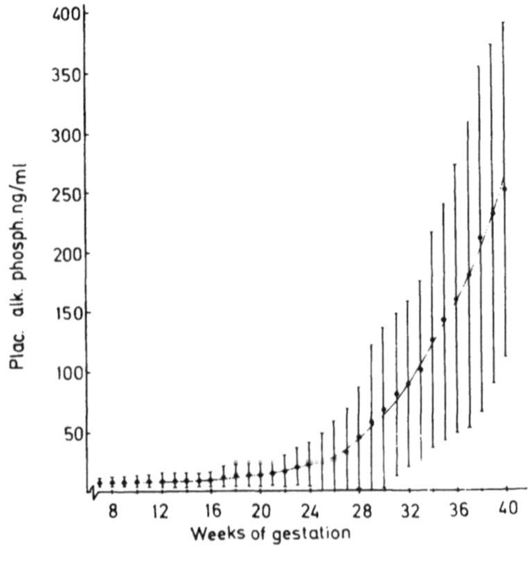

Fig. 2. Mean serum concentration of placental alkaline phosphatase during normal (within shaded area mean ±2SD) pregnancies.

radioimmunoassay techniques (26, 27). The rate of enzyme enrichment in the maternal blood follows a logarithmic course. Despite its convenience in sampling and measurement, however, PLAP levels during normal pregnancy are considered too widely distributed to define discriminating working ranges. The introduction of "prediction curves" by Yamaguchi and Shimozato (28) indicates that placental function tests can be used as screening tests during the third trimester of pregnancy. Several complications like "small for date" children, severe and mild toxemia and diabetes are associated with significantly low levels of placental alkaline phosphatase, but the future use of this parameter is still to be established (27, 28).

Simultaneous determinations of urinary estrogens and PLAP levels in maternal serum seem to give a clear differentiation between pregnancy complications like intrauterine growth retardaton and other high risk pregnancies (27). Combinations of tests might thus yield more information than single parameter assays.

Previous reports on PLAP levels during high risk pregnancies are, however, controversial. Both high (29, 30) and low (31, 32) levels have been reported. The high levels might be explained by release of enzymes from placentae with multiple infarctions. Although low levels of placental alkaline phosphatase mostly are observed in high risk pregnancies, the wide range of values found during the last trimester of pregnancy limits the clinical usefulness at the individual case level.

Tumor associated forms of alkaline phosphatases

The re-expression of embryonic or fetal antigens in tumors has generated a great deal of interest from the observation by Fishman et al. (33, 34) of circulating PLAP in a male patient with an oat cell carcinoma of the lung. This discovery occurred shortly after the description of other "oncofetal" antigens, i.e. the carcinoembryonic antigen (CEA) (35) and α-fetoprotein (AFP) (36). In these three findings, the concept of oncodevelopmental biology was elaborated and the re-expression of placental and fetal genes during malignancy is recognized as a misregulation of normal embryonic genes or developmental phase-specific genes (37). During the last fifteen years, many reports have been made on the occurrence of PLAP and PLAP-like APs in sera and in tumors, often with greatly discordant results. Besides the initial observation of re-expression of PLAP, several other forms of ectopically synthesized enzyme have been reported. Nakayama _et al_. (38) described in 1970 a PLAP-like enzyme (Nagao), sensitive to L-leucine in a tumor patient, and soon after Inglis _et al_. (39) re-examined 39 cancer patient sera by electrophoretic and catalytic methods known to be PLAP. Up to half of them were found to be inhibitable by L-leucine, i.e. to be PLAP like AP.

Elevated levels of PLAP or PLAP-like AP have been demonstrated for a number of different malignancies including those from the pancreas, lung, breast, colon, lymph nodes, kidney, stomach and bladder. (For review, see 21). The highest incidence of elevation seems to comprise germ cell tumors of the testis (40, 41) and ovarian and uterine cancer (40, 42, 43, 44). In the case of seminomas, PLAP or PLAP-like enzymes now seem to be established as clinically useful tumor markers (45, 46). Recently, successful tumor immunolocalization in nude mice, by use of poly- and monoclonal anti-PLAP antibodies against HeLa cell tumors expressing PLAP, have been performed (46).

Acknowledgements

The present article was prepared with support of the Swedish Cancer Research Council, including a grant to K. Hirano, the Medical Faculty, University of Umeå, and Lion's Foundation, Umeå University.

References

1. McComb, R. B., G. N. Bowers, S. Posen. "Alkaline Phosphatases" Plenum Publ. Corp. 1979.

2. Fishman, W. H., H. G. Sie. 1971. Enzymologia _41_. 141.

3. Mulivor, R. A., L. I. Plothin, H. Harris. 1978. Ann. Hum. Genet. (London) 42, 1.

4. McKenna, M. J., T. A. Hamilton, H. H. Sussman. 1979. Biochem. J. 181, 67.

5. Seargeant, L. E., R. A. Stinson. 1979. Nature 281, 152.

6. Badger, K. S., H. H. Sussman. 1976. Proc. Nat. Acad. Sci. (USA) 73, 2201.

7. Sussman, H. H., P. A. Small, E. Cotlove. 1968. J. Biol. Chem. 243, 160.

8. Doellgast, G. J., W. H. Fishman. 1976. Nature 259, 49.

9. Lehmann, F. G. 1976. In: Onco-Developmental Gene Expression (W. H. Fishman and S Sell, eds.) New York: Academic Press, p. 731.

10. McComb, R. B., G. N. Bowers, S. Posen. 1979. "Alkaline Phosphatases". New York: Plenum.

11. Manning, J. P., N. R. Inglis, S. Green, W. H. Fishman. 1970. Enzymologia 39, 307.

12. Goldstein, D. J., H. Harris. 1979. Nature 280, 602.

13. Goldstein, D. J., C. E. Rogers, H. Harris. 1980. Proc. Natl. Acad. Sci. USA 77, 2857.

14. Moak, G., H. Harris. 1979. Proc. Natl. Acad. Sci. USA 76,1948.

15. Fishman, L., H. Miyayama, S. G. Driscoll, W. H. Fishman. 1976. Cancer Res. 36, 2268.

16. Manning, J. P., N. R. Inglis, S. Green, S. H. Fishman. 1969. Enzymologia 37, 251.

17. Doellgast, G. J., K. Benirshke. 1979. Nature 280, 601.

18. Chang, C. H., D. Angellis, W. H. Fishman. 1980. Cancer Res. 40, 1506.

19. Nozawa, S., M. Ohta, S. Izumi, S. Kayashi, F. Tsutsni, S. Kurihara, K. Watanabe. 1980. Acta Histochem. Cytochem. 13,521.

20. Goldstein, D. J., C. E. Rogers, H. Harris. 1982. Clin. Chem. Acta 125, 63.

21. Stigbrand, T., W. H. Fishman. 1984. In: Human Alkaline Phosphatases. Alan Liss Inc. New York.

22. Slaughter, C. A., M. C. Coseo, M. P. Cancro, H. Harris. 1981. Proc. Natl. Acad. Sci. USA 78, 1124.

23. Millan, J., T. Stigbrand. 1983. Europ. J. Biochem. 136, 1.

24. McLaughlin, P. J., P. J. Travers, I. W. McDicken, P. M. Johnson. 1984. Clin. Chim. Acta 137, 341.

25. Harris, H. 1982. Multilocus enzyme systems and the evolution of gene expression: The alkaline phosphatases as a model example. The Harvey lectures, series 76, p. 95.

26. Holmgren, P. A., T. Stigbrand, M. G. Damber. B. v.Schoultz. 1978. Clin. Chim. Acta 83, 205.

27. Holmgren, P. A., T. Stigbrand, M. G. Damber, B. v.Schoultz. 1979. Obstet. Gynaecol. 54, 631.

28. Yamaguchi, R., N. Shimozato. 1978. Tohoku J. Exp. Med. 124,73.

29. Oesterling, J. J., S. E. Cox, E. R. Carrington. 1977. Am. J. Clin. Nutr. 30, 182.

30. Hunter, R. J., J. H. M. Pinkerton, M. Johnston. 1970. Obstet. Gynaecol. 36, 536.

31. Spellacy, W. N., M. Usategui-Gomez, A. Fernandez de Castro. 1977. Am. J. Obstet. Gynaecol. 127, 10.

32. Kunz, J., P. J. Keller. 1976. J. Perinat. Med. 4, 85.

33. Fishman, W. H., N. R. Inglis, L. L. Stolbach, M. J. Krant. 1968. Cancer Res. 28, 150.

34. Fishman, W. H., N. R. Inglis, S. Green, L. L. Anstiss, N. K. Ghosh, A. E. Reif, R. Rustigan, M. J. Krant, L. L. Stolbach. 1968. Nature 219, 697.

35. Gold, P., S. O. Freeman. 1965. J. Exp. Med. 122, 467.

36. Abelev, G. I., S. D. Perova, N. I. Khramkova, Z. A. Postnikova, I. S. Irlin. 1963. Transplantation I, 174.

37. Manes, C. 1974. Cancer Res. 34, 2044.

38. Nakayama, T., M. Yoshida, M. Kitamura. 1970. Clin. Chim. Acta 30, 546.

39. Inglis, N. R., S. Kirley, L. L. Stolbach, W. H. Fishman. 1973. Cancer Res. 33, 1657.

40. Nathanson, L., W. H. Fishman. 1971. Cancer 27, 1388.

41. Wahren, D., P. A. Holmgren, T. Stigbrand. 1979. Int. J. Cancer 24, 749.

42. Sasaki, M., W. H. Fishman. 1973. Cancer Res. 33, 3008.

43. Usategui-Gomez, M., F. M. yeager, A. Fernandez de Castro. 1973. Cancer Res. 33, 1574.

44. Nishiyama, T. L. L. Stolbach, A. H. Rule, R. A. de Lellis, N. R. Inglis, W. H. Fishman. 1980. Acta Histochem. Cytochem. 13, 245.

45. Lange, P. H., J. L. Millan, T. Stigbrand, R. L. Vesella, E. Ruoslahti, W. H. Fishman. 1982. Cancer Res. 42, 3244.

46. Jeppsson, A., B. Wahren, J. L. Millan, T. Stigbrand. 1984. Brit. J. Cancer 49, 123.

NEW PREGNANCY RELATED PROTEINS IN HUMANS AND THEIR ANALOGUES IN ANIMALS

Hans Bohn
Research Laboratories, Behringwerke AG, D-3550 Marburg, Federal Republic of Germany

Biologically Active Proteins in Humans

Pregnancy proteins are proteins, which do not occur or occur only in trace amounts in normal sera, and which during gravidity appear in the maternal bloodstream or are strongly elevated in their concentration. A number of pregnancy proteins are functional proteins which have been detected by their biological activity and named accordingly.

To this category belong the pregnancy-specific hormones human chorionic gonadotropin (hCG) and human placental lactogen (hPL), as well as the pregnancy-specific enzymes heat-stable alkaline phosphatase (HSAP), diamine oxidase (DAO) or histaminase and cystine aminopeptidase (CAP) or oxytocinase.

The pregnancy specific hormones and enzymes are synthesized by the trophoblast, with the exception of DAO which appears to be mainly produced by the decidua.

The physicochemical properties of these biologically active pregnancy proteins are summarized in Table 1. The amounts in which these proteins occur in term placentas, in normal sera and in pregnancy sera are shown in Table 2.

Table 1. Pregnancy Specific Hormones and Enzymes and Their Characteristics

	Physicochemical properties			
	Electrophoretic mobility	Molecular weight	Carbohydrate content (%)	Reference
hCG	ß	47,000	31	(1)
hPL	α_1 - α_2	22,000	0	(2)
HSAP	β_1	125,000	~10	(3)
DAO	α_2	180,000	?	(4)
CAP	?	290,000	44	(5)

Table 2. Concentrations of Pregnancy Specific Hormones and Enzymes

	Term placentas mg per placenta (mean values)	Normal sera mg per liter (mean values)	Pregnancy sera mg per liter (mean values at maximum)	Reference
hCG	1	$<$ 0.001	10	(6)
hPL	100	$<$ 0.005	7	(7)
HSAP	40	~ 0.01	~ 0.2	(3, 8)
DAO	20	$<$ 0.01	~ 0.2	(9, 10)
CAP	?	~0.12	~ 5.7	(11, 12)

hPL is the most abundant hormone in term placentas. The most abundant of the placental specific enzymes in the tissue appears to be HSAP. During pregnancy especially hCG, hPL and CAP are secreted in larger amounts into the maternal bloodstream.

The pregnancy specific hormones and enzymes have attracted much interest as indicators of pregnancy or placental function during pregnancy. In former times most of the functional proteins have been quantitated by bioassays, but nowadays usually immunochemically methods are used for their detection and determination.

Detection of hCG in serum or urine is used for the early diagnosis of pregnancy. hCG becomes detectable in the maternal blood already 8 - 12 days after conception i.e. at the time of implantation. Levels then rise rapidly and reach a maximum at the 10th week of pregnancy after which the concentration declines rapidly and remains almost constant from the 4th month onwards. The concentrations of hCG secreted in urine of pregnant women show a similar distribution.

hPL reaches measurable levels in serum only by the sixth week of pregnancy. The levels then rise steadily reaching a plateau during the third trimester. A close correlation was found to exist between placental mass and circulating hPL values. Serial determinations therefore are of diagnostic value in the assessment of placental function, especially in cases complicated by threatened abortion. The pregnancy specific enzymes also appear in the maternal bloodstream in amounts increasing with advancing pregnancy. Determination of these proteins in serum also has a certain diagnostic significance in the surveillance of pregnancy.

Of the known functional proteins of the placenta we have isolated in our laboratory hPL and HSAP. hPL could be obtained in highly purified form by crystallization (13). HSAP was prepared from soluble as well as from solubilized protein fractions of the placenta using an immunoadsorption technique.

New Pregnancy Related Proteins in Humans Detected by Immunochemical Methods

During the last 20 years a number of new pregnancy proteins have been detected in placental extracts or in sera from pregnant women by immunochemical methods.

A systematic investigation of placental extracts for new pregnancy proteins performed in our laboratory revealed the presence of three proteins which were present in high concentrations in sera from pregnant women but did not occur or occurred only in trace amounts in normal sera. These proteins were designated as "Schwangerschaftsprotein" 1, 2 and 3 and abbreviated to SP1, SP2, SP3 (14). The antisera used for their detection had been prepared by immunizing rabbits with protein fractions from the placenta. Table 3 summarizes the names and synonyms of these proteins as well as some of their physicochemical properties.

SP1 turned out to be a pregnancy-specific protein different from all pregnancy-specific hormones and enzymes known thus far. This protein was shown to have the electrophoretic mobility of a ß1-globulin and a high carbohydrate content; it therefore was designated as pregnancy-specific ß1-glycoprotein. Other names given to this pregnancy protein are pregnancy-associated plasma protein C (PAPP-C) and trophoblast-specific ß1-glycoprotein (TSG). The molecular weight of SP1 was found to be 90 000 daltons. Later in addition other variants of SP1 have been described having α2-electrophoretic mobilities and higher molecular weights of 200 000 and 400 000 daltons respectively.

SP2 and SP3 on the other hand, are pregnancy-associated proteins. They occur in trace amounts in all normal sera. Their concentrations in serum are strongly elevated, not only during pregnancy, but also in women taking hormonal contraceptives. SP2 turned out to be identical with the steroid-binding ß-globulin or sex hormone-binding globulin and SP3 was found to be identical with the so-called pregnancy-zone protein which is now generally designated as pregnancy-associated α2-glycoprotein (α2-PAG). SP2 and SP3 are not apparently derived from the placenta: their occurrence in the placental extract appears to be only due to a certain content of maternal blood which is always present in the placenta. SP1 on the other hand is known to be synthesized by the trophoblast (30).

The biological function of SP2 (SHBG) is to bind and transport certain steroid hormones. SP1 and SP3 (α2-PAG) appear to have immunosuppressive properties, at least in vitro. Other functions of these pregnancy proteins are not yet known thus far.

In addition to the pregnancy proteins SP1 to SP3, a number of other soluble placental proteins have been detected, isolated and characterized in our laboratory; they were found to occur mainly in the placental tissue with very little secreted into the maternal bloodstream (29, 31). Their concentration in sera of pregnant individuals is usually less than 1 mg/l. These proteins have been designated placental tissue proteins or simply placental proteins, abbreviated PP and numbered consecutively. Some of these proteins show a significant increase in serum during gestation and therefore may be ranged with the pregnancy proteins. To this group belong PP5, PP10, PP11, PP12 and PP14; these pregnancy related soluble placental tissue proteins are also listed in Table 3.

PP12 and PP14 were found to be immunochemically identical or at least related to chorionic α1-microglobulin (CAG-1) and to chorionic α2-microglobulin (CAG-2), respectively, which were first described by Petrunin et al. Later these proteins were designated also as placenta-specific α1-microglobulin (PAMG-1) and placenta-specific α2-microglobulin (PAMG-2) respectively by Tatarinov et al. However our data on the physicochemical properties of PP12 and PP14 are somewhat different from those of these authors.

The soluble placental tissue proteins listed in Table 3 have molecular weights in the range of 25 000 to 50 000 daltons. They are glycoproteins - the carbohydrate content being highest in PP5 and PP14.

Table 3. Immunochemically Detected Pregnancy Proteins and Their Characteristics

Abbrev.	Name and synonyms	Physicochemical properties: Electrophoretic mobility	Molecular weight	Carbohydrate content (%)	Reference
SP_1	pregnancy-specific β_1-glycoprotein	β_1 (α_2)	90,000(400,000)	29,3	(14, 16)
	pregnancy-associated plasma protein C (PAPP-C)				(17)
	trophoblast-specific β_1-glycoprotein (TSG)				(18)
SP_2	steroid-binding β-globulin (SBβG)	β_1	65,000	12.6	(19)
	sex-hormone-binding globulin (SHBG)				
SP_3	pregnancy-associated α_2-glycoprotein (α_2PAG)	α_2	360,000	12.2	(20)
	pregnancy-zone protein (PZ)				(21)
	pregnancy-associated macroglobulin (PAM)				(22)
PP_5	placental protein 5	β_1	36,600	19.8	(23)
PP_{10}	placental protein 10	α_1	48,000	6.6	(24)
PP_{11}	placental protein 11	α_1	44,300	3,9	(25)
PP_{12}	placental protein 12	α_1	25,200	4.3	(26)
	chorionic α_1-microglobulin (CAG-1)				(27)
	placenta-specific α_1-microglobulin (PAMG-1)				(28)
PP_{14}	placental protein 14	α_2	43,000	17.5	(29)
	chorionic α_2-microglobulin (CAG-2)				(27)
	placenta-specific α_2-microglobulin (PAMG-2)				(28)

Immunohistochemical studies revealed that all these placental proteins could be localized in the syncytiotrophoblast (32, 33). In the case of PP10 and PP12 a strong staining was also observed in histiocytes occuring in amniotic, villous and decidual tissue. In addition it was found that granulocytes in normal blood also stained strongly positive for PP10 and PP12; this indicated that these two proteins are not really specific to the placenta. PP12 and PP14 could be detected in the endometrium. PP14 in addition was found to occur in high concentrations in seminal plasma. PP12 was recently shown by Rutanen et al. to be mainly produced by the decidua during pregnancy, and the same appears to be true for PP14. Thus only PP5 and PP11 are relatively specific to the placenta and appear to be synthesized by the syncytio-trophoblast.

Biological functions are known thus far only for PP5. This protein was found to inhibit the function of trypsin and plasmin and thus turned out to be a protease inhibitor. PP5 also was shown to form complexes with heparin and to be involved with the coagulation system.

For the other soluble placental tissue proteins no biological activities have yet been found. But there are indications that the endometrial proteins PP12 and PP14 may play an important role in the early events of human reproduction: They appear in the female genital tract before ovulation and implantation, and the release of PP12 from secretory endometrium was found to be stimulated by progesterone (Rutanen et al., personal communciations).

Table 4. Concentrations of the Immunochemically Detected Pregnancy Proteins

	Term placentas mg per placenta (mean values)	Normal sera mg per liter (mean values)	Pregnancy sera mg per liter (mean values at max.)	Amniotic fluid mg per liter at term (16th week)	Reference
SP_1	30	< 0.001	180	1.2	(32)
SP_2 (SHBG)	6.5	~ 6	75	?	(19, 34)
SP_3 (α_2PAG)	<7	~ 20	860	?	(35)
PP_5	1.5	not detectable	0.035	0.004	(36, 37)
PP_{10}	20	0.002	0.026	~ 0.002	(38)
PP_{11}	11	< 0.001	0.004	?	(39)
PP_{12}	5	0.024	0.169	(30)	(40, 41)
PP_{14}	4.8	~ 0.03	1.0	(124)	(42)

Table 4 summarizes the concentrations in which the new pregnancy related proteins occur in human term placentas and in body fluids . The amounts in which the pregnancy related proteins occur in term placentas have been quantitated by an electro-immunoassy. Usually more sensitive methods, such as radioimmuno-assay or enzyme immunoassays, were necessary to measure these proteins in body fluids.

In normal sera these proteins could not be detected or were found to occur only in trace amounts. In sera from pregnant women the so-called "pregnancy proteins" reach levels much higher than 1 mg/l, whereas the concentrations of the placental tissue proteins usually remain below this level. But compared to the normal values in serum the placental tissue proteins also show an increase in concentration during pregnancy.

In amniotic fluid the concentrations of the pregnancy related proteins are usually low. Exceptions are PP12 and PP14 which reach very high levels in this body fluid - the maximal values being found around the 16th week of pregnancy.

The patterns of the concentration of the pregnancy related proteins in sera from normal pregnant women throughout pregnancy are as follows: SP1 can be detected in the maternal blood as early as 7 - 10 days after conception. Measurement of this protein, therefore, was found to be useful in the early detection of pregnancy. Levels of SP1 rise steadily during gestation and tend to reach a plateau near term. They were found to correlate well with th stage of gestation. Measurement of SP1, therefore, can be used as an indicator of placental function. After delivery SP1 disappears from the maternal circulation with a mean half-life or 30 hr (32).

SP2 (SHBG) shows a main increase during the first half of pregnancy; in the second half, the concentration is only slightly more elevated (34).

SP3 (α2-PAG) levels reach a maximum by week 31 - 34 and there-after appear to fall slowly until parturition. Concentrations of SP2 and SP3 do not correlate with the stage of gestation (34). Post partum, the SP2 and SP3 concentrations return to normal values within 5 - 10 weeks (apparent half-life 6 - 7 days).

The concentrations of PP5, PP10 and PP11 during pregnancy, similar to SP1, increase to reach a maximum or plateau near term (36, 37, 38, 39).

In case of PP12 the maximal values in serum are found already around week 20 of pregnancy; then there is a minimum around week 30 and in late pregnancy PP12 levels rise again (32, 38, 40). This pattern of PP12 levels is unique amongst this group of proteins.

The levels of PP14 have recently been measured in serum of women throughout pregnancy with use of radioimmunoassays by Julkunen et al. (42) and by Chapman et al. (personal communications): They found that in nonpregnant sera, the PP14 levels are low, and that a significant rise occurs in early pregnancy. Serum levels rise rapidly following the establishment of pregnancy and reach a peak at 9 weeks gestation. From that point they decline towards term. This pattern of PP14 levels bears a striking resemblance with hCG.

Animal Analogues of the New Pregnancy Related Proteins

To almost all pregnancy related proteins mentioned above immunochemically identical or related proteins could be detected in placentas or pregnancy sera of apes and/or monkeys. On the one hand these proteins could be localized in monkey placentas by immunohistochemical methods using antisera to the corresponding human antigens. The findings in cynomolgus monkeys were similar to those in the human (43 - 45). On the other hand the presence of immunochemically related proteins in placental extracts or pregnancy sera of subhuman primates was also demonstrated using immunodiffusion tests. First we have shown using the gel diffusion test of Ouchterlony that pregnancy sera from cynomolgus monkeys contain proteins immunochemically related to SP1, SP2 and SP3, respectively (46). Then an electroimmunoassay was used to investigate the concentrations of these three pregnancy proteins in sera of cynomolgus monkeys before pregnancy, during pregnancy and after delivery (47). As in the human, monkey SP1 appeared to be specific for pregnancy, not being detectable in nonpregnant animals. Its concentration in serum increased with advancing pregnancy; after delivery it disappeared from the maternal circulation with a half-life of around 40 hr. On the other hand SP2 and SP3 were found to be present in quite high levels in serum of nonpregnant animals. During pregnancy these proteins decreased in concentration progressively to reach a minimum at the time of delivery or immediately afterwards; following that they slowly increased to return to normal values. This behaviour of SP2 and SP3 levels in monkeys is in striking contrast to the situation observed in humans, where the concentration of these proteins is very low in normal serum and strongly increased during pregnancy. With regard to SP3 (pregnancy zone protein) the same observation in monkeys has been reported by Lin et al. (48).

With the gel diffusion test of Ouchterlony we also have demonstrated that the placental extracts from chimpanzee, baboon, rhesus and cynomolgus monkey contain proteins immunochemically related to hPL, SP1 and PP5 (49, 50): hPL and SP1 were found to be only partly identical with the corresponding proteins in apes and monkeys as indicated by a spur formation. No spur was formed in the case of PP5; the corresponding antigen in monkeys therefore seems to be identical or very closely related to the human protein. In the gel diffusion test no material cross-reacting with antibodies to HSAP could be detected in placental fractions from subhuman primates. But immunofluorescence studies revealed that an antigen immunochemically related to HSAP is present in the tissue of monkey placentas (43, 51). An explanation for the absence of this antigen in the placental extract could well be that the corresponding protein in monkeys is very firmly associated with membranes, therefore insoluble in aqueous solution and not detectable in the aqueous extract.

In a similar way we have demonstrated that placental extracts of apes and / or monkeys contained proteins immunochemically closely related to human PP10, PP11 and PP12 (29). In case of PP14 no cross-reacting proteins could be detected in subhuman primates. May be that the concentration of the corresponding proteins in monkeys is too low to be detected with the gel diffusion test.

The amounts in which the new pregnancy related proteins occur in extracts from term placentas of man and cynomolgus monkeys (Macaca fascicularis) have been quantitated by an electroimmunoassay. The results are shown in Table 5. SP1 was found to be around tenfold and PP12 even more than hundred-fold higher in the cynomolgus placenta than in human placenta. PP10 on the other hand was significantly lower in the monkey. The concentrations of PP5 and PP11, respectively, were of the same order of magnitude in the two species.

Table 5: Amounts of Pregnancy Related Proteins Found in Extracts from Term Placentas of Man and Monkey (mg per 100 g placenta)

Proteins	Human placenta	Cynomolgus placenta
SP1	5.0	47
SP2 (SHBG)	1.0	?
SP3 (α_2-PAG)	7.8	?
PP5	0.2	$<$ 0.2
PP10	3.3	$<$ 0.2
PP11	1.8	2.8
PP12	0.8	120
PP14	0.8	-

In our laboratory a method was developed to isolate SP1 from placental extracts in good yield and highly purified form by using an immunoadsorption technique in combination with hydroxy-apatite chromatography. The same immunoadsorption technique we applied to purify the proteins antigenically related to SP1 from placental extracts of apes and monkeys (52). Comparative electrophoretic studies of these preparations revealed that the corresponding proteins of chimpanzee, rhesus and cynomolgus monkeys are more acid than human SP1 and migrate faster to the anode than the human protein. Two-dimensional electrophoresis in addition showed that the related monkey proteins are much more heterogeneous than human SP1. Rhesus SP1 has been prepared in larger amounts; its amino acid and carbohydrate composition was determined and compared with that of human SP1. Both proteins were found to be similar in their chemical composition.

In a similar way we have isolated rhesus PP12 and cynomolgus PP12 from the corresponding monkey placentas using an immunoadsorbent which has been prepared by coupling rabbit antibodies to human PP12 to Sepharose.

Human PP12 was found to contain two immunochemically completely different epitopes. When rabbits were immunized with purified PP12, antibodies were usually obtained which reacted with two different antigenic determinants. These different antigenic determinants have been designated A and B. We still do not know whether PP12 is composed of two different subunits having different antigenic determinants, or whether the two determinants A and B are present in one and the same polypeptide chain.

The PP12 analogous isolated from rhesus and cynomolgus monkey placentas also were found to contain two different epitopes which were immunochemically closely related to determinants A and B, respectively, of human PP12. In case of cynomolgus PP12 we could separate the A from the B determinants by euglobulin precipitation, the A determinants being found in the precipitate. These findings indicate that in monkey PP12 the A and B determinants are located in different subunits which are held together by noncovalent bonds. Whereas in case of human PP12 we did not succeed in separating the two epitopes under the same condition.

Human PP12 has the electrophoretic mobility of an α1-globulin. The monkey analogues of PP12 were found to be less acidic and to have α2-mobility. Sometimes a much more basic component migrating in the range between ß2- and γ-globulins could be detected in placental extracts especially of cynomolgus monkeys (26). This component which appeared to be a split product of PP12 contained only the B determinants and did not react with anti-PP12 (A) i.e. an antiserum reacting only with the A determinants of PP12. Such an antiserum was obtained by immunizing rabbits with the euglobulin precipitate of cynomolgus PP12.

Thus far I have shown that proteins immunochemically identical or at least related to the human pregnancy proteins could be detected in pregnancy sera or placentas of subhuman primates. In animals of lower phylogenetic levels usually no crossreacting proteins could be found in the gel diffusion test of Ouchterlony with antisera to the human pregnancy proteins. Only in case of anti-SP3 rabbit serum a precipitin reaction was obtained with sera from pregnant dogs and guinea pigs. This antiserum to human SP3 we used to prepare an immunoadsorbent which enabled us to isolate the SP3-related protein from pregnancy sera of beagle dogs. The dog analogue thus obtained was used to prepare a rabbit antiserum to beagle dog SP3. With such an antiserum it was possible to detect immunochemically related proteins not only in humans and guinea pigs but also in pregnancy serum of rat, cow, horse and sheep. These findings had been first reported by Stigbrand et al. (53) who had prepared an "anti-dog-PZ" serum by immunizing rabbits with the precipitate formed by pregnancy dog serum against anti-human PZ (SP3) serum.

With our anti-dog-SP3 serum we investigated the concentrations of dog-SP3 before, during and after pregnancy in beagle dogs. A characteristic pattern is shown in Fig. 1. During pregnancy the dog-SP3 levels increased to reach a maximum at the 6th week of pregnancy; then they declined towards term.

Fig. 1

Serum levels of dog-SP_3

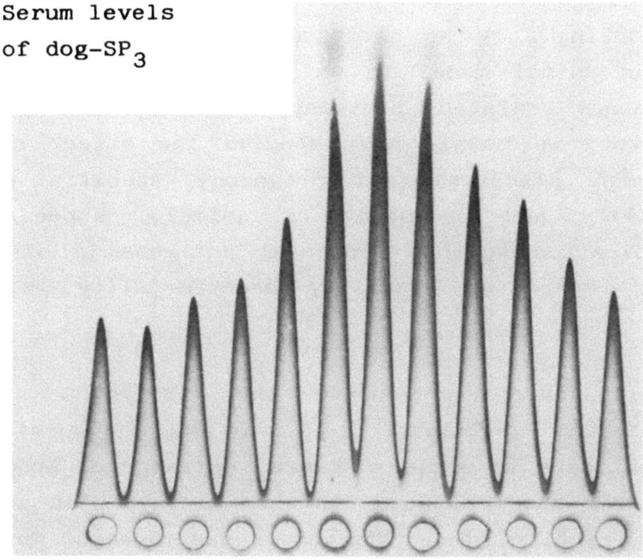

before / 1 2 3 4 5 6 7 8 9 / after

weeks of pregnancy

A mouse analogue to SP3 was described by Hau (54) who investigated the pregnancy-associated proteins in mouse. He could detect 4 such proteins which he called PAMP-1 to PAMP-4. PAMP-1 was found to be immunochemically related to SP3 or pregnancy-zone protein. PAMP-4 turned out to be the mouse analogue of AFP (alpha-fetoprotein), PAMP-3 was supposed to be the mouse analogue to SP2, whereas PAMP-2 was shown to crossreact with SP1. These investigations were performed by Hau using antisera raised against murine and human pregnancy proteins in hens.

Tatarinov et al. (55) using a special technique detected that human SP1 is immunochemically related to pregnancy specific ß-globulins of rabbit, rat and guinea pig. They used a special technique combining counter current immunoelectrophoresis and immunochemical adsorption by heterologous antibodies.

Antifertility Testing in Animals Using Human Placenta-"Specific" Proteins

The aim of looking for the corresponding proteins in animals was to find an animal model in which the biological function of the pregnancy related proteins could be investigated. In our laboratory we mainly have studied the effect of antibodies to human placental and pregnancy specific proteins on the fertility and pregnancy of animals, especially of monkeys. Immunization against placental antigens is of special interest with regard to sterility and fertility regulation (50, 56).

Numerous studies have demonstrated that it is possible to prevent or disrupt gestation in a variety of species with heterologous antisera to whole placental extracts or trophoblast cells of the same species. Such experiments have been performed in guinea pigs, rabbits, mice, rats and rhesus monkeys. Antifertility testing in animals also has been conducted using antibodies to purified human placental proteins: O. Förster et al. (57) reported interruption of pregnancy in rats by administering rabbit antisera to hCG. El Tomi et al. (58) and later Gusdon (59) demonstrated that abortion could be produced in rats by application of antisera to highly purified hPL.

In a similar manner we have tested the effect of rabbit antibodies to SP1 in pregnant rats and monkeys. In rats passive immunization with rabbit antibodies to SP1 proved to be unsuccessful in terminating pregnancy in rats (46). But the same antibodies induced abortion when intravenously injected into pregnant cynomolgus monkeys (47). In this experiment a total of 750 mg of the IgG fraction of the rabbit antiserum were injected into pregnant cynomolgus monkeys on three consecutive days between days 19 - 55 of gestation. Eight of ten pregnant monkeys aborted.

A similar antifertility effect was obtained in passively immunizing pregnant cynomolgus monkeys with rabbit antibodies to PP5. In this experiment 5 of 7 animals aborted between days 33 and 48 of pregnancy and one monkey had a still-birth. Only in one of the seven monkeys pregnancy continued and resulted in normal delivery.

Table 6. Antifertility Effect of Active Immunization of Cynomolgus Monkeys with "Placenta-Specific" Proteins

Antigens used for immunization	No. of animals	No. of matings	Abortions observed	Pregnancies resulting in normal birth	Matings per pregnancy
SP_1-DSA	6	22	5	5	4.4
PP_5-DSA	6	22	1	6	3.7
HSAP-DSA	6	13	0	5	2.6
PP_{11}-DSA	6	12	0	6	2.0
PP_{12}-DSA	5	10	0	4	2.5
Adjuvant alone (Controls)	10	19	1	9	2.1

DSA, diazotized sulphanilic acid

In addition we have studied the ability of an active immunization of fertile female monkeys with SP1, PP5, HSAP, PP11 and PP12 to induce contraception in the monkeys (61, 62). The purified human antigens were modified by coupling diazotized sulfanilic acid to the proteins to make them more antigenic. Fertile female cynomolgus monkeys then were actively immunized with the modified antigen preparations prior to mating. Per animal 3 mg protein were injected intravenously and subcutaneously over a six weeksperiod using silica gel as an adjuvant. Following immunization the female monkeys were mated with males of proven fertility for three days at midcycle starting two days prior to expected ovulation. The outcome of the mating was checked by intrarectal palpation of the uterus. Those monkeys who did not conceive or aborted were mated again at the next midcycle.

The effect of active immunization with the above mentioned antigens on reproduction in cynomolgus monkeys is summarized in Table 6. The ratio of matings to pregnancies resulting in normal birth can be used as an indicator of the antifertility effect. Ten animals receiving adjuvant alone served as controls. The ratio of matings to pregnancies resulting in birth with the control animals was 2.1. This ratio was much higher in the monkeys immunized with SP1 or PP5; these monkeys showed a significant reduction in fertility. Immunization with HSAP, PP11 and PP12 appeared to be less effective.

References

1. Bahl, O.P. 1969. Human chorionic gonadotropin I. Purification physicochemical properties. J. Biol. Chem. 244, 567.

2. Li , C.H., J.S. Dixon, D. Chung 1973. Amino acid sequence of human chorionic somatomammotropin. Arch. Biochem. Biophys. 155, 95.

3. Harkness, D.R. 1968. Studies on human placental alkaline phosphatase. Arch. Biochem. Biophys. 126, 503.

4. Baylin, S.B., S. Margolis 1975. Purification of histaminase (Diamine oxidase) from human pregnancy plasma by affinity chromatography. Biochem. Biophys. Acta 397, 294.

5. Yman, L. 1970. Composition of oxytocinase (cystine aminopeptidase). Acta Pharm. Suecia 7, 29.

6. Braunstein, G.D. 1982. Chorionic gonadotropin (hCG) and hCG-like substances in human tissues and bacteria. In: Pregnancy Proteins (J.G. Grudzinskas, B. Teisner and M. Seppälä, eds.) Academic Press, Sydney, pp. 39-49.

7. Letchworth, A.T. 1976. Human placental lactogen assay as a guide to fetal wellbeing. In: Plasma Hormone Assays in Evaluation of Fetal Wellbeing (A. Klopper, ed.), Churchill Livingstone, Edinburgh, p. 147.

8. Leroux, M., W. F. Perry. 1970. Serum heat-stable alkaline phosphatase in pregnancy. Am. J. Obstet. Gyn. 108, 235-239.

9. Lin, V.W., S.D. Kirley, M.S. Pierre 1981. Tumour and placental histaminase I. Oncodev. Biol. Med. 2, 267-280.

10. Lorenz, W., J. Kusche, F. Werle. Diamine oxidase. In: Methods of Enzymatic Analysis, Vol. 2 (H.V. Bergmeyer ed.) Academic Press, Inc. Second English Edition, pp. 660-665.

11. Ryden, G. 1971. Cystine aminopeptidase activity in pregnancy. Acta Obstet. Gynec. Scand. 50, 253-257.

12. Watkins, W.B. , C. W. Small 1972. Immunologic inactivation of human pregnancy serum oxytocinase activity. Am. J. Obstet. Gyn. 113, 973-978

13. Bohn, H. 1971. Kristallisation und Hochreinigung des humanen Plazenta-Laktogens. Experientia 27, 1223.

14. Bohn, H. 1971. Nachweis und Charakterisierung von Schwangerschaftsproteinen in der menschlichen Plazenta sowie ihre quantitative immunologische Bestimmung im Serum schwangerer Frauen. Arch. Gynäkol. 210, 440.

15. Bohn, H. 1972. Isolierung und Charakterisierung des schwangerschaftsspezifischen ß1-Glykoproteins. Blut 24, 292.

16. Teisner, B., J. Folkersen, P. Hindersson, J. Jensenius, J. Westergaard. 1979. Quantification of the pregnancy-specific ß1-glycoprotein (SP1) by immunoprecipitation techniques: the influence of a cross-reacting high molecular weight α2 protein. Scand J. Immunol. 9, 409-417

17. Lin, T.M., S.P. Halbert, D.J. Kiefer, W.N. Spellacy, S. Gall. 1974. Characterization of four human pregnancy-associated plasma proteins. Am. J. Obstet. Gynecol. 118, 223.

18. Tatarinov, Y.S. 1978. Trophoblast-specific ß1-glycoprotein as a marker for pregnancy and malignancies. Gynecol. obstet. Invest. 9, 65-97

19. Bohn, H. 1974. Isolierung, Charakterisierung und quantitative immunologische Bestimmung des Steroid-bindenden ß-Globulins. Blut 29, 17.

20. Bohn, H., W. Winckler 1976. Isolierung und Charakterisierung des schwangerschaftsassoziierten α 2-Glykoproteins (α 2-PAG). Blut 33, 377.

21. Schoultz, B., T. Stigbrand 1974. Characterization of the "pregnancy zone protein" in relation to other α2-globulins of pregnancy. Biochim. Biophys. Acta 359, 303.

22. Stimson, W.H., J.M. Sinclair 1974. An immunoassay for a pregnancy-associated α-macroglobulin using antibody-enzyme conjugates. Febs Letters 47, 190.

23. Bohn, H., W. Winckler 1977. Isolierung und Charakterisierung des Plazentaproteins PP5. Arch. Gynäkol. 223, 179.

24. Bohn, H., W. Winckler 1979. Isolierung und Charakterisierung eines neuen plazentaspezifischen Proteins (PP10). Arch. Gynäkol. 227, 125.

25. Bohn, H., W. Winckler 1980. Isolierung und Charakterisierung eines neuen Plazenta-Gewebsproteins (PP11). Arch. Gynecol. 229, 293.

26. Bohn, H., W. Kraus (1980). Isolierung und Charakterisierung eines neuen plazentaspezifischen Proteins (PP12). Arch. Gynecol. 229, 279.

27. Petrunin, D.D., I.M. Gryaznova, Y.A. Petrunina, Y.S. Tatarinov (1978). Comparative immunochemical and physico-chemical characteristics of human chorionic α1- and α2-microglobulins. Byull. Eksp. Biol. Med. 5, 600.

28. Tatarinov, Y.S., G.A. Kozljaeva, D.D. Petrunin, Y.A. Petrunina (1980). Two new human placenta-specific α-globulins: identification, purification, characteristics, cellular localisation and clinical investigation. In: (Klopper, A., A. Genazzani and P.G. Crosignani, eds.) The human placenta: Proteins and hormones. London Academic Press, p. 35-46

29. Bohn, H., W. Kraus, W. Winckler 1982. New soluble placental tissue proteins. In: (Klopper, A. ed.) Immunology of Human Placental Proteins. Placenta (Suppl. 4). New York: Praeger, p. 67-81.

30. Horne, C.H.W., C.M. Towler, R.G.P., A. Q. Thomson, H. Bohn 1976. Pregnancy-specific ß1-glycoprotein, a product of the syncytiotrophoblast. Experientia 32, 1197-1199.

31. Bohn, H. 1972. Nachweis und Charakterisierung von löslichen Antigenen in der menschlichen Plazenta. Arch. Gynaek. 212, 165-175.

32. Bohn, H., F. Dati, G. Lüben (1983). Human trophoblast specific products other than hormones. In: (Loke, Y.W., Whyte, A. eds.) Biology of trophoblast. Amsterdam: Elsevier; p. 317-352

33. Wahlström, T., H. Bohn, M. Seppälä 1982. Immunohistochemical studies on pregnancy proteins. In: (Grudzinskas, J.G., Teisner, B., Seppälä, M. eds.) Pregnancy proteins. Sydney: Academic Press, p. 415-422.

34. Towler, C.M., C.H.W. Horne, V. Jandial, H. Bohn. 1976. A serial study of pregnancy proteins in primigravidae. Br. J. Obstet. Gynaecol. 83, 368.

35. Than, G.N., I.F. Csaba, D.G. Szabo 1974. An antigen associated with pregnancy. Lancet 2, 1578.

36. Obiekwe, B.C., D.J. Pendlebury, Y.B. Gordon, J.G. Grudszinskas, T. Chard, H. Bohn 1979. The radioimmunoassay of placental protein 5 and circulating levels in maternal blood in the third trimester of normal pregnancy. Clin. Chim. Acta 95, 509.

37. Seppälä, M. T. Wahlström, H. Bohn. 1979. Circulating levels and tissue localization of placental protein five (PP5) in pregnancy and trophoblast disease. Absence of PP5 expression in the malignant trophoblast. Int. J. Cancer 24, 6.

38. Than, G.N., I.F. Csaba, D.G. Szabo 1983. Circulating levels of new placenta-specific tissue proteins PP10 and PP12 during healthy pregnancy. IRCS Med. Sci. 11, 627-628.

39. Lüben, G., G. Tatra, H. Bohn, F. Nasr 1985. Radioimmunoassay of circulating placental protein PP11 in normal individuals and in patients with ovarian tumours (in preparation).

40. Rutanen, E.-M., H. Bohn, M. Seppälä 1982. Radioimmunoassay of placental protein 12: Levels in amniotic fluid, cord blood, and serum of healthy adults, pregnant women, and patients with trophoblastic disease. Am. J. Obstet. Gynecol. 144, 460-463.

41. Than, G.N., I.F. Csaba, D.G. Szabo, Z. Bognar, A. Arany, H. Bohn 1983. Levels of placenta-specific tissue protein 12 (PP12) in serum during normal pregnancy and in patients with trophoblastic tumour. Arch. Gynecol. 234, 39-46.

42. Julkunen, M., E.-M. Rutanen, R. Koistinen, A. Koskimies, T. Ranta, H. Bohn, M. Seppälä 1985. Distribution of placental protein 14 in tissues and body fluids during pregnancy. Brit. J. Obstet. Gynaecol., submitted.

43. Sedlacek, H.H., R. Rehkopf, H. Bohn 1976. Immunofluorescence histological localization of human pregnancy and placenta proteins in the placenta of man and monkeys (cynomolgus). Behring Inst. Mitt. 59, 81.

44. Inaba, N., T. Renk, E. Weinmann, H. Bohn. 1981. Immunohistochemical location of proteins related to pregnancy-specific ß1-glycoprotein (SP1) and placental tissue proteins (PP5, PP8, PP9, PP10, PP11, PP12) in the afterbirth of cynomolgus monkeys (Macaca fascicularis). Arch. Gynecol. 230, 195-204

45. Inaba et al. 1981. Blut 43, 315-323

46. Bohn, H., H. Ronneberger 1973. Immunologischer Nachweis von Schwangerschaftsproteinen des Menschen im Serum trächtiger Tiere. Arch. Gynäk. 215, 277-284.

47. Bohn, H., E. Weinmann 1974. Immunologische Unterbrechung der Schwangerschaft bei Affen mit Antikörpern gegen das menschliche schwangerschafts-spezifische ß1-Glykoprotein (SP1). Arch. Gynäk. 217, 209-218.

48. Lin, T.M., S.P. Halbert, R. Plasencia 1976. Pregnancy zone pprotein analogue inpregnant and non-pregnant primates, and its decrease during pregnancy in some monkey species. Clin. exp. Immunol. 26, 609-622.

49. Bohn, H., H. Sedlacek 1975. Eine vergleichende Untersuchung von plazentaspezifischen Proteinen bei Mensch und subhumanen Primaten. Arch. Gynäk. 220, 105-121.

50. Bohn, H. 1976. The protein antigens of human placenta as a basis for the development of contraceptive vaccine. In "Development of Vaccines for Fertility regulation"

(World Health Organization, ed.) Scriptor, Copenhagen, p. 111.

51. Inaba, N., N. Sato, M. Ijichi, I. Fukazawa, A. Nito, H. Takamizawa, G. Lüben, H. Bohn 1984. The immuno-cytochemical location of two membrane-associated placental tissue proteins in human and cynomolgus monkey placentae. Tumour Biology 5, 75-85

52. Bohn, H., R. Schmidtberger, H. Zilg 1976. Isolierung des schwangerschaftsspezifischen ß1-Glycoproteins (SP1) und antigenverwandter Proteine durch Immunadsorption. Blut 32, 103-113.

53. Stigbrand, T., M.-G. Damber, B. Schoultz 1976. Biochemistry and properties of the immunosuppressive pregnancy zone protein (PZ). In Protides of Biological fluids (Peeters H. ed.) Vol. 24, Pergamon Press, Oxford, New York, pp. 181-188.

54. Hau, J. 1982. A murine model for the study of pregnancy proteins of fetal, placental and maternal origin. In: Pregnancy Proteins. Biology, Chemistry and Clinical Application (Grudzinskas, J.G., B. Teisner, M. Seppälä eds.) Academic Press, Sydney, p. 435.

55. Tatarinov, Y.S., D.D. Petrunin, O.P. Shevchenko 1980. Immunochemical study of human trophoblast-specific ß1-globulin and its analogues in animals. Bull. Exp. Biol. & Med. U.S.S.R. 89, 294.

56. Stevens, V.C. 1976. Perspectives of development of a fertility control vaccine from hormonal antigens of the trophoblast. in: Development of Vaccines for Fertility Regulation (World Health Organization, ed.) Scriptor, Copenhagen, pp. 93-110.

57. Förster, O., F. Friedrich, H. Holzner and A. Kratochwil 1969. Schwangerschaftsunterbrechung durch Anti-HCG bei der Ratte. Zeitschrift Geburtsh. Gynäk. 170, 47-55.

58. El Tomi, A.E.F., L. Boots, V.C. Stevens 1971. Effects of antibodies to human placental lactogen on reproduction and contraception. Endocrinology 88, 805-809.

59. Gusdon, J.P. 1972. An immunologic method of pregnancy destruction and contraception. Amer. J. Obstet. Gynec. 112, 472.

60. Bohn, H., E. Weinmann 1976. Antifertilitätswirkung einer aktiven Immunisierung von Affen mit dem schwangerschaftsspezifischen ß1-Glykoprotein (SP1) des Menschen. Arch. Gynäk. 221, 305-312.

61. Bohn, H. 1980. Protein antigens of the human placenta. in: The Human Placenta (Klopper, A., A. Genazzani and P.G. Crosignani, eds.) Academic Press, London, New York, pp. 23-34.

EVOLUTION OF PREGNANCY-ASSOCIATED PLASMA PROTEIN-A

M.J. Sinosich

Department of Obstetrics and Gynaecology, Royal North Shore Hospital, St Leonards NSW 2065 Australia

Introduction

Pregnancy-associated plasma protein-A (PAPP-A), a large glycoprotein of trophoblastic origin (1), was first detected in the circulation of pregnant women, 28 days after conception (2). Thereafter, circulating PAPP-A levels increase expotentially, demonstrating a close relationship to trophoblastic development (3). Depressed serum PAPP-A levels offered poor pregnancy prognosis, even in the presence of normal ultrasonic findings (3 - 6). *In-vitro* functional studies have shown PAPP-A to reversibly bind heparin (7,8), contain zinc (9,10) and to specifically inhibit leucocyte elastase (11). In view of these findings and immunohistochemical localisation of PAPP-A at the feto-maternal interface (12), PAPP-A was suggested to participate in a systemic and local barrier, protecting vital placental structural proteins against proteolytic attack originating from maternal leucocytes (13).

More recently PAPP-A was detected in human seminal plasma (14), ovarian follicular fluid (15,16), pre- and post-coital cervical mucus and peritoneal fluid (1). The occurrence of PAPP-A, in conjunction with many other protease inhibitors, within the female reproductive tract suggested that PAPP-A formed part of a local barrier protecting cellular elements of paternal genetic origin from maternal phagocytic-proteolytic degredation (17). To further study the biological role of PAPP-A in species propagation it was necessary to identify a suitable model. However, since haemochorial placentation is an essential criterion for selection of a model, this study was restricted to placental mammals.

Pregnancy Proteins in Animals

Materials and Methods

Samples

Pregnacy serum and placental tissue were collected from a variety of mammals including primates, rodents, lagomorphs, horses and ungulates (Table 1). All serum samples were stored at $-20^{o}C$ until required for assay. Placental tissue was sliced, rinsed in 50mM Tris-HCl buffer (pH 7.8), containing 150 mM NaCl (TBS) and 15 mM NaN_3, and patted dry. After tissue weighing, an equal volume of TBS (W/V) was added and the tissue homogenised with an Ultra-Turax homogeniser (Kika-Werk, John Morris Pty Ltd, Sydney). After centrifugation 30 min at 15000 rpm, the supernates were aspirated and stored at $-20^{o}C$ until required.

Assays

PAPP-A was quantified by a specific and sensitive radioimmunoassay (RIA) based on human derived primary reagents (2,15). Pure PAPP-A for radioiodination was isolated from pooled serum collected from pregnant women at term of gestation (11). Primary antibody, rabbit anti-PAPP-A (Code A 230; Dako, Bioscientific, Sydney, Australia) was used at an initial dilution of 1/20,000. Dilutions of pooled pregnancy serum, calibrated against International Reference WHO 78/640 (designated 100 IU/l), were used as standards.

Characterisation

Heparin affinity chromatography was performed on a 15 ml column of Heparin-Sepharose (Pharmacia South Seas, Sydney Australia) equilibrated with TBS. Samples (5ml) and column control (5ml TBS) were applied at 1 ml/min. Elution was achieved with a linear 40 ml gradient in increasing NaCl concentration from 0.15 M to 1.0 M. Fractions (1 ml) were collected and assayed for PAPP-A and protein elution monitored by fused immunoelectrophoresis (15).

Zinc affinity chromatography was performed on a 5 ml column of zinc chelate Sepharose (Pharmacia South Seas, Sydney, Australia) equilibrated with 50 mM phosphate buffer (pH 7.0) containing 0.8 M NaCl (PBS). Samples (1 ml) and buffer control (1 ml PBS) were applied at 1 ml/min with 1 min fractions being collected. Elution of bound proteins was achieved with 50 mM EDTA. As before, fractions were assayed for PAPP-A and protein elution monitored by immunoelectrophoresis.

Size characterisation was performed by gel filtration on a K16/70 column (Pharmacia South Seas, Sydney, Australia) packed with Ultrogel AcA34 (L'Industrie Biologique, LKB, Sydney, Australia) and equilibrated with TBS. Heparin bound fractions containing PAPP-A were applied at 5.0 ml/h. PAPP-A elution profile was monitored by RIA and the column calibrated with internal markers by rocket immunoelectrophoresis (15).

Heparin bound fractions with peak PAPP-A levels were analysed for elastase inhibitory activity and compared against buffer control (11). Each preparation was arbitrarily designated 100 AU PAPP-A to eliminate the problem of partial cross-reactivity.

Serially diluted heparin-Sepharose bound proteins and 2M NaCl were subjected to PAPP-A RIA. The resultant binding data was logit-log transformed and lines of best fit were statistically determined by linear regression analysis (15).

Results

PAPP-A immunoreactivity was readily detected in placental homogenate and serum samples obtained from pregnant high order primates (macaques, baboons, chimpanzee, gorilla and human). No PAPP-A was detected in any serum samples collected from male or non-pregnant female animals or from pregnant marmosets, rats, sheep, horses or cattle (Table 1) Following sample concentration on heparin-Sepharose, of the non-primates, low levels of PAPP-A were detected only in the guinea pig. Although serum from pregnant rabbits demonstrated apparent PAPP-A immunoreactivity, this entity did not interact with either heparin or zinc chelate Sepharose (Table 1). All primate PAPP-A was heterogeneous with respect to heparin binding, with the major proportion interacting reversibly with the immobilised polyanion. All heparin bound PAPP-A, independent of animal origin, eluted at 0.45 - 0.5 M NaCl.

Table 1 Distribution and Physicochemical Characterisation of PAPP-A Immunoreactivity

Source	Affinity Chromatography[a] Heparin Sepharose	Zinc Chelate	Molecular Size (k)
Primate			
Human	+	+	820
Gorilla	+	+	820
Chimpanzee	+	+	820
Baboon	+	+	820
Macaque	+	+	820
Marmoset	-	-	-
Non Primate			
Guinea Pig	+	+	≤100
Rat	-	-	-
Mouse	-	-	-
Rabbit	-	-	≃100
Sheep	-	-	-
Cow	-	-	-
Horse	-	-	-

[a], where + = positive interaction, - = no detectable PAPP-A

Logit-log transformed binding data of serially diluted heparin bound PAPP-A demonstrated complete immunological identity between all primate derived PAPP-A (Table 2). Partial cross-reactivity was demonstrated between guinea pig ($m = -2.391$) and human circulating (m = -2.144) and placental (m = -2.179) PAPP-A. By contrast, apparent rabbit PAPP-A immunoreactivity (m = -3.802) was immunologically distinct. A regression line of similar slope (m = -3.788) was determined for heparin binding rat placental proteins. However, this apparent PAPP-A immunoreactivity was completely accounted for by the NaCl concentration in the sample (Table 2).

Table 2 Immunological Characterisation of PAPP-A

Source	Slope (-m)	Intercept (b)	Correlation Coefficient (r)
Human			
Pregnancy serum	2.144	-1.461	0.994
Placental	2.179	0.956	0.996
Gorilla	2.207	0.445	0.998
Macaque	2.118	4.642	0.98
Guinea Pig	2.391	4.877	0.992
Rabbit	3.802	-0.433	0.995
Rat	3.788	9.188	0.999
2M NaCl	5.319	10.861	0.998

Fused rocket immunoelectrophoretic analysis of protein elution following chromatography on zinc chelate Sepharose showed the majority of proteins to elute in the column breakthrough (Figure 1). All primate and guinea pig derived PAPP-A interacted reversibly with elution achieved by application of 50 mM EDTA (Figure 1 ↓). By contrast, following chromatography of rabbit pregnancy serum, apparent PAPP-A immunoreactivity was detected only in the column void volume.

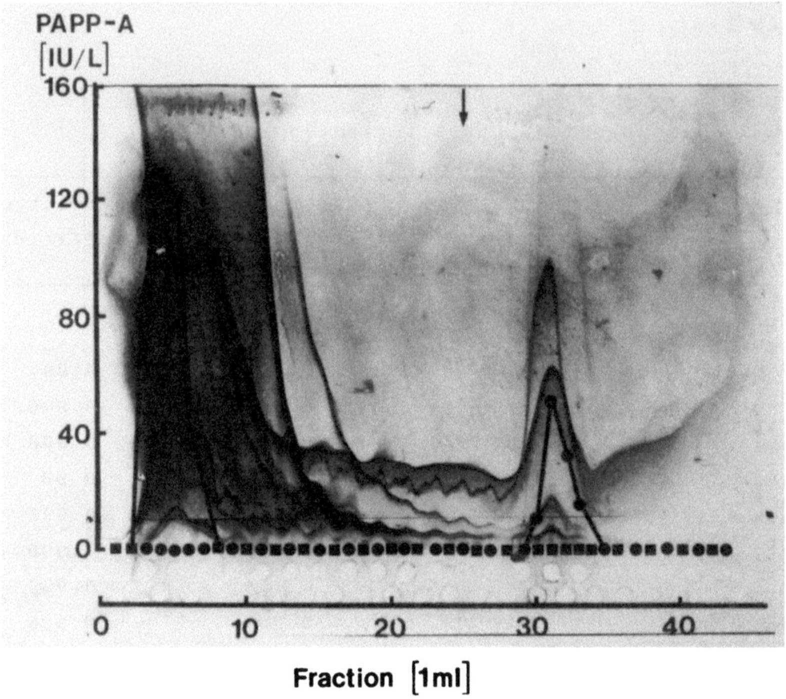

Figure 1: Fused rocket immunoelectrophoretic analysis of protein elution profile after serum (1 ml) chromatography on a 5 ml column of zinc chelate Sepharose. Human (●) and rabbit (■) PAPP-A immunoreactivity were monitored by RIA. Application of 50 mM EDTA (↓) quantitatively eluted matrix bound proteins.

Size characterisation on Ultrogel Aca34 showed all primate derived PAPP-A to coelute with an Mr = 820 k (Table 1). By contrast, non-primate PAPP-A immunoreactivity eluted with an Mr $\leq$ 110 k.

Human, gorilla, guinea pig and rat heparin binding placental proteins inhibited human granulocyte elastase activity in a dose dependent relationship (Figure 2). The gorilla preparation was least active (m = 0.112), possibly due to delay in sample processing. The other tissues were prepared immediately and proved more potent inhibitors. Human (m = 1.056) and guinea pig (m = 1.272) preparations were almost equipotent, whereas, rat the (m = 1.68) preparation was 50% more potent.

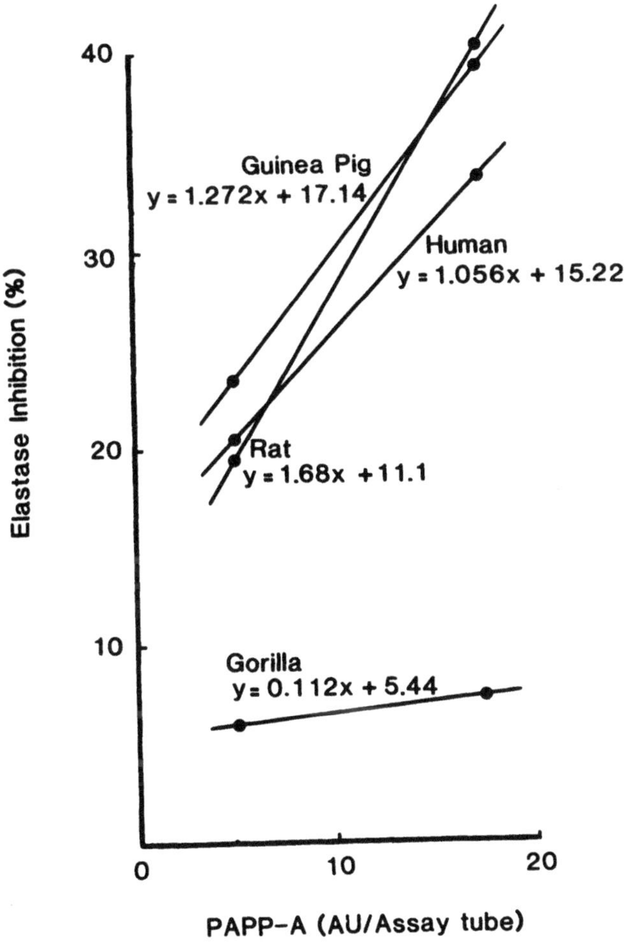

Figure 2: Linear regression analysis of dose dependent inhibition of granulocyte elastase activity by placental heparin binding proteins.

Evolution of placental mammalia shows that many of the modern families have diverged from the placental stem early in cenozoic era (Figure 3). Human (primate) PAPP-A was detected only in the more recently evolved anthropoids, but not in the platyrrhini (New World Monkeys). Of the non-primate species, PAPP-A like immunoreactivity was detected only in rodents with hystricomorphic placental development. No PAPP-A was detected in lagomorphs, artiodactyls, perissodactyls or myomorphic rodents (Figure 3).

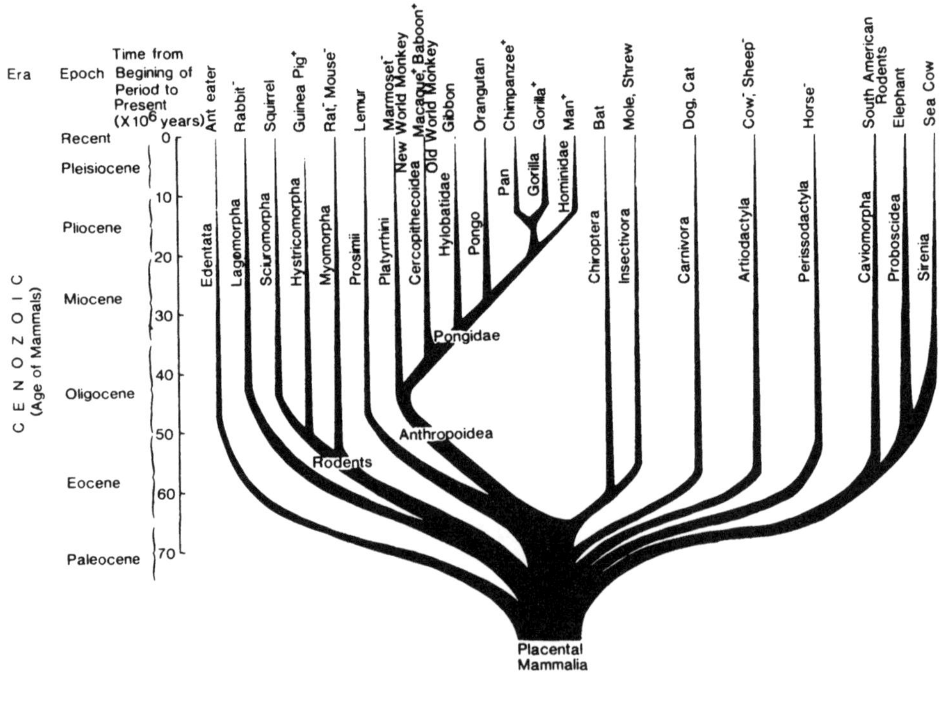

Figure 3: Evolutionary tree for human PAPP-A. + = PAPP-A positive, - = PAPP-A negative

Discussion

This study was based on measurements determined by RIA established on human derived primary reagents (2). Dako anti-human PAPP-A antibody was used as a primary antibody (15) because of polyspecificities (including α_2-macroglobulin, β-lipoprotein and pregnancy zone protein) detected in the antiserum provided by Dr P Bischof. Clearly, studies and assays employing such a polyspecific antiserum in conjunction with a contaminated antigen would produce erroneous or misleading findings. By contrast, the author's assay is based on a monospecific antiserum and pure human PAPP-A antigen, as judged by a variety of physicochemical and immunological techniques (1,8).

PAPP-A immunoreactivity was readily detected in placental tissue and serum obtained from pregnant anthropoids, but not from non-pregnant animals or New World Monkeys. Primate PAPP-A was immunologically, physicochemically and functionally indistinguishable. Although serum from pregnant rabbits inhibited tracer binding in the PAPP-A RIA this factor was immunologically, physicochemically and functionally distinct from primate PAPP-A. In addition, with the exception of hystricomorphic rodents (guinea pigs), all other pregnant and non-pregnant non-primate species were negative for PAPP-A. Guinea pig PAPP-A shared partial immunological identity with the human analogue. Furthermore, guinea pig PAPP-A reversibly interacted with heparin and zinc chelate affinity matrices and inhibited elastase activity with a similar dose dependent relationship as determined for human PAPP-A. However, non-primate PAPP-A is significantly smaller ($M_r \leqslant 110$ k) than the primate analogue ($M_r = 820$ k).

Evolution of PAPP-A, a specific and potent elastase inhibitor, appears restricted to very recent times. No PAPP-A was demonstrated in those species which diverged from ancestral stock to develop epitheliochorial placentation. Only within the cercopithecoidea, pongidae and hominidae of the anthropoids was PAPP-A immunologically, physicochemically and functionally indistinguishable. For within these groups, haemochorial placentation is most intimate. No PAPP-A was detected in the more primitive primates. Of the non-primates, PAPP-A was detected only in the guinea pig. Unlike other rodents which have localised trophoblastic growth, guinea placentation is hystricomorphic, acquiring the character of interstitial trophoblast invasion, as seen in the human (18). Although other rodents (rat) with myomorphic haemochorial placentation do not produce an immunological PAPP-A analogue, a functional analogue (heparin binding inhibitor of elastase activity) was readily detected.

This study suggests that PAPP-A evolved in parallel with those species where contact between fetal and maternal tissues is most intimate, implying the need to evolve a protective mechanism against the maternal phagocytic-proteolytic defense system. Of the anthropoids, PAPP-A was not detected in the platyrrhini, suggesting no relationship to menstruation. Thus, primate PAPP-A evolved recently, perhaps during the oligocene epoch (about 40 x 10^6 years ago), as part of a placental protective mechanism. Functional PAPP-A analogues may have evolved earlier in rodents, during the late eocene period. In conclusion, the most suitable animals to further study the biological role of PAPP-A in reproduction are the primates, particularly Old World Monkeys. However, guinea pigs may prove a useful and cheaper alternative.

Acknowledgement

The author is grateful to Mr M D Bonifacio, Mr A Zaker and Miss C Cobb for technical assistance and to Ms L E Frizell and Mrs E Sinosich for preparation of the manuscript.

References

1. Sinosich, M.J. In: Proteins of the Placenta; Biochemistry, Biology and Clinical Application (P. Bischof, ed.), S Karger, Basel (in press).

2. Sinosich, M.J., B. Teisner, J. Folkersen, D. Saunders, J.G. Grudzinskas. 1982. Clin. Chem. 28, 50.

3. Sinosich, M.J., A. Ferrier, B. Tesiner, R. Porter, J.G. Westergaard, D.M. Saunders, J.G. Grudzinskas: Clin. Reprod. Feril. (in press).

4. Westergaard, J.G., M.J. Sinosich, M. Bugge, L. Masden, B. Teisner, J.G. Grudzinskas. 1983. Am. J. Obstet. Gynecol. 145, 67.

5. Westergaard, J.G., B. Teisner, M.J. Sinosich, L.T. Madsen, J.G. Grudzinskas. 1985. Br. J. Obstet. Gynaecol. 92, 77.

6. Sinosich, M.J., A. Ferrier, D.M. Saunders: Fertil. Steril. (in press).

7. Sinosich, M.J., B. Teisner, M. Davey, J.G. Grudzinskas. 1981. Aust. N.Z. J. Med. 11, 429.

8. Davey, M., B. Teisner, M.J. Sinosich, J.G. Grudzinskas. 1983. Ann. Biochem. 131, 18.

9. Sinosich, M.J., M. Davey, B. Teisner, J.G. Grudzinskas. 1983. Biochem. Int. 7, 33.

10. Sinosich, M.J., M. Davey, J.G. Grudzinskas, 1984. In: Protides of Biological Fluids, XXXIst Colloquium (H. Peeters, ed). Pergamon Press, Oxford, p 19.

11. Sinosich, M.J., M. Davey, P. Ghosh, J.G. Grudzinskas. 1982. Biochem. Int. 5, 777.

12. Wahlstrom, T., B. Teisner, J. Folkersen. 1981. Placenta 2, 253.

13. Sinosich, M.J., D.M. Saunders. J.G. Grudzinskas. 1984. In: In-Vitro Fertilization, Embryo Transfer and Early Pregnancy (R.F. Harrison, J. Bonnar and W.T. Thompson, eds). MTP Press, Lancaster, p 209.

14. Sinosich, M.J., M. Seppala, D.M. Saunders, J.G. Grudzinskas. 1984. Ann. N.Y. Acad. Sci. 442, 287.

15. Sinosich, M.J., R. Porter, P. Sloss, M.D. Bonifacio, D.M. Saunders. 1984. J Clin. Endocrinol. Metab. 58, 500.

16. Sinosich, M.J., L.W. Westergaard, B. Teisner, J.G. Westergaard, J.G. Grudzinskas, D.M. Saunders. 1984. In: Protides of Biological Fluids, XXXIInd Colloquium (H. Peeters, ed) Pergamon Press, Oxford, p.259.

17. Sinosich, M.J., D.M. Saunders, J.G. Grudzinskas. 1984. Ann. N.Y. Acad. Sci. 442, 269.

18. Pijnenborg, R., W.B. Robertson, I. Brosens, G. Dixon. 1981. Placenta 2, 71.

STRUCTURAL AND FUNCTIONAL RELATIONSHIPS BETWEEN MACROGLOBULIN INHIBITORS FROM DIFFERENT MAMMALIAN SPECIES.

Björn R. Weström, Cynthia A. de Wit, and Börje W. Karlsson.
Department of Zoophysiology, University of Lund, Helgonavägen 3B, S-223 62 Lund, Sweden

Introduction

Proteinase inhibitors are present in all vertebrates investigated. When studied in blood, the gross electrophoretic pattern of these inhibitors appears to be similar in different species, with trypsin and chymotrypsin inhibition predominantly found in the α_1- and α_2-regions, and occasionally also in the β-region upon gel electrophoresis (1,2). The individual plasma/serum proteinase inhibitors have been thoroughly studied mostly in mammals and it appears as if one or two macroglobulin inhibitors, an α_1-proteinase inhibitor (α_1- antitrypsin) and an antithrombin III are present in all mammals that have been investigated (3). In the different species each one of these inhibitors shows similar properties, e.g., proteinase inhibition spectra, physiochemical properties, immunological cross-reactivity or sequence homologies.

The high molecular weight proteinase inhibitors (~725-800 kDa), the α-macroglobulins (αM:s), have been shown to play important roles in the regulation of proteinases by inhibiting and eliminating these from blood or other body fluids (4,5). The αM:s are unique among the proteinase inhibitors for their broad specifity for all four catalytic classes of endopeptidases and in their reaction mechanism with these enzymes. Binding of the proteinase is initiated by cleavage of the αM:s polypeptide chain in the "bait" region, leading to a conformational change and an entrapment of the proteinase within the αM molecule. The active site of the proteinase

Pregnancy Proteins in Animals

is only involved in this initial interaction with the αM, but not in the following binding of the enzyme to the inhibitor. Thus, upon interaction with the αM the proteinase retains, at least partly, it´s enzymatic activity against low-molecular substrates.

αM:s showing these properties have been detected, isolated and charcterized from several vertebrate species, mainly mammals, i.e., man, pig, cow, horse, dog, rabbit, rat, mouse, hamster, hedgehog and plaice (for reveiw 3,6). Furthermore, comparative investigations on either structural homologies by immunological cross-reactivity studies (7-10) or functional homologies by screening for proteinase binding properties (6) have shown that α-macroglobulins appear in sera/plasma from all vertebrates studied.

Two αM:s with similar physiochemical and inhibition properties, but which are antigenically distict, are found in many of the species, including the pig, dog, rat and rabbit. The reasons for the presence of two αM:s with similar properties in some species seems obscure, but differences in their ontogenetic development and their presence during various physicopathological conditions, like pregnancy and acute inflamation, might indicate differences in their functions. In humans only one αM inhibitor, α_2-macroglobulin (α_2M), has been found (4,5). However, during pregnancy increasing amounts of another αM, pregnancy-associated α_2-glycoprotein (pregnancy zone protein), appears which in addition to similar electrophoretic mobility also shares other physico-chemical properties with α_2M (11-14). Both consist of two or four identical subunits with a Mw of about 180 kDa, making up a dimeric or tetrameric molecule.

In this presentation we would like to give a survey of our work on the detection and characterization of the macroglobulin inhibitors in the pig and the phylogenetically old mammal, the European hedgehog, and their antigenic relationship with αM:s of some other species, including human pregnancy-associated α_2-glycoprotein. Functional homology was studied by specific staining for proteinase inhibiting properties after agarose gel electrophoresis. Antigenic relationships between the macroglobulins in human, pig, dog, rat and hedgehog blood serum/plasma were investigated using immunoelectrophoresis. The results imply that the αM:s from the different species are functionally similar and more or less antigenically related. The latter is true also for human pregnancy-associated α_2-glycoprotein.

Material and Methods

Sera

Porcine serum, dog serum, rat serum from rats with acute-phase inflammation, human plasma from late pregnant women and hedgehog (Erinaceus europaeus) plasma were obtained as previously described (11,15). The macroglobulin fraction was obtained from these sera after gel filtration on Sephacryl S-200 (Pharmacia Fine Chemicals, Uppsala, Sweden) as described (15).

Antisera

Rabbit antisera to human α_2-macroglobulin (α_2M) and pregnancy-associated α_2-glycoprotein were purchased from Dakopatts A/S (Copenhagen, Denmark). Rabbit antisera to dog α_1M and α_2M, rat α_1M and α_2-acute phase globulin (α_2-macroglobulin, α_2M) and swine α_2Mf (fast α_2-moving upon electrophoresis) and α_2-Ms (slow) were prepared as previously described (10,16).

Analysis

Agarose gel electrophoresis of the different sera and/or macroglobulin fractions was carried out using 1 % gels (HSA, Litex, Glostrup, Denmark) in 0,075 M barbital-Na barbital buffer, pH=8,6 containing 2 mM Ca-lactate. After electrophoresis at 20 V/cm for 40 min, the gels were either tested for proteinase inhibition or immunological cross-reactions by immunoelectrophoresis.

The proteinase (trypsin, chymotrypsin and elastase) inhibiting properties of the αM:s were localized in the gels using both low-molecular weight substrates, Ac-Phe-β-Naphtyl ester and CBZ-Ala-β-Naphtyl ester, and the high molecular weight substrate casein (15,17).

Antigenic cross-reactivity between the αM:s from the various species was established using ordinary immunoelectrophoresis (16), and line immunoelectrophoresis (17). In the latter method, the cross-reactivity was established as an absorption of an identified immunoprecipitation line by the various antisera to the αM:s (10).

Results

The vast and complex immunological cross-reactivity patterns obtained between the αM:s from the species investigated are summarized in Table 1. Where cross-reactivity was shown

Table 1. Summary of the immunological cross-reactivities between the αM:s from different species studied by using ordinary immunoelectrophoresis and line immunoelectrophoresis.

		Antisera to							
		Human		Porcine		Dog		Rat	
		α_2PAG	α_2M	α_2Mf	α_2Ms	α_1M	α_2M	α_1M	α_2M
Human	α_2PAG	+++	++	++	+	++	+		+
	α_2M		+++	++	++				++
Porcine	α_2Mf	+	+++	+++		+			
	α_2Ms		++		+++		+		
Dog	α_1M	+	++	++	++	+++			++
	α_2M		++		+++		+++		+
Rat	α_1M		+		+		+	+++	
	α_2M		+++	++	+		+		+++
Hedgehog	α_2M		+++	+++					++
	α_2-βM								
	βM				++				

+, ++, +++ indicates weak, medium and strong cross-reactivity.

using ordinary immunoelectrophoresis, the results were verified using line immunoelectrophoresis. In some cases, however, cross-reactivity could only be demonstrated with the more sensitive technique, line immunoelectrophoresis. Reciprocal cross-reactions i.e., two αM:s which both cross-reacted with their corresponding antisera, were shown for 8 macroglobulin pairs: human α_2M - swine α_2Mf; human α_2M - swine α_2Ms; human α_2M - rat α_2M; human α_2PAG - swine α_2Mf; human α_2-PAG - dog α_1M; swine α_2Mf - dog α_1M; swine α_2Ms - dog α_2M and dog α_2M - rat α_2M.

Table 2 shows the inhibiting properties of the macroglobulins from the species studied in comparison to the information found in the literature for other species macroglobulins. The macroglobulins all function as proteinase inhibitors, and their specificity appears to be broad in most cases since several proteinases are inhibited both within a functional class, the serine proteinases, and between different functional classes when tested.

Table 2. Summary of the proteinase inhibiting properties of the macroglobulins in serum/plasma from different species.

		Inhibition of				
		Trypsin	Chymotrypsin	Elastase	Papain	Ref.
Human	α_2PAG	(-)[1]	+[1]	+[1]		18
	α_2M	+	+	+	+	4
	PAPP-A	-	-	+[2]		19
Porcine	α_2Mf	+	+	+		20
	α_2Ms	+	+	+		
Dog	α_1M	+	+	+[3]		21,22
	α_2M	+	+	+[3]		
Rat	α_1M	+	+			23
	α_2M	+	+			
Hedgehog	α_2M	+	+	+	+[4]	
	α_2-βM	+	+	+	+[4]	
	βM	+	+	+	+[4]	

[1]proteinase binding properties
[2]human granulocyte elastase
[3]Ohlsson, K., personal commun.
[4]de Wit, C.A., Weström, B.R., unpubl.

Discussion

The great number and the complexity of the immuno-cross-reactions between the macroglobulins from the five mammalian species studied, show that all the macroglobulins are more or less antigenically related and thus structurally related. One must, however bear in mind that in a study like this the results, in additon to showing antigenic similarities, also reflect the special properties of the antisera used, which in this case all were of rabbit origin. The structural homology between the αM:s is further strengthened by the apparent similarities in molecular size and stucture and in other physiochemical characteristics. However, the macroglobulins within a particular species also show strong non-cross-reacting antigens, indicating a difference in specific structures.

It is notable also that human α_2PAG, the pregnancy-associated human macroglobulin, showed antigenic relationships to the other macroglobulins. Even the antiserum to human α_2M cross-reacted with human α_2PAG, indicating the existence of common determinants. Though such cross-reactivity has not previously been reported (4, 12), our results have now been confirmed using monoclonal antibodies (26). Furthermore, recently partial sequence determination of human α_2PAG has shown a close homology between human α_2M and α_2PAG (27).

The structural homology between the αM:s is also accompanied with a functional homology, since all the macroglobulins studied can function as proteinase inhibitors or binders, most with broad specifity. Since two functional macroglobulin inhibitors have been found in many mammalian species, it seemed interesting to study the phylogenetically old mammal, the European hedgehog, for macroglobulins. To our surprise, these studies showed what appear to be three hedgehog macroglobulin inhibitors with slightly different electrophoretic mobilities in the α-to β-region.

Recently, also a third human macroglobulin with inhibiting properties has been identified (Table 2), namely Pregnancy Associated Plasma Protein A (PAPP-A) (21). This second pregnancy associated macroglobulin, besides α_2PAG, shares many physiochemical similarities with the αM:s, like α_2-mobility in agarose gels, large Mw of 820 kDa and a tetrameric structure. No immuno-cross-reactivity has been reported between PAPP-A and human α_2M, but antigenically related proteins to PAPP-A have been found in the higher apes and in the guinea pig (28).

The physiological reason for two or even three macroglobulin inhibitors with very similar properties within a species is still largely unknown. But their different appearance during ontogenetic development and various pathophysiological conditions might indicate different functions. For example, high serum levels, above the adult ones, are found for human α_2M, rat α_2M and rabbit α_1M during the late fetal period (3,29). Only low serum levels of both swine αM:s are found during fetal development but especially swine α_2Ms increases very rapidly after birth (30). Also increased serum levels of rat α_2M, human α_2PAG and PAPP-A are found during pregnancy (3). Furthermore, rat α_2M, rabbit α_1M and human α_2PAG may all function as acute-phase reactants (3,27). Recently, structural difference has been shown for one of the αM:s in the rat, α_1M, and for mouse α_1M where the "normally" found 200 kDa subunit instead appears to be composed of two subunits of 160 kDa and 40 kDa covalently linked to each other (31,32).

An attempt was made to group the αM:s according to their cross-reactivities as shown in the figure (10) Interestingly, the electrophoretically more fast-moving αM:s appeared to form one group (human α_2PAG, dog α_1M and swine α_2Mf) while the more slow-moving αM:s formed another group (dog α_2M, swine α_2Ms and rat α_2M) with human α_2M as a common dominator. The hedgehog αM:s also appear to fit in this grouping, since both the fast-moving (hedgehog α_2M and

Figure. Schematic figure of the immunological cross-reactivities between the αM:s.
Simple antigenic cross-reactivity, i.e. one αM interacts with an antiserum towards another αM, is indicated by the symbol ——— and reciprocal cross-reactivity, i.e. a pair of αM:s which both cross-react with their corresponding antisera is indicated by the symbol ⟷. (Modified from ref. 10).

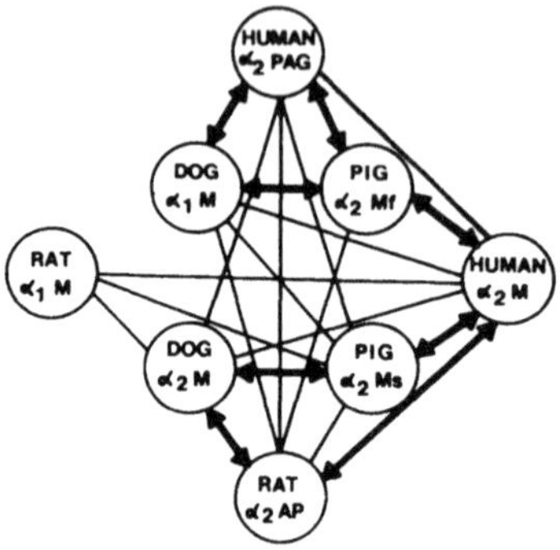

swine α_2Mf) and the slower-moving (hedgehog βM and swine α_2Ms) hedgehog and swine αM:s cross-reacted. The relatively poor cross-reactivity of rat α_1M with the other αM:s might be explained by it´s different subunit construction.

Acknowledgements

This work was supported by grants from the Swedish Natural Science Research Concil.

References

1. Fossum, K. 1970. Acta Path. Microbiol. Scand. Section B 78, 605-618.

2. Nakamura, S., H. Ogata, R. Suzuno. 1972. Comp. Biochem. Physiol. 41, 201-215.

3. Weström, B.R. 1981. Protease Inhibitors in Mammalian Blood Serum and Mammary Secretions with Special Reference to their Function in the Growing Pig. Dr. Disseration, Lund, Sweden (LUNBDS/(NBZF/1010)).

4. Starkey, P.M., A.J. Barrett. 1977. in: Barrett (Ed.) Proteinases in Mammalian Cells and Tissues. pp. 663-696 Elsevier-North Holland Biomedical Press.

5. Travis, J., G.S. Salvesen. 1983. Ann. Rev. Biochem. 52, 655-709.

6. Starkey, P.M., A.J. Barrett. 1982. Biochem. J. 205, 91-95.

7. James, K.A. 1965. Immunology 8, 55-61.

8. Berne, B.H., S. Dray, K.L. Knight. 1971. Proc. Soc. Exp. Biol. Med. 138, 531-535.

9. Shortridge, K.F., G. Belyavin, D.E. Bidwell. 1976. Comp. Biochem. Physiol. 54, 319-321.

10. Weström, B.R., B.W. Karlsson, K. Ohlsson. 1983. Hoppe-Seyler´s S. Physiol. Chem. 364, 375-381.

11. Berne, B., H. Bohn, R. Hofmann, B. Klausch, C.H.W. Horne, R. Kasukawa, C. Rittner, W.H. Stimson, G.N. Than. 1975. Lancet ii 367-368.

12. Stimson, W.H., L. Eubank-Scott. 1972. FEBS Lett. 23, 298-302.

13. von Schoultz, B., T. Stigbrand. 1973. Acta Obstet. Gynecol. Scand. 52, 51-57.

14. Stimson, W.H., D.M. Farquharson. 1978. Int. J. Biochem. 9, 839-843.

15. de Wit, C.A., B.R. Weström. 1985. Comp. Biochem. Physiol. in press.

16. Weström, B.R. 1979. Hoppe-Seyler´s Z. Physiol. Chem. 360, 1861-1867.

17. Weström, B.R., L.C.T. Carlsson. 1976. Int. J. Biochem. 7, 41-47.

18. Krøll, L., M. Andersen. 1975. J. Immunol. Methods. 9, 141-146.

19. Scheidegger, J. 1955. Int. Arch. Allergy 7, 103-110. 141-146.

20. Sottrup-Jensen, L., O. Sand, J. Folkerson. This volume.

21. Sinosich, M.J., M.W. Davey, P. Ghosh, J.G. Grundzinskas. 1983. Biochem. Internat. 5, 777.

22. Weström, B.R. 1979. Hoppe-Seyler's Z. Physiol. Chem. 360, 1869-1878.

23. Ohlsson, K. 1971a. Biochim. Biophys. Acta. 236, 84-91.

24. Ohlsson, K. 1971b. Scand. J. Clin. Lab. Invest. 28, 5-11.

25. Gauthier, F., H. Mouray. 1975. Protides Biol. Fluids 23, 139-143.

26. Carlsson, L., J. Folkersen, T. Stigbrand. 1985. Molec. Immunol. in press.

27. Sottrup-Jensen, L., J. Folkersen, T. Kristensen, B.F. Tack. 1984. Proc. Natl. Acad. Sci. USA 81, 7353-7357.

28. Sinosich, M.J., P.M. Saunders. This volume.

29. Versavel, C., A. Feve, F. Esnard, T. Lebreton de Vonne, H. Mouray. 1983. Comp. Biochem. Physiol. 75B, 701-702.

30. Weström, B.R., B.W. Karlsson, J. Svendsen. 1982. Biol. Neonate 41, 22-31.

31. Nelles, L.P., H.P. Schnebli. 1982. Hoppe-Seyler's Z. Physiol. Chem. 363, 677-682.

32. Hudson, N.W., P.H. Koo. 1982. Biochim. Biophys. Acta 704, 290-303.

MODIFICATION OF THE PROPERTIES OF MURINE ALPHA-FETOPROTEIN AND HUMAN SEX BINDING PROTEIN BY NON ESTERIFIED FATTY ACIDS

M.E. Martin, R. Vranckx, G. Vallette, C. Benassayag, E.A. Nunez
U. 224, INSERM - Faculté de Médecine Xavier Bichat - 16, rue Henri Huchard, 75018 Paris

Introduction

Several lines of research suggest that non esterified fatty acids (NEFAs) could play a major role in the mechanisms of hormonal action.

Data from our laboratory have shown that in the developing rat the NEFAs specifically interfere with the binding of estrogen to serum and tissue proteins. Unsaturated NEFAs inhibit the binding of estrogens to uterine receptors (1, 2, 3) and serum alpha-fetoprotein (AFP) (4, 5, 6). The inhibition level varies with both the NEFA concentration and the degree of NEFA unsaturation. AFP is found in high concentrations (5 g/l) in serum during fetal and early post-natal life. This alpha-glycoprotein is monomeric and displays a high binding affinity for polyunsaturated fatty acids (Ka = $10^7 M^{-1}$) in several species, and for estrogens (Ka = $10^8 M^{-1}$) specifically in rat and mouse (7).

The aim of this study was to investigate the influence of different classes of NEFAs on various properties of a human pregnancy steroid binding protein, the sex binding protein (hSBP) as compared to their effects on mAFP. The hSBP is a dimeric glycoprotein (9, 10)which binds the steroids E_2, T and DHT with increasing affinities (Ka E_2 = $0{,}6 \times 10^9 M^{-1}$; Ka T = $1 \times 10^9 M^{-1}$; Ka DHT = $2 \times 10^9 M^{-1}$) (11). This protein is found in five to ten times higher concentrations (50 mg/l) during pregnancy.

Pregnancy Proteins in Animals

We report here that the fatty acids induced modifications of the binding properties, the immunoelectrophoretic behaviour and the immuno-reactivity of both serum proteins : mAFP and hSBP. All our studies has been carried out with NEFA/protein ratios similar to these found in pregnancy sera.

Material and Methods

Proteins :

Rat AFP from 19-day embryo sera was purified by immunoaffinity chromatography on anti-AFP as previously described (12).
Human SBP was purified from third trimester pregnancy sera in 3 steps :
1) 25-55% ammonium sulfate precipitation - **2)** phenyl-sepharose chromatography (13) - **3)** cortisol hemisuccinate affinity chromatography (13).
The first two steps eliminated the albumin and CBG (transcortin) : major contaminants of SBP.
The protein concentration was measured by method of Lowry (14) or Bradford (15).

Crossed immunoelectrophoresis :

The method of Laurell (16) as modified by Clarke and Freeman (17) was used.

Protein - hormone interaction :

Protein sample previously incubated for 2 h. at 4° with tritiated labelled steroid with or without NEFA were separated by polyacrylamide electrophoresis (5).
Equilibrium dialysis analysis was carried out using the batch gel technique of Pearlman and Crépy (18).

Results

I - Modification of steroid binding properties of murine AFP and human SBP by NEFAS.

The influence of several classes of NEFAs on mAFP-E_2, hSBP-E_2, hSBP-T, hSBP-DHT interactions were analyzed by polyacrylamide electrophoresis and equilibrium dialysis techniques.
Figure 1 shows that the binding of ^3H-E_2 to purified murine AFP (Fig. 1A) and of ^3H-T to purified human SBP (Fig. 1B) is strongly inhibited by unsaturated fatty acids while saturated fatty acids had no effect or slightly potentiate these bindings.
The results obtained by equilibrium dialysis with total rat embryo serum and purified AFP (Fig. 2) or with pregnant human serum and purified hSBP (Fig. 3) confirmed and extended the differential effect of each classes of fatty acids on the steroid binding. The saturated NEFAs such as stearic acid (C18:0) did not prevent the binding of E_2 to serum AFP or the binding of E_2, T and DHT to serum SBP. The binding of T and E_2 to purified hSBP was significantly increased. Unsaturated NEFAs such as oleic acid (C18:1), arachidonic acid (C20:4) and docosahexaenoic acid (C22:6), inhibited steroid binding to both the purified proteins and to sera containing these proteins.
The polyunsaturated compounds are the strongest competitors for the AFP estrogen interactions (Fig. 4).The displacement of the steroid bound to hSBP by unsaturated fatty acid is dose dependent (not shown) and inversely correlated to the association constant of each steroid.

Thus, the quantity of NEFA producing 50% binding inhibition to hSBP was smaller for E_2 than for T and DHT.

Fig. 1 – Electrophoresis in 7% polyacrylamide gelof tritiated purified mAFP and hSBP in the presence of fatty acids.

(A)– AFP incubated with ^3H-E_2 (—●—) and in the presence of ^3H-E_2 and saturated (C18:0) (—o—) or unsaturated (C22:6) (·-·-·) NEFA. The ratio were 20 μg of AFP/5×10^4 cpm (0,4 ng) ^3H-E_2/5 μg fatty acid.

(B)– SBP incubated with ^3H-T (—●—) and in the presence of ^3H-T and saturated (C18:0) (-----) or unsaturated (C22:6) (·-·-·) NEFA. The ratio were 2,5 μg of SBP/5×10^4 cpm ^3H-T (175 pg)/50 μg fatty acid.

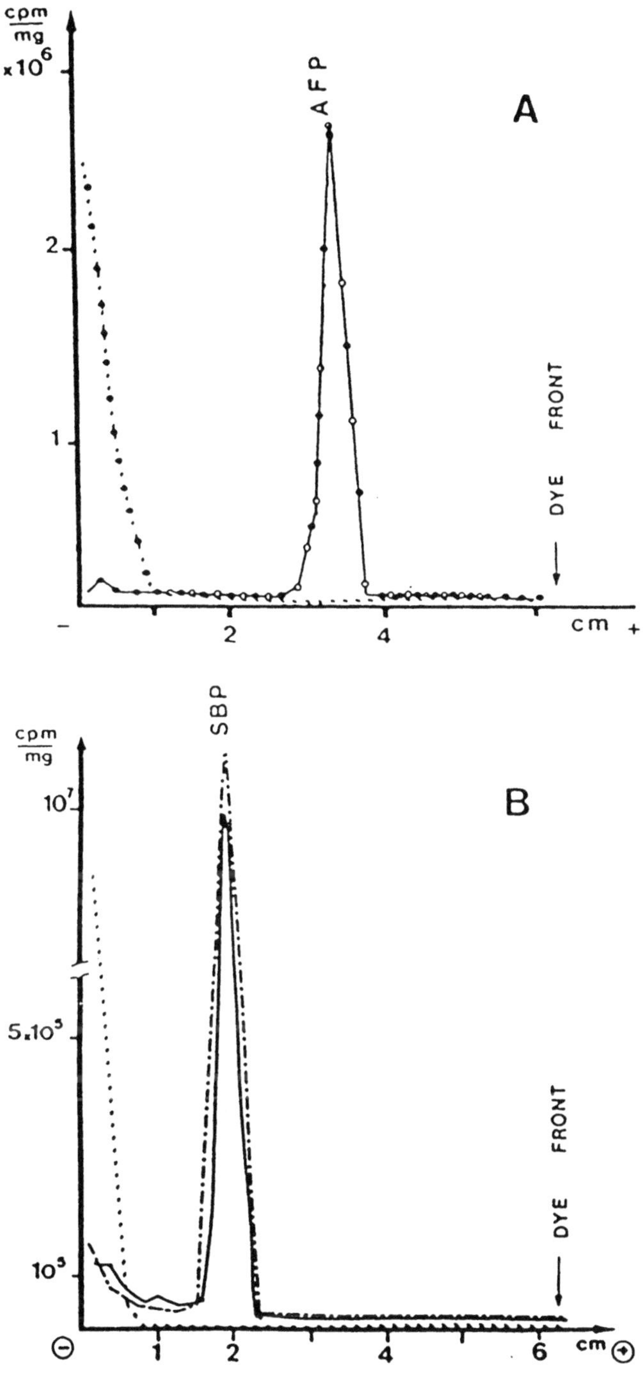
cpm/mg x10^6
AFP
A
2
1
DYE FRONT
cm
2
4
cpm/mg
SBP
10^7
B
5.10^5
10^5
DYE FRONT
1
2
4
6
cm

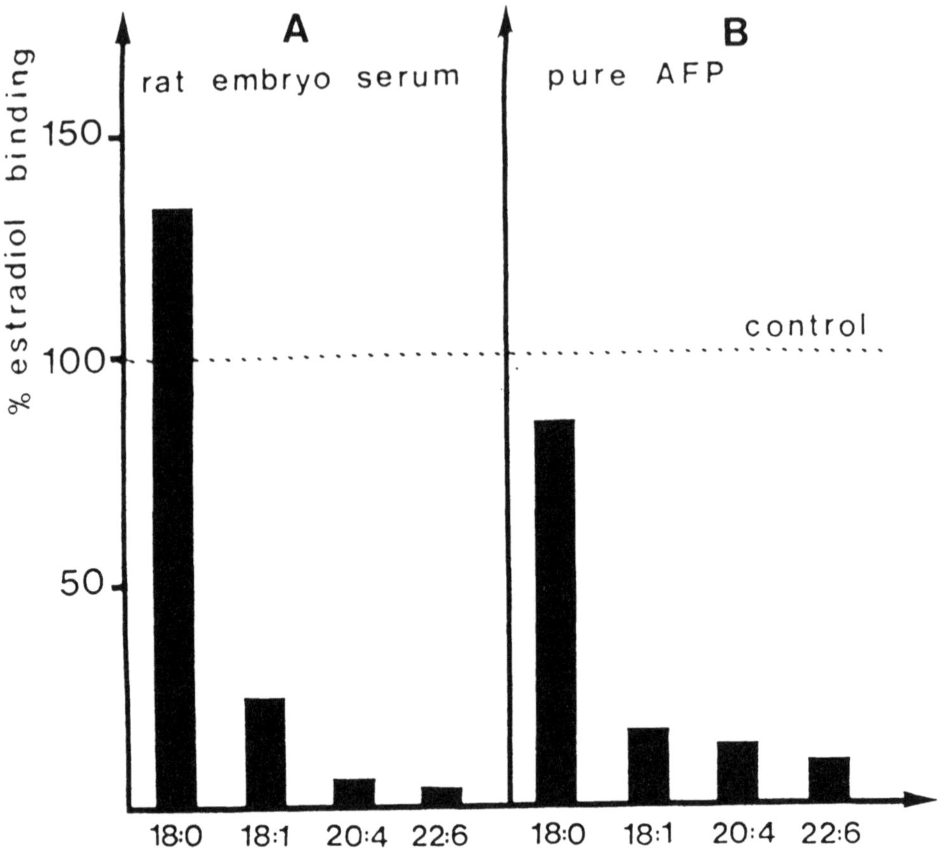

Fig. 2 – Effect of different classes of NEFA on estrogen binding of mAFP.

E_2 binding was determined by equilibrium dialysis and the results plotted as percent of $\frac{Sbo}{Suo}$ i.e. of bound versus unbound radioactive ligand in the absence of NEFA .

Incubation mixture contained in (A) 20 µg of embryo serum proteins, 10^5 cpm (0,8 ng) of ^3H-E_2 and 25 µg of each fatty acids ; in (B) 4 µg of pure AFP, ^3H-E_2 and 2,5 µg of each fatty acids.

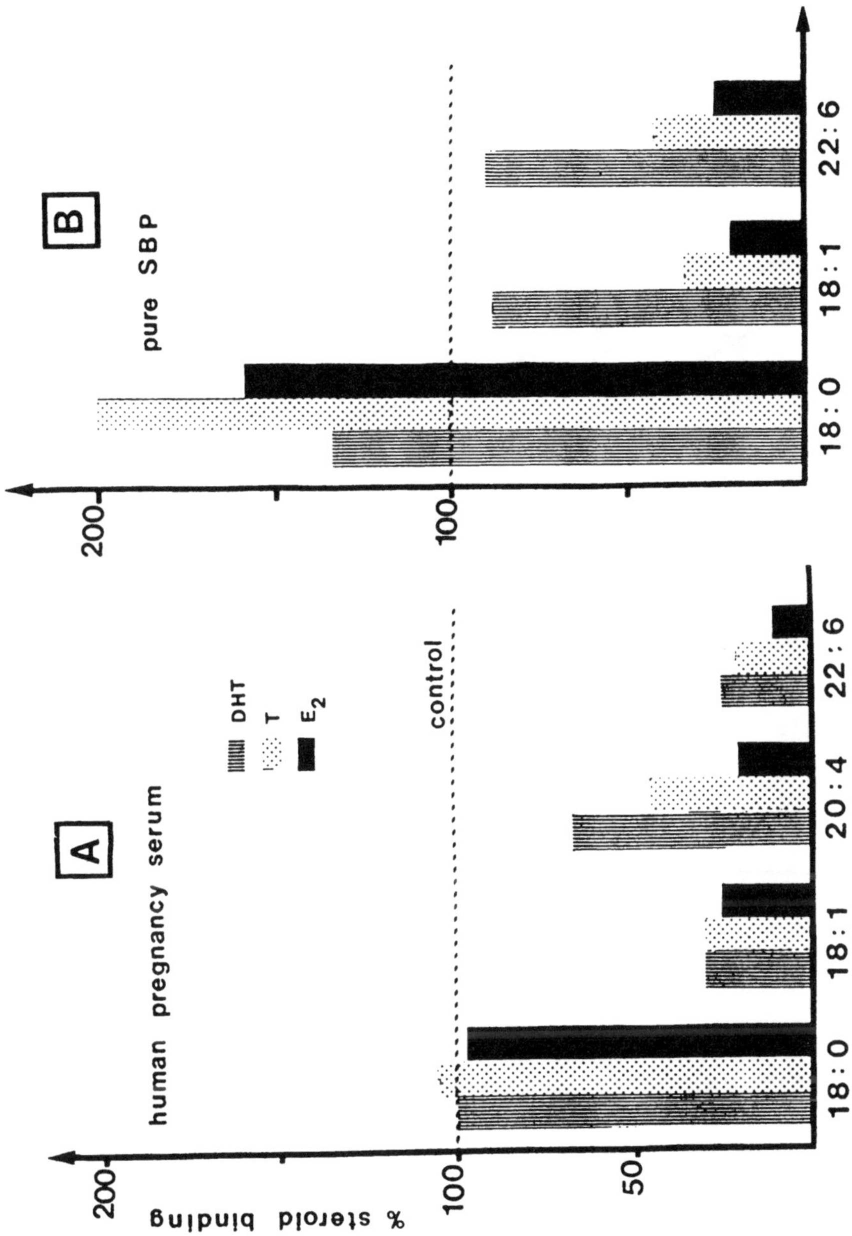

Fig. 3 - Effect of different classes of NEFAs on DHT, T, E_2 binding of hSβP. Steroid binding was determined as described in Fig. 2. Incubation mixtures contained in 300 ug of pregnancy human serum proteins and in 2,5 ug of pure SβP with 10^4 cpm (350 pg) or each ^{3}H-steroid and 25 ug of each fatty acids.

Incubation mixtures contained in (A) 300 µg of pregnancy human serum proteins and in (B) 2,5 µg of pure SBP with 10^4 cpm (350 pg) or each ^{3}H-steroid and 25 µg of each fatty acids.

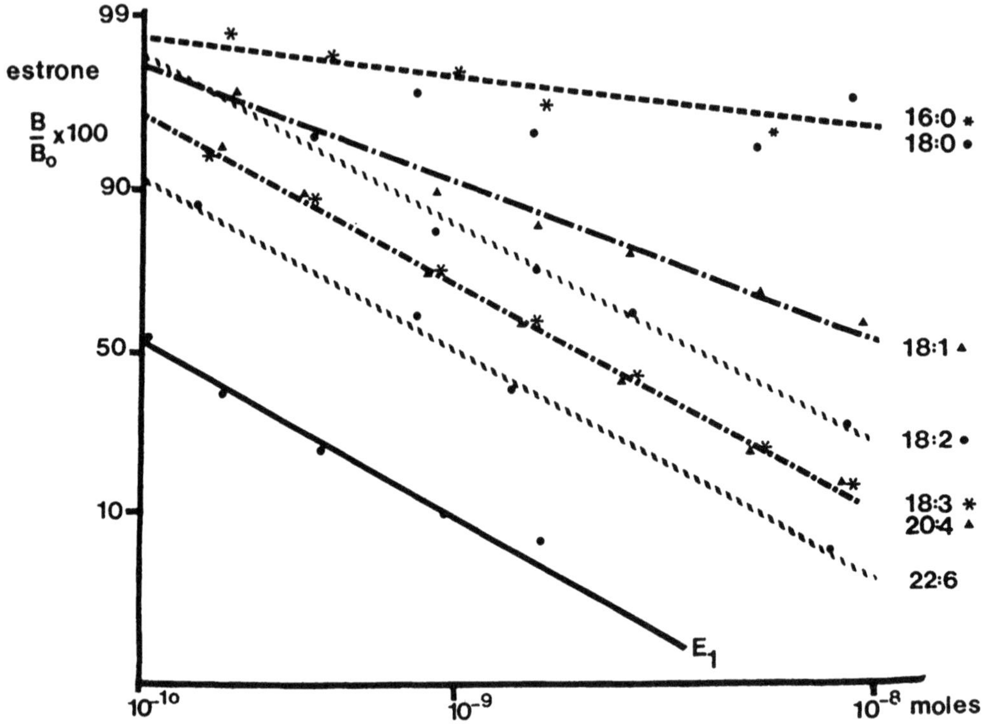

Fig. 4 – Comparative displacements of (^{3}H)-E_1 from AFP by increasing amounts of non labelled E_1 (control) or different individual fatty acids.

5 µg AFP were incubated with 0,35 pmole (10^4 cpm) ^{3}H-E_1 and 0,1-10 nmole estrone or fatty acids in phosphate buffer pH 7,4 with 1% gelatin. The binding of ^{3}H-estrone was measured subsequently with the charcoal/Dextran technique (from Benassayag and al - réf. 5).

II - Fatty acid induced modifications of immunoelectrophoretic behaviour of mAFP and hSBP.

Immunoelectrophoresis (Fig. 5) clearly demonstrated a dramatic unsaturated NEFA-induced shift in hSBP from a β migrating steroid-binding form (Fig. 5 Ba) to a α non steroid binding one (Fig. 5Bb). This shift

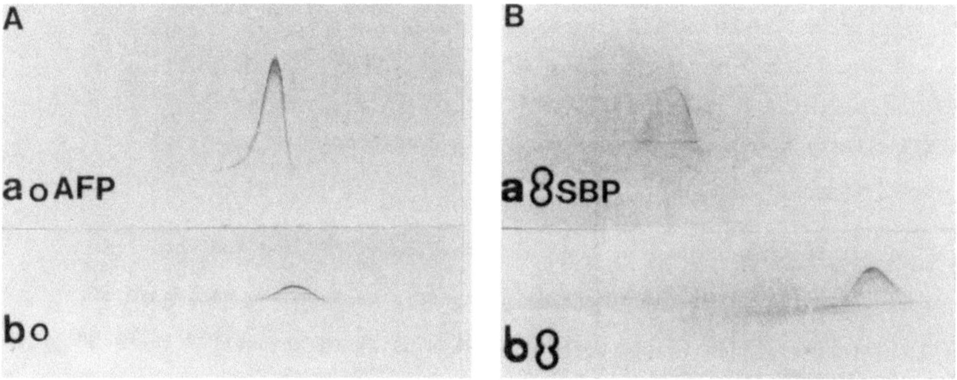

Fig. 5 - Effect of unsaturated NEFAs on the crossed immunoelectrophoretic behaviour of purified mAFP and hSBP.

(A) Antiserum = rabbit anti mAFP (1%)

a. - Pure AFP (250 ng)

b. - AFP preincubated with 5.5 nanomole C22:6 per picomole of AFP.

(B) Antiserum = rabbit anti hSBP (5%)

a. - Pure SBP (1 μg)

b. - Pure SBP preincubated with 5,5 nanomole per picomole of SBP.

was dependent upon unsaturated NEFA concentration. Several transitory forms were seen at 2.2 nanomole unsaturated NEFA/pmole protein (results not shown) while a single peak in the α migrating zone was seen at 5.5 nanomole NEFA/pmole protein. In the same way it may be observed that the non-steroid binding mAFP (Fig. 5Ab) after incubation with unsaturated fatty acid is faster moving than the native steroid-binding form (Fig. 5Aa).

III - Fatty acid induced modification in quantitation of mAFP and hSBP.

It is noteworthy that the shifted forms of AFP and SBP were less immunoreactive against anti-AFP or anti-SBP sera than the native forms. Quantitative immunoelectrophoretic studies using the Laurell " rocket " system confirmed the changes in AFP and SBP immunological behaviour in the presence of unsaturated NEFAs. The apparent amounts of mAFP and of hSBP (Fig. 6) are significantly lower in the presence of unsaturated NEFAs as compared to controls.

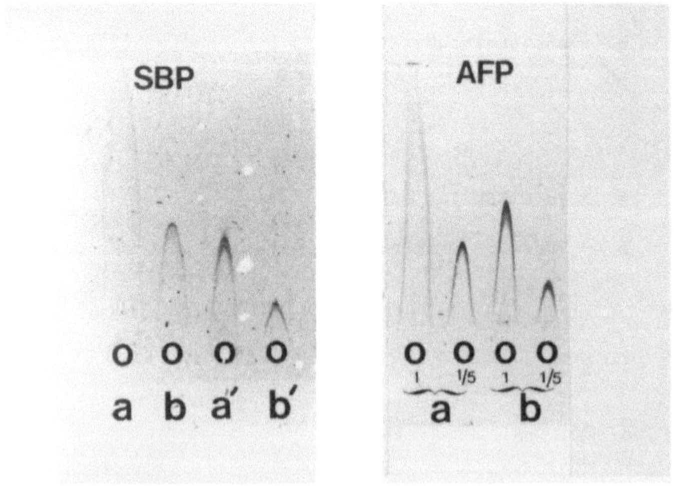

Fig. 6 - Laurell electroimmunodiffusion " rocket " of pure AFP and SBP in the presence of unsaturated NEFA (C22:6)

a. - Purified proteins

b. - Purified proteins after incubation with 5.5 nmole of C22:6/pmole of protein.

(**a'**, **b'** = Same analysis after charcoal treatment which removes the free fatty acids 10 μg charcoal/1 μg Dextran/μg protein).

Conclusion

The present report describes fatty acids induced changes in binding, electrophoretic and immunological properties of murine AFP or human SBP.

Saturated NEFAs do not prevent steroid binding but seem rather to stabilize the steroid protein interactions. By contrast, unsaturated fatty acids, in a well defined range of concentrations, inhibit E_2 binding to AFP and the E_2, T and DHT binding to SBP. The inhibitory effect of unsaturated NEFAs occurs with total sera or with purified proteins. But in the sera the quantities required to produce 50% inhibition of binding were ten times greater than with purified proteins, suggesting that other proteins may interfere and/or bind unsaturated NEFAs as has been shown for albumin and lipoproteins (6). The inhibitory effect of unsaturated fatty acids on E_2-mAFP and steroid-hSBP interactions is dose-dependent and seems to be correlated with each specific steroid-protein association constants. These modifications in binding properties of mAFP and hSBP could be explained by a fatty acid-induced conformational change in the proteins. The electrophoretic and immunologic behaviour of mAFP and hSBP in the presence of unsaturated NEFA indicates that changes in charge and/or structure of these proteins occur. Thus unsaturated NEFAs induce a shift of the mAFP and hSBP from a slow migrating active native form to a fast moving inactive and less immunoreactive one.

The modifications observed may occur as a result of unsaturated NEFAs interacting with the proteins to reduce steroid accessibility. For example the dimeric form of hSBP (β form) would be destabilized by unsaturated NEFAs leading to production of an " open dimer " (transitory forms) or even its dissociation into subunits (α forms).

One of the implications of these results is that, in pathological conditions related to lipid disorders, the measurements of mAFP and hSBP concentrations and functional activity of these proteins might be influenced by endogenous serum NEFAs levels.

Taken together, these results suggest how a critical fatty acid concentration appearing in the vicinity of the binding proteins could release the steroids from their binding sites and thus induce a specific cellular response.

References

1. Nunez, E.A., Benassayag, C., Vallette, G., Savu, L., Clerc-Hofmann, F., Delorme, J., Christeff, N. (1983) Oncodevelopmental Biology and Medicine Ann. Ac. N.Y. Sc., 417, 137-148
2. Clerc-Hofmann, F., Vallette, G., Secco-Millet, C., Christeff, N., Benassayag, C., Nunez, E.A. (1983) C.R. Acad. Sci. D. 296, 53-58
3. Vallette, G., Benassayag, C., Christeff, N., Clerc, F., Nunez, E.A. (1984) Dev. Pharmacol. Ther. 7 : Suppl. 1, 30-37

4. Benassayag, C., Savu, L., Vallette, G., Delorme, J., Nunez, E.A. (1979) Biochim. Biophys. Acta 587, 227-237

5. Benassayag, C., Vallette, G., Delorme, J., Savu, L., Nunez, E.A. (1980) Oncodevelopmental Biology and Medicine 1, 27-36

6. Savu, L., Benassayag, C., Vallette, G., Christeff, N. and Nunez, E.A. (1981) J. Biol. Chem. 256, 18, 9414-9418

7. Nunez, E.A., Vallette, G., Benassayag, C. and Jayle, M.F. (1974) Biochem. Biophys. Res. Comm. 57, 1, 126-133.

8. Mercier-Bodard C., Renoir, J.M., Baulieu, E.E. (1979) J. Steroid Biochem. 11, 253-259.

9. Petra, P.H., Stanczyk, F.Z., Senear, D.F., Namkung, P.C., Novy, M.J., ALexander Ross, J.P., Turner, E., Brown, J.A. (1983) J. Steroid Biochem. 19 (1), 699-706

10. Cheng, C.Y., Musto, N.A., GUnsalus, G.L., Bardin, C.W. (1983) J. Steroid Biochem. 19 (4), 1379-1389

11. Rosner, W., Smith, R.N. (1975) Biochemistry 14, 4815-4820

12. Nishi, S. and Hirai, M. (1974) Protides of the Biological Fluids 18, 43-47

13. Ferlund, P. and Laurell, C.B. (1981) J. Steroid Biochem. 14, 545-552

14. Lowry, O.H., Rosebrough, N.H., Farr, A.L. and Randall, R.J. (1951) J. Biol. Chem. 193, 265-275

15. Bradford, M.M. (1976) Analyt. Biochem. 72, 248-254

16. Laurell, C.B. (1965) Anal. Biochem. 10, 358-361

17. Clarke and Freeman (1968) Clin. Sci. 35, 403-413

18. Pearlman, W.H. and Crepy, O. (1967) J. Biol. Chem. 242, 182-189.

TRANSCORTIN ACTIVITY DURING PREGNANCY

B. Martin
Laboratoire de Physiologie de la Reproduction (Groupe stéroides)
Université P. et M. Curie, 9, Quai St Bernard, 75005 Paris, France

M. Bonnin
Laboratoire d'Endocrinologie expérimentale, Université Bordeaux II,
146 rue Léo Saignat, 33076 Bordeaux Cedex.

P. Silberzahn
Laboratoire de Biochimie, ERA CNRS 958, Université, 14920 Caen,
France

F. Xavier
Laboratoire de Zoologie, Ecole Normale Supérieure, UA 686 CNRS,
46, rue d'Ulm, 75230 Paris Cedex 05

Introduction

Pregnancy-related variations of steroid-binding activities have been demonstrated in the plasma of a variety of mammalian species (1, 2, 3, 4, 5). The results obtained to date show important differences among the studied species with respect to the nature and the ligand specificities of the binding proteins and the placental vascular relationships of pregnancy. It has been established that elevated levels of transcortin (the major circulating specific plasma binder for C_{21} steroids, named also Corticosteroid Binding Globulin or CBG) occurred during pregnancy in conjunction with hemochorial placentation (2). The major biological function of transcortin is to modulate the hormonal activity of C_{21} steroids in blood plasma. It is generally accepted that the fraction of steroids which is bound with high affinity to transcortin is physiologically inactive, i.e. not available for uptake by target tissues, and only the free or unbound steroids can cross the target tissues and initiate actions (3, 6, 7). In this way, the steroid-transcortin complex has a regulatory in-

Pregnancy Proteins in Animals

fluence on the hormonal action. Association with transcortin protects the C_{21} steroids from enzymatic metabolism. Since the complex dissociate, the active form of the hormone is available at any target tissue (Fig. 1). The biological action of transcortin results from an equilibrium between its binding and dissociation with steroid hormones.

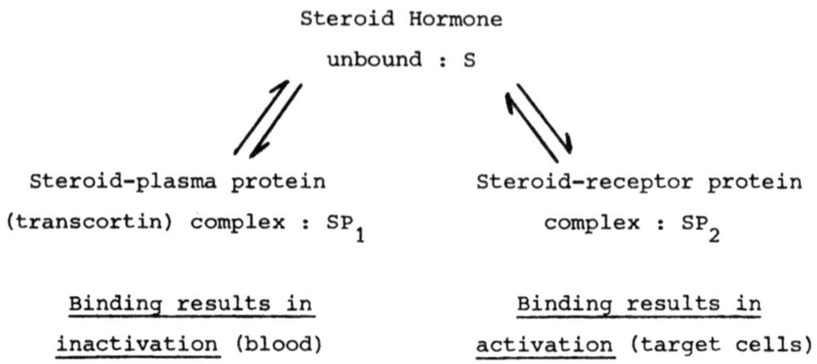

Figure 1

Effects of pregnancy on CBG were studies in 3 species which exhibit different pattern of embryonic development and type of placentation: the European badger (*Meles meles* L.), the viviparous lizard (*Lacerta vivipara*) and the French trotter mares.

Pregnancy occurs in the European badger in February-March just after parturition. Ovulation and fertilization are followed by a prolonged period of embryonic diapause during which the blastocysts are retained unattached within the uterine lumen lasting for 300-310 days. Implantation is delayed until mid-December or early January (8). The gestation phase (with endotheliochorial placental vascular relationships) lasts for 45 days. From fertilization in February-March to parturition in the following February-March of the next year, 30 female badgers were bled on arrival at the laboratory just after capture. Plasma samples from these animals gave data for monthly means.

60 adult females of the viviparous lizard were collected from May og September (period of activity) in the Massif Central (France) at 1200 m above sea level. During autumn and winter lizards hibernate. Vitellogenesis, gestation (2 months) and parturition take place in spring and summer. In the lizard, two types of placenta, the chorio-vitelline placenta and the chorio-allantoic placenta are found during the period of pregnancy. The passage of nutritive substances from maternal to the fetal tissues, via the placenta, is unneccessary in the lizard; the mass of yolk localized in the yolk sac appears to be sufficient to feed the embryo.

In the pregnant French trotter mares (12 animals studied) the maternal blood is separated from the fetal tissues by maternal tissue layers (epithelio-chorial placenta). This type of placenta is quite different from the hemochorial placenta found in primates. In the case of the hemochorial placenta the cavities of the maternal blood vessels are opened up and the villi of the chorion become bathed in maternal blood.

All species studied by Seal and Doe (2) exhibiting significant elevated CBG levels during pregnancy were of the hemochorial type of placentation. It was first reported by Sandberg and Slaunwhite (9) that during human pregnancy CBG plasma concentrations were two to three times that of non pregnant women. The biological significance of this placentation-related variation in CBG remains to be elucidated.

CBG-type proteins were previously identified and characterized in 2 of the 3 species studies in this paper: the badger (10) and the lizard (11, 12). Evidence for the presence of a CBG type protein, in low concentration (3.3 10^{-7}M) with high affinity for cortisol ($K_{A4°C} = 3 \times 10^{9} M^{-1}$) (Fig. 2) and corticosterone ($K_{A4°C} = 3 \times 10^{8} M^{-1}$) (not shown) was obtained for mare plasma. A number of properties are common to these different CBG type proteins and those characterized in the human CBG: high affinity for C_{21}steroids ($K_{A4°C} > 10^{8} - 10^{9} M^{-1}$) and saturable binding sites.

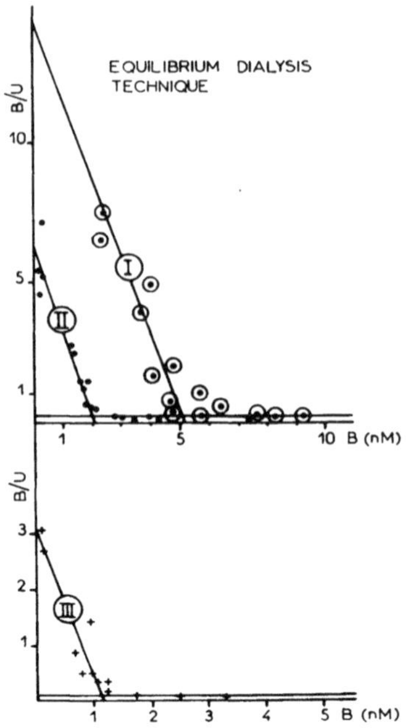

Fig. 2. Scatchard plots for the binding of corticol with plasma proteins of non pregnant (I) and pregnant (II and III) mares. Equilibrium dialysis experiments with 50-fold diluted plasma. B: Concentration of bound cortisol; U: concentration of unbound cortisol. II: 250 days of pregnancy; III: 330 days of pregnancy.

The following studies were conducted to quantify plasma CBG concentrations at multiple time points throughout badger, mare and lizard pregnancies. Plasma progesterone and corticosteroid levels were also measured at various times during pregnancy. The physiological meaning of the measured total plasma C_{21} steroid concentrations will be discussed in regard to the CBG concentrations. Comparison of our results with those obtained in the literature and concerning the human will be made.

Materials and Methods

Materials and experimented procedures were essentially the same as those previously described (10, 12, 13).

Two methods for the quantitative measurement of CBG in whole plasma were assayed: cortisol, corticosterone and progesterone binding experiments were performed by equilibrium dialysis (14) and gel equilibration methods (15). Experimental data were computed using a C II HB Iris computer program (16) and by graphic analysis according to Scatchard (17). Progesterone and corticosteroids were assayed respectively by radioimmunoassay and a radio-competitive assay.

Results

I - CBG Concentration in pregnant badger

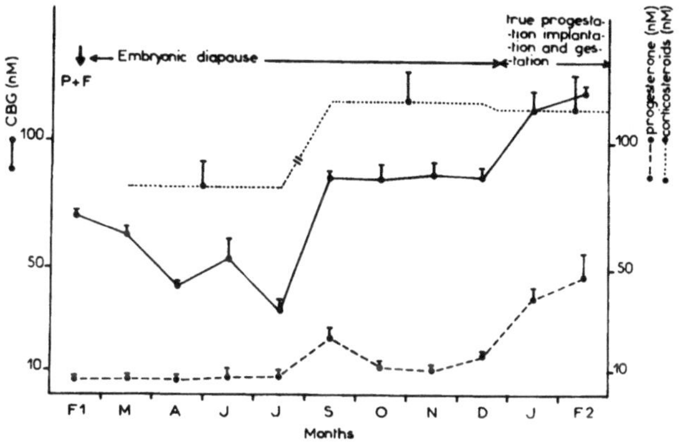

Fig. 3. Plasma concentration of CBG, progesterone and corticosteroids during delayed implantation, true progestation and gestation in female badger. Results are presented as the mean ± SEM. For CBG and progesterone concentrations monthly means are given. For corticosteroid concentration, means are calculated for February (F_1), from March to July, from September to December, January-February (F_2). Because pregnancy lasts for one year two mean values are given for February: $February_1$ = females at the beginning of pregnancy. $February_2$ = females in gestation.
P + F: parturition and fertilization.

The results reported here (Fig. 3) for pregnant badger show two significant changes in CBG concentration during the year. The first increase, during the embryonic diapause from September to December, must be considered a seasonal variation since it has also been observed during the same period in non pregnant females. The second increase, in January and February$_2$, seems to be related to true progestation and gestation. The two mean values obtained for February (February$_1$ and February$_2$) were significantly different ($P < 0.025$). CBG concentration was 1.7 times higher in February$_2$ than in February$_1$. Data given in Fig. 4 are normalized to the day of implantation (day 0). They confirm an increase in CBG concentrations just before implantation.

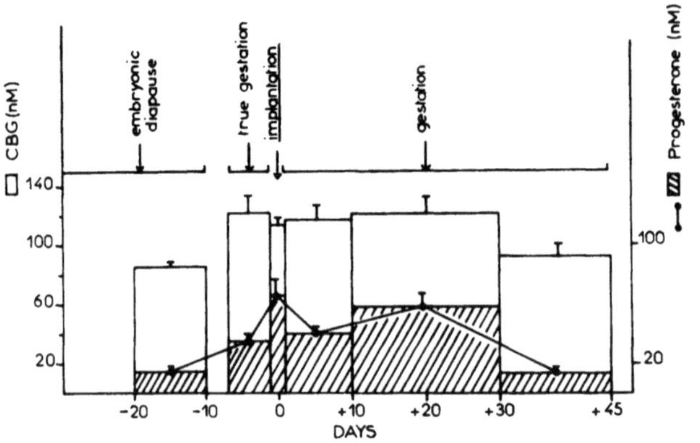

Fig. 4. Plasma concentration of CBG and progesterone during the end of embryonic diapause, true progestation and gestation in badger.
Results are normalized to the implantation day (day 0) determined by the first P peak in animals bled twice a week. Mean values ± SEM.

Plasma progesterone concentration fluctuates during the delayed implantation period (February$_1$ to December) (Fig. 3), increases at the end of the diapause (peak coinciding with implantation), and remains high during gestation (Fig. 4).

There was no significant variation in the plasma corticosteroid level during the first half of the diapause (Fig. 3). A significant increase of corticosteroids is observed during the second half of the embryonic diapause. No change in total corticosteroid concentration is observed during the gestative phase.

II - CBG Concentration in pregnant mare

Table I presents CBG concentrations measured on plasma from normal and pregnant mares. Mean concentration were not significantly different between non pregnant and 60-120 days pregnant females, but they were significantly different between non pregnant and 8-9 months pregnant mares ($P < 0.01$): lower levels of CBG were found in plasma after 4 months of pregnancy. Scatchard plots for the interaction of cortisol with plasma proteins from individual non-pregnant and pregnant mares are given in Fig. 2. The level of total progesterone of the pregnant mare increases up to the 3rd month and then decreases progressively reaching low levels in the 5-6 month (Table I).

TABLE I

	CBG (nM)	[1]P (nM)
Non pregnant mares (in the estrous phase)	330 ± 80†* (n = 7)	< 1.6
Pregnant mares		
. 60-120 days of pregnancy	240 ± 40† (n = 5)	38-44
. 180 days	160	8
. 240-270 days	140 ± 20* (n = 4)	3
. 330 days	64 ± 5 (n = 2)	9

Table I. Plasma concentration of CBG and progesterone in pregnant and non pregnant mares. Results for CBG are presented as the mean ± SD; n = number of animals; non significantly different, $P > 0.05$; *significantly different, $P < 0.01$; 1 results obtained from the literature (18); P:Progesterone.

III - CBG Concentration in pregnant lizard

A high concentration of CBG was found in the course of the year, approximately 10^{-5}M with a significant twice higher level in pregnant lizards. During pregnancy, the binding capacity of CBG for progesterone was first characterized by a slow increase up to the stage 30 of the embryonic development, then an important augmentation between stages 30 and 39 and finally a rapid and regular descrease up to 2 days after parturition (Fig. 5). During these periods, the plasma progesterone concentrations increased progressively to reach maximal values and followed by a rapid decrease before parturition.

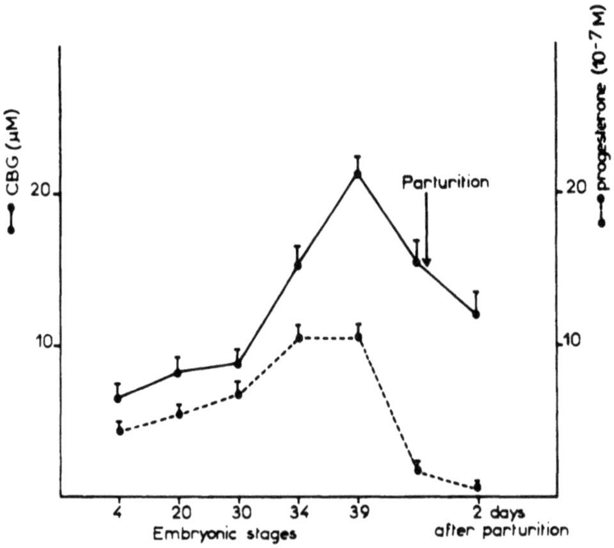

Fig. 5. CBG and progesterone concentrations in plasma of pregnant lizard.
Values are expressed as means; vertical bars are SD

Discussion

Values for both progesterone and CBG plasma concentrations measured in the 3 species studied during the pregnancy period are presented in Table II; human data are included for comparison.

The elevation of plasma CBG seen previously only in human and species with hemochorial placentation type (2) is seen here in a species with a different type of placentation, the badger (with endotheliochorial placentation type). The temporal sequence of events shown in Fig. 3, where data for both progesterone and CBG are presented, during badger pregnancy appeared to go through two stages, as described during human pregnancy (19): an initial and a further step increase.

High estrogen levels are known to be effective in inducing an increase in CBG concentration in man (3, 5, 9, 20) and in some primates (2). Since in the peripheral plasma of female badgers the estrogen concentration is generally low, except for episodic release occurring in late May and late July (21) when we have shown no increase in CBG, endogenous estrogen activity cannot be related to the CBG increase observed during true progestation and gestation. During lizard pregnancy a 2 fold rise of CBG concentration was found; in this case also no rise of estrogen çould be detected (22). On the other hand, a 2 fold decrease in CBG concentration was seen in the mare during the 2nd half of gestation. At this time blood estrogen content reaches its highest value. Hence, in the mare the rise and fall of blood unconjugated and conjugated estrogens during gestation follow a pattern which exhibits peak levels around the seventh and eighth month respectively, and then declines slowly to reach basal value within a few hours after parturition (23). From our results, it seems that the CBG elevation during pregnancy seen in 2 of the 3 species studied is independent of endogenous estrogen levels. For the other species (the mare) the drop of CBG during the 2nd half of pregnancy cannot be related to a decrease of estrogen levels. A biosynthetic stimulus other than estrogen must be responsible for the rise in CBG concentration, as has already been suggested to explain the relatively high CBG level before puberty in man since at this time steroid production is minimal (5).

Alterations in the concentration of CBG cause major changes in the plasma distribution of C_{21} steroids. Computer analysis is the only method available for solving the complex equilibrium interaction of multiple ligand-binding systems which most closely

TABLE II

	WOMEN[1]	BADGER	MARE	LIZARD
CBG (nM)				
Non pregnancy	700	70	330	11000
Pregnancy[2]	1700	120	140	21400
Effect of pregnancy	x2.5	x2	$x\frac{1}{2}$	x2
[3]P (nM)				
Non pregnancy	38	7	1.6[a]	7
Pregnancy[2]	480	47	3[a]	1032
Effect of pregnancy	x12	x7	x2	x147

Table II. Plasma concentration of CBG and progesterone in various species; effect of pregnancy.
[1]: results obtained from the literature (24, 25)
[a]: results obtained from the literature (18)
[2]pregnancy: women (3rd trimester); badger (gestation stage); mare (8-9th month of pregnancy); lizard (stage 39-40).
[3]P: progesterone.

approximates the in vivo situation. Recently Dunn et al. (25) have used a computer model based on the law of mass action to characterize C_{21}-steroid transport during human pregnancy. Thus, in spite of a 2 fold increase in the total cortisol (0.4 to 0.7 µM), the unbound cortisol concentration in pregnant women is essentially the same as that in non pregnant women (15 and 14 nM). This results from the 2.5 fold increase in CBG concentration (Table II). Only higher elevated steroid levels during human pregnancy [as for progesterone (0.038 to 0.48 µM)] cause marked changes in the plasma distribution: the unbound progesterone concentration is significantly enhanced (0.9 to 11 nM). Such evaluation of the unbound steroid concentration (the biologicially active fraction) in the plasma of the 3 species studied here would be hazardous since a computer simulation needs all data concerning (1) the plasma concentration of the different endogenous steroids, (2) the plasma concentration of the different steroid binders, and

(3) the different binding affinities of each steroid to the diferent steroid binders. Nevertheless the important elevated total progesterone levels during the pregnancy period in badger (x 7) and lizard (x 147), and the relatively low increase of CBG (x 2) during the same time, leads us, according to the law of mass action, to consider that the unbound active progesterone fraction during pregnancy must be enhanced. The 50% CBG decrease during the second half of mare pregnancy must also emphasize the biologically active form of progesterone.

References

1. Seal, U.S. and Doe, R.P. 1966. In: Steroid dynamics. Academic Press, N.Y. and London, 63.
2. Seal, U.S. and Doe, R.P. 1967. Proc. 2nd Intern. Congr. on Hormonal Steroids. Excerpta Med.Inter.Congr.Series n° 132, 697.
3. Westphal, U. 1971. "Steroid-Protein interactions". Springer-Verlag, Berlin.
4. Savu, L., Nunez, E.A. and Jayle, M.F. 1977. Endocrinology 101, 369.
5. Wagner, R.K. 1978. Acta Endocrinol. Suppl. 218, 73.
6. Westphal, U. 1983. J. Steroid Biochem. 19, 1, 1.
7. Funder, J.W. 1977. Proceedings of the Vth International Congress of Endocrinology. pp. 458-462. Excerpta Medica, Hamburg/Amsterdam/Oxford.
8. Canivenc, R. and Laffargue, M. 1956. C.R.Soc.Biol. 150, 1193.
9. Sandberg, A.A. and Slaunwhite, W.R. 1959. J.Clin.Invest. 38, 1290.
10. Bonnin, M., Martin, B., Charron, G., Audy, M.C. and Canivenc, R. 1984. J.Steroid Biochem. 20, 2, 575.
11. Martin, B. 1980. Steroid-protein interactions in non mammalian vertebrates: distribution, origin, regulation and physiological significance of plasma steroid binding pro- In: Steroids and their mechanisms of action in non mammalian vertebrates. edited by D. Delrio and J. Brachet. Raven Press, N.Y.
12. Martin, B. and Xavier, F. 1981. Gen. and Comp.Endocr. 43, 229.
13. Silberzahn, P., Zwain, I. and Martin, B. Endocrinology 115, no. 1.

14. Martin, B., Rotten, D., Jolivet, A. and Gautray, J.P. 1981. J.Clin.Endocrinol.Metab. 53, 443-447.

15. Pearlman, W.H. and Crepy, O. 1967. J.Biol.Chem. 242, 128.

16. Raynaud, J.P. 1973. Comput.Program.Biomed. 3, 63.

17. Scatchard, G. 1949. Ann.N.Y.Acad.Sci. 51, 660.

18. Holtan, D.W., Nett, T.M. and Estergreen, V.L. 1975. J.Reprod. Fert.Suppl. 23, 419.

19. Doe, R.P., Fernandez, R. and Seal, U.S. 1964. J.Clin.Endocr. 24, 1029.

20. Musa, B.U., Seal, U.S. and Doe, R.P. 1965. J.Clin.Endocr. 25, 1163.

21. Modain-Monval, M., Bonnin, M., Canivenc, R. and Scholler, R. 1980. Gen.Comp.Endocr. 41, 143.

22. Gavaud, J. 1984. Role de l'hibernation dans le déterminisme de la vitellogenèse de Lacerta vivipara J.(reptile squamate): étude morphologique, endocrine et biochemique. Thèse de Doctorat d'Etat, Paris.

23. Pashen, R.L. and Allen, W.R. 1979. J.Reprod.Fertil (Suppl.) 27, 499.

24. Rosenthal, H.E., Slaunwhite, W.R. and Sandberg, A.A. 1969. J.Clin.Endocr. 29, 352.

25. Dunn, J.F., Nisula, B.C. and Rodbard, D. 1981. J.Clin.Endocrinol.Metab. 53, 58.

REPRODUCTION SPECIFIC VITAMIN CARRIER PROTEINS INVOLVED IN TRANSPLACENTAL VITAMIN TRANSPORT IN MAMMALS INCLUDING PRIMATES

P.R.Adiga, P.B. Seshagiri, P.V. Malathy & S.V. Sandhya
ICMR Centre for Advanced Research in Reproductive Biology, Department of Biochemistry, Indian Institute of Science, Bangalore - 560 012, India

Introduction

In egg laying lower vertebrates estrogen assumes an important function in preparing the maternal system to provide adequate nutritional support to the developing embryos. Thus, in many fishes during the reproductive phase, estrogen of ovarian origin induces the elaboration in the maternal liver of several yolk proteins (exemplified by vitellogin) in massive amounts to be deposited in developing oocyte. In the birds, the steroid hormone additionally stimulates the *de novo* production of several oviducal proteins destined to be incorporated into egg white (1,2). Now the question arises whether the dominant role played by this sex steroid in terms of maternal nutrient mobilization in egg laying species is conserved atleast in part during evolutionary transition to viviparity. Obviously, there is no compelling reason to believe so since in mammals, the blastocyst with its limited nutritional reserves and following implantation it is under continuous maternal survillence for nutritional support and growth *in utero*. However, the recent demonstration in this laboratory of the occurrence and functional importance in transplacental vitamin transport of reproduction-specific high-affinity carrier proteins for riboflavin and thiamin in mammals and that they are estrogen induced gene products of hepatic origin testifies to the fact that one of important biochemical roles of this hormone in pregnant mammals is to ensure adequate embryonic vitamin nutrition and survival. This paper describes the salient

Pregnancy Proteins in Animals

features of these exciting findings and emphasizes the evolutionary conservation of these vitamin carriers not only during transition from oviparity to viviparity, but all the way to primates in terms of the structure and function as well as the site of synthesis and hormonal modulation thereof.

Vitamin Carrier-Proteins for Riboflavin and Thiamin in the Chicken Oocyte

The major functional attributes of vitamin carrier-(or binding) proteins, which interact reversibly and selectively (in 1:1 molar ratio) with fat- and water- soluble vitamins with high affinity and avidity, are dietary absorption, circulatory transport and conservation from metabolic degradation/ urinary excretion of these micronutrients (3). Among this class of important proteins, retinol binding proteins, transcobalmines and folate binding proteins have an ubiquitous distribution as normal physiological entities throughout the animal kingdom. Neither the natural occurrence nor the possible physiological relevance of carrier-proteins specific for the major water-soluble vitamins (like riboflavin and thiamin) were known until the discovery of a unique flavoprotein in the chicken egg white and the yolk (4), which interacts preferentially with free riboflavin rather than FAD and FMN. That the elaboration of this unique flavoprotein by the egg-laying hen solely as a reproductive stratagem to ensure adequate flavin deposition in the developing oocyte was suggested by (a) the absence of such a protein in male as well as rapidly growing birds of either sex, (b) the unhampered growth rates, sexual maturity and egg-laying capacity of a mutant strain of the chickens lacking the functional gene coding for this protein (5) and (c) the finding that the fertilized eggs from these mutant birds failed to hatch since flavin deficiency precipitated by the maternal failure to deposit adequate vitamin in the egg led to embryonic mortality.

These initial observations on the chicken egg RCP were exploited in our laboratory to investigate in some detail its physicochemical properties, affinity and specificity for the ligand and the characteristics of its active site interacting with the vitamin (6). Carrier protein-mediated vitamin deposition in the egg yolk is clearly not unique to riboflavin alone, but is applicable to thiamin as well. A thiamin carrier protein (TCP) was isolated and characterized from the egg yolk (7) and white (8) as a distinct protein with high affinity for its ligand. Immunologically and chemically, TCP could be clearly distinguished from RCP as a different gene product despite similarities alluded to above.

Hormonal Regulation of RCP and TCP in the Chickens

Since both RCP and TCP are undetectable in either rapidly growing immature pullets or the adult male birds but are encountered in measurable amounts in the plasma of egg-laying hens, hormonal regulation of the expression of their genes was presumed. It was hypothesised that in analogy with the production of other yolk and white proteins in the avian and amphibian systems, E may be the specific trigger for the expression of their genes. To investigate this, the genetic potentiality of the hepatocytes from the male birds to respond to E administration in terms of yolk protein production (2) was exploited and using sensitive and specific radioimmunoassays (RIA), the kinetics of elaboration of these two proteins were investigated (8,9). Following a single injection of estradiol-17β to the male chicks, the plasma levels of both these proteins rapidly accumulated attaining the maximum concentrations at different times and declining to undetectable levels at the same time. Administration of a second but identical dose of the hormone elicited significant amplification of the responses. It is interesting that the extents of response was E_2 dose-dependent but the durations of ascending and descending phases remained constant. The absolute E specificity of these induction phenomena was established by the finding that (a) the antiestrogens completely countered the estrogenic stimuli and (b) progesterone (P) by itself or in combination with E failed to modulate the inductive responses (8,9).

Relevance to Mammalian Reproduction

In higher animals and humans it has been repeatedly observed that the rapidly developing fetoplacental unit with its parasitic existence <u>in utero</u>, demands a continuous supply of nutrients from the mother and consequently accumulates, often against concentration gradients, several essential nutrients (10,11). Water-soluble vitamins belong to this category of nutrients which exhibit high fetal maternal ratios and the molecular mechanisms underlying this preferential fetal vitamin accumulation have remained enigmatic for a long time particularly since the placenta prevents the transit of the free vitamins or their co-enzyme forms. In view of this, we hypothesized that this placental barrier function may be similar to that of the plasma membrane of the avian oocyte and hence, the carrier-protein mediated vitamin delivery mechanisms may also be operative here. It was also attractive to visualize that the genes coding for these vitamin carriers are conserved during evolution to viviparity and expressed under instruction of the same hormonal signal (i.e., E). If some aspects of structure and conformation are retained in the carrier proteins, then detection of such proteins in pregnant mammals should be possible by immunological means.

This working hypothesis was clearly borne out by the observation that the sera from pregnant, but not adult male or immature rats contained proteins which clearly exhibited gross immunological cross-reactivity with the specific antisera (a/s) to both chick RCP and TCP (12,13). Direct biochemical evidence for the existence of these proteins was provided by the isolation in a homogeneous form, of both these proteins from either pregnant or estrogenized rat sera by affinity chromatography on immobilized vitamins (14,17). Their high affinity interaction specifically with the respective free vitamins in preference to their co-enzyme forms and their existence in the rodent circulation during gestation clearly favoured the view that their proposed function was that of transplacental vitamin transport to subserve embryonic growth as in the chickens.

Estrogen Regulation of Plasma Concentration of RCP and TCP in the Rat

When the plasma concentrations of these proteins in the rat were assessed during different stages of gestation and different periods of estrous cycle, it was clear that their levels fluctuated in concert with corresponding changes in the circulatory levels of E (12,15). In pregnant rat significant amounts of both RCP and TCP were encountered as early as Day 4, i.e., before implantation, showing that the proteins are of maternal origin. Between Days 10-14, coinciding with intense embryonic growth and organogenesis, the levels of both the proteins were maintained high. After Day 15, TCP level appears to decrease while RCP levels remained elevated upto Day 18. The reasons for this abrupt fall in TCP levels in the absence of corresponding change in E levels are currently not known and may be related to such diverse factors as the faster rate of fetal sequestration than that of synthesis and/or rapid clearance from circulation (15). During the 4-day estrous cycle, however, a good correlation between the plasma E levels on one hand and those of RCP and TCP were clearly evident with proestrous E surge coinciding with the maximum levels of the proteins (12,17).

Unequivocal proof that E is the primary signal for induction of these vitamin carriers in the rodent liver was provided by experiments in which the male rats were administered a pharmacological dose (10 mg/kg body wt.) of the hormone (16,17). Detailed kinetic analysis of RCP/TCP accumulation in the plasma have brought out several features of overall similarity in the induction phenomena between the rat and the chicken systems. These include an initial lag phase, an ascending phase of rapid accumulation followed by a decay phase. The 'memory effect' characterized by significant amplification of inductive responses during secondary stimulation with the steroid hormone was also clearly discernible. Again, as in the chicken system, P alone or in combination with E failed to modulate the E response whereas the antiestrogens and cycloheximide severely curtailed the production of both the vitamin carriers.

The only major difference between the two systems was quantitative, i.e., the response was generally more sluggish and the amplification during secondary estrogenic stimulation was of lesser magnitude in the rat <u>vis</u>-<u>a</u>-<u>vis</u> the avian system (15,16). It was thus abundantly clear that both the vitamin-carriers were highly conserved not only in terms of ligand-binding and other conformational characteristics, but also with regard to the specificity and the pattern of hormonal induction in the maternal liver.

Functional Significance in Fetal Development and Survival

Our first approach to ascertain the functions of these proteins was by administration of rabbit antibodies to either the chicken or the homologous vitamin carriers to pregnant rats and to follow the progress of their pregnancies by assaying the maternal serum P (12,17). Following passive immunoneutralization, the slowly increasing maternal P in circulation precipitously dropped to pregestation levels indicating acute fetal distress and pregnancy termination. Laparotomy of such animals at early stages of gestation revealed fetal resorption, whereas abortion occurred in those of advanced pregnancy. The effective period of pregnancy termination extended from Days 6 to 16 and the specificity of the a/s as abortifacients was confirmed by the failure of the a/s to other egg proteins like ovalbumin and ovomucoid to terminate pregnancy.

This acute fetal wastage culminating in abrupt pregnancy termination could be traced to drastic curtailment of transport of respective vitamin across the placenta with the resultant precipitation of severe fetal vitamin deficiencies (18,19). In the case of RCP a/s treated animals, inhibited flavin uptake (>90%) seems to set in motion a series of intra-embryonic events characterized by significant disturbance in relative contents and concentrations of flavin co-enzymes with depletion in FAD being most dramatic (19). On exploring the enzymic basis of these changes in fetal flavin co-enzyme levels, fetal FAD pyrophosphorylase was found to be markedly depressed due to flavin deficiency, which coupled with intrinsically higher activities of embryonic FAD degrading phosphatases, presumably depleted fetal FAD levels beyond the critical levels required to sustain development (20). Similarly, marked reduction in total thiamin content of fetuses from the animals treated with TCP a/s led to disturbances in relative levels of various thiamin derivatives, the decrease in TPP levels being most marked (17). It would appear therefore that the fetoplacental unit under conditions of deprivation of carrier protein-mediated vitamin delivery mechanism was unable to maintain its homeostasis culminating in pregnancy termination.

These remarkable findings prompted us to investigate the possibility that active immunization with heterologous vitamin carriers could be exploited as a novel approach to pregnancy suppression. Fertile female rats were actively immunized with either chick RCP (21) or TCP (17), mated with fertile males and ensuing pregnancies were monitored by assaying circulatory P levels. It was observed that the initial linear increases in serum P levels were abruptly reversed on Day 9 (RCP) and Day 6 (TCP) respectively signalling fetal death/resorption. This was in marked contrast to non-immunized control and ovomucoid immunized animals which carried pregnancies to term to deliver normal pups. Concomitantly, however, the general well-being of such actively-immunized animals, including their vitamin status, wt. gain, cyclicity and hormonal profiles were unimpaired testifying the selectivity of the adverse effects on the fetoplacental unit. This pattern of early pregnancy suppression could be repeatedly observed as long as high antibody titers were maintained by repeated boosting of their immunological memory; if however, the titers were allowed to wane with time, the animals carried their pregnancies to term thereby showing the reversibility of this approach to pregnancy termination.

Further Evolutionary Conservation

If the carrier protein-mediated vitamin delivery mechanisms to ensure embryonic development are conserved from the avian species to rodents, it may not be presumptous to expect that during further evolution, to primates these basic processes could have been modified only marginally. In conformity with the above, similar molecular species which immunologically cross-react in radioimmunoassays with specific antibodies to either the chicken or the rat vitamin carriers have been encountered in pregnancy sera from mouse, guinea pig, sheep, cow and sub-human primates (M. radiata). In fact, it has been possible to detect and purify similar cross-reacting proteins from pooled human pregnancy and cord sera (17,22). It is however not known at present whether during transplacental vitamin transport, the intact carrier protein-vitamin complexes themselves traverse the placental barrier or a separate sets of fetal vitamin carriers are elaborated to sequester the free vitamins delivered at the placental plasma membrane.

Experiments with the Primate Models

The hormonal modulation and functionality of these vitamin carriers in the sub-human primates were investigated to extrapolate these basic observations made in the rodent model. Using the South Indian bonnet monkey (M. radiata) as a model, it could be clearly shown that in a heterologous RIA, only the sera from either the cycling adult or pregnant monkeys harboured a cross-reacting protein (RCP) which could displace in a dose dependent manner the binding of ^{125}I-labelled chicken RCP to the respective a/s. Using this heterologous RIA, fluctuation in the circulatory levels of RCP-like protein during the menstrual cycle in the adult female bonnets were examined. It was clear that the relative concentrations of the protein in circulation bear a significant correlation with corresponding levels of E, the maximum amount of the protein being measured approximately 3-4 days after preovulatory surge of E_2 in the cycling monkeys. This is reminiscent of the situation in the rat in which the highest concentrations of the protein were encountered coinciding with proestrous E surge (16). Our preliminary data (23) also reveal that even in male, the protein can be indeed induced <u>de novo</u> by E administration, its level fluctuating in synchrony with alterations in circulatory levels of administered hormone. Requirement for repeated doses of E to elicit sustained protein production seems to be directly related to very short half-life of the exogenous hormone administered.

Encouraged by the above findings, attempts were made to interfere with the functioning of these proteins during gestation in the primates. We directly resorted to active immunization and used the purified chicken RCP as the antigen in view of its availability in a pure form in larger quantities as well as the prospects of eliciting high titre antibody because of the evolutionary divergence between the avian and the primate system. This is notwithstanding the rather restricted immunological cross-reactivity between the proteins from the two animal species. The immune response in the five animals chosen revealed that the sera prepared from these monkeys could neutralize 300-400 μg of cRCP/ml. During the course of either active immunization (2-3 months) or subsequent gestation, the general well-being, wt. gain. cyclicity and gonadal hormonal levels as well as vitamin status of these animals were normal, showing that the immunization had no adverse effect on their health and endocrine profiles. When such animals were mated with fertile males and their pregnancies were monitored by measuring serum P and CG (between 28-33 days), it could be shown that their pregnancies were terminated between 30-45 days after ovulation provided the antibody titers were high. Low antibody titers led to the animals carrying their pregnancies to term with the delivery of normal babies without any detectable abnormalities. Confirmatory evidence that RCP exists in the primates and passive immunoneutralization of the carrier protein by administering rabbit a/s to chick RCP would result in embryonic resorption has recently and independently been demonstrated in a study with marmosets at Institute for Research in Reproduction, Bombay (Drs. Usha Natraj and T.C. AnandKumar personal communication). These observations with the primate model testify to the specificity of physiological functions of the vitamin carrier as well as the reversibility of the approach of immunoneutralization of the protein as a method of pregnancy suppression.

Conclusions and Future Prospects

From the foregoing, it is abundantly clear that the vitamin carrier proteins utilized by the avian system for yolk deposition of the water-soluble vitamin for use by the prospective embryo are remarkably conserved not only during evolutionary transition from oviparity to viviparity but also during further evolution of the species all the way to humans. The fact that mammals have evolved special mechanisms for embryonic nutrition and yet have not eliminated the carrier-mediated vitamin delivery mechanisms to support fetal growth emphasizes the importance of these proteins in survival of the species. It is remarkable that not only the basic aspects of the structure and functions, but also the sites of synthesis, hormonal induction and modulation are scrupulously conserved throughout. The absolute E specificity of protein induction bestows on the steroid hormone an important functional significance during gestation viz., the production of these carrier proteins for adequate vitamin nutrition of fetuses unperturbed by competing demands of the maternal system. Thus it is clear that the ever-increasing E production is clearly then one of the endocrine communications originating from the fetoplacental unit to command the mother to provide adequate nutrients to sustain its growth. Since immunoneutralization of the carrier proteins terminated early pregnancy, the above findings imply that genetic/endocrine defects in the functionality/production of these proteins may explain some of yet ill-defined causes of habitual/spontaneous abortions/intrauterine growth retardation as a maternal error after fertilization. Finally, these findings suggest a novel immunological yet reversible approach to pregnancy suppression post-coitally in mammals since the maternal well-being and hormonal profile remain unimpaired under these conditions. It may be also mentioned at this junction, that our recent preliminary observations (15) show that the mammalian uterine tissue is also capable of independently secreting the carrier proteins under E stimulation. It is possible that these locally produced vitamin carrier may have some relevance to early blastocyst nutrition. It is a moot point whether E requirement for implantation observed in some mammalian species may be explicable in part to the requirement of uterine secretion of the carrier proteins. Work in progress presently in our laboratory should clarify some of these findings.

Acknowledgement

The financial assistance from Indian Council of Medical Research and the Family Planning Foundation, New Delhi is gratefully acknowledged.

References

1. Chan, L., Means, R.A. and O'Malley, W.B. Steroid Hormone Regulation of Specific Gene Expression. Vitamins and Hormones, 86, 259-295, (1978).

2. Tata, J.R. and Smith, D.F. Vitellogenesis, a Versatile Model for Hormonal Regulation of Gene. Recent Prog. Horm. Res. 35, 47-90 (1979).

3. Glover, J. Retinol Binding Proteins. Vitamins and Hormones. 31, 1-42, (1973).

4. Feeney, R.E. and Allison, R.G. Evolutionary Biochemistry of Proteins, Wiley, New York (1969).

5. Clagett, C.O. Genetic Control of the Riboflavin Carrier Protein. Fed. Proc. 30, 127-129, (1971).

6. Murthy, U.S., Podder, S.K. and Adiga, P.R. The Interaction of Riboflavin with a Protein Isolated from Hen's Egg White: a Spectrofluorimetric Study. Biochim. Biophys. Acta. 434, 68-81, (1976).

7. Muniyappa, K. and Adiga, P.R. Nature of Thiamin Binding Protein from Chicken Egg Yolk, Biochem. J. 193, 679-685, (1981).

8. Muniyappa, K. and Adiga, P.R. Isolation and Characterization of Thiamin Binding Protein from Chicken Egg White. Biochem. J. 177, 887-894, (1979).

9. Murthy, U.S. and Adiga, P.R. Estrogen Induced Synthesis of Riboflavin Binding Protein in Immature Chicks. Biochim. Biophys. Acta. 538, 364-375, (1978).

10. Dancis, J. and Schneider, H. in Placenta and Maternal Supply Line (Ed. Grunewald, P). MTP Press. Lancester. p 99-124, (1975).

11. Miller, R.K., Koszald, T.R. and Brent, R.L. Transport of Molecules across Placental Membrane, Cell. Surface. Reviews. 1, 145-207, (1976).

12. Muniyappa, K. and Adiga, P.R. Occurrence and Functional Importance of a Riboflavin Carrier Protein in Pregnant Rat. FEBS Letters 110, 209-212, (1980).

13. Adiga, P.R. and Murty, C.V.R. in Molecular Biology of Egg Maturation. Ciba Foundation Symposium no. 98. Pitman, London. p 111-136, (1983).

14. Muniyappa, K. and Adiga, P.R. Isolation and Characterization of Riboflavin-Binding Protein from Pregnant Rat Serum. Biochem. J. 187, 537-540, (1980).

15. Malathy, P.V. and Adiga, P.R. Hormonal Modulation of Gestation Specific Thiamin Carrier Protein in the Rat. J. Bioscience. (In Press), (1985).

16. Murty, C.V.R. and Adiga, P.R. Induction of Riboflavin Carrier Protein in the Immature Male Rat by Estrogen: Kinetics and Hormonal Specificity. J. Biosci. 4, 227-237, (1982).

17. Malathy, P.V. Studies on Thiamin Carrier Proteins: Physicochemical, Biosynthetic and Functional Aspects. Ph.D thesis, Indian Institute of Science, Bangalore (1984).

18. Murty, C.V.R. and Adiga, P.R. Mechanism of Fetal Wastage following Immunoneutralization of Riboflavin Carrier Protein in the Pregnant Rat. FEBS Letters. 135, 281-284, (1981).

19. Krishnamurthy, K. Surolia, N. and Adiga, P.R. Mechanism of Foetal Wastage Following Immunoneutralization of Riboflavin Carrier Protein in the Pregnant Rat: Disturbances in Flavin Co-enzyme Levels. FEBS Letters. 178, 87-91, (1984).

20. Krishnamurthy, K. Studies on Vitamin Carrier Proteins: Physicochemical and Functional Aspects. Ph.D thesis, Indian Institute of Science, Bangalore. (1984).

21. Murty, C.V.R. and Adiga, P.R. Pregnancy Suppression by Active Immunization against Gestation Specific Riboflavin Carrier Protein. Science. 216, 191-193, (1982).

22. Murty, C.V.R. and Adiga, P.R. Isolation and Characterization of a Riboflavin Carrier Protein from Human Pregnancy Serum. Biochem. International. 5, 289-296, (1982).

23. Sandhya, S.V., Seshagiri, P.B. and Adiga, P.R. Unpublished Observations, (1984).

PROGESTERONE-BINDING GLOBULIN FROM PREGNANT GUINEA PIG SERUM: IMPROVED PURIFICATION PROCEDURE AND COMPARISON WITH OTHER STEROID BINDING PROTEINS

Oscar A. Lea

Hormone Laboratory and Department of Pharcacology, University of Bergen, Norway

Introduction

The progesterone-binding globulin ,PBG, in guinea pig serum is a unique binding-protein produced by the placenta. PBG has found application as a binding agent in competitive isotope displacement analysis for progesterone(1) and 5-alpha-dihydroprogesterone.(2). It also serves as a model protein for studies of steroid-protein interactions in general (3,4). Antibodies against PBG has been used for immunocytochemical localization of PBG (5) and is of potential value for studying placental evolution and differentiation. PBG is however, a poor immunogen, as judged by the poor immunological response it induces in rabbits. This may not be surprising considering the fact that PBG is a strange molecule to the guinea pig itself being present only during pregnancy .Low immunogeisity could be due to a very high carbohydrate content (40-70 %) with a unique structure which renders it unable to bind to lechtins like concanavalin A (6). PBG is a polydisperse protein and exhibits both size and charge heterogeneity probably due to variations in carbohydrate content between PBG molecules produced within the same animal. (7,8). This charge heterogeneity makes purification by conventionaol ion-exchange procedures a demanding task and results in PBG preparations of varying chemical composition. The low immunogenisity of PBG also puts strick limits of the level of impurities tolerated in the preparation used for immunization if a monospecific antiserum is to be obtained. The following isolation procedure copurifies all molecular forms of PBG disrespective of size

Pregnancy Proteins in Animals

and charge configuration and requires only commercially available chromatography media. Antibodies to PBG has been used to look for immunological crossreactivity with other animal sera and progesterone-binding proteins isolated from various sources.

Materials and Methods

Pregnant guinea pig serum from the last trimester (days 40-60) was obtained following cardiac puncture and was stored at -20 ^{0}C until processed. Dialysis was performed in 5/8 inch dialysis tubing (A.H. Thomas Co. PA. USA). Hydroxyl apatite

Figure 1

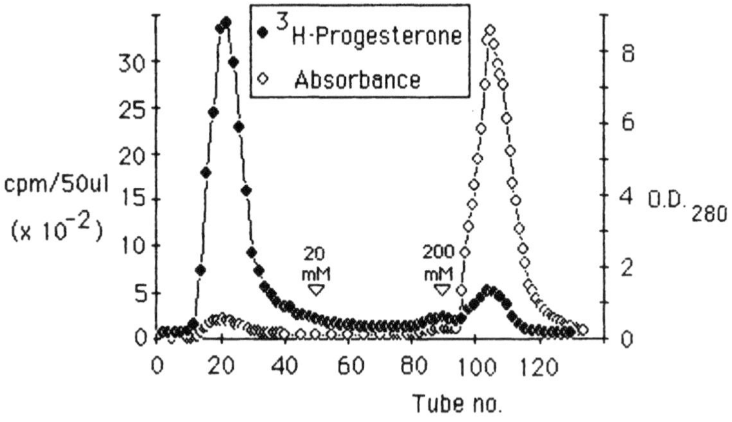

Chromatography of 50 ml pregnant guinea pig serum on Hydroxyl apatite. Serum was dialysed against starting buffer and incubated with ^{3}H-progesterone prior to fractionation. Column dimensions:5x15 cm, flow: ≈ 20 ml/h, temperature : 4 ^{0}C. Arrows indicate buffer changes to 20 and 200 mM phosphate buffer, pH 7.0.

chromatography was carried out in short columns (Pharmacia K 50/30) . Aproximately 100 g Bio-Gel HTP Hydroxylapatite (Bio-Rad, Richmond, Ca, USA.) was equilibrated with starting buffer (10 mM phosphate buffer pH 7.0) and packed according to the manufacturers recommendations. Prior to fractionation serum was dialyzed against 40 volumes of water, then against 20 volumes of 10 mM phosphate buffer pH 7.4 and clarified by centrifugation. During application of sample it was sometimes necessary to stir the upper third of the column to avoid cracking of the gel surface. Approximately 50 ml serum could be loaded to 100 g adsorbent. Protein was eluted from the column using increasing concentrations of phosphate buffer in a stepwise fashion.

Affinity chromatography on concanavalin A was carried out in columns 2.5cm x 12 cm packed with Con-A-sepharose 4B (Pharmacia fine chem. Sweden) as previously described (4). Sample (3-4mg protein/ml gel) was slowly (30 ml/h) drained through the gel and the effluent collected. Adsorbed protein was eluted using a 2% solution of α-methyl-D-mannopyranoside in phophate buffer.

Figure 2

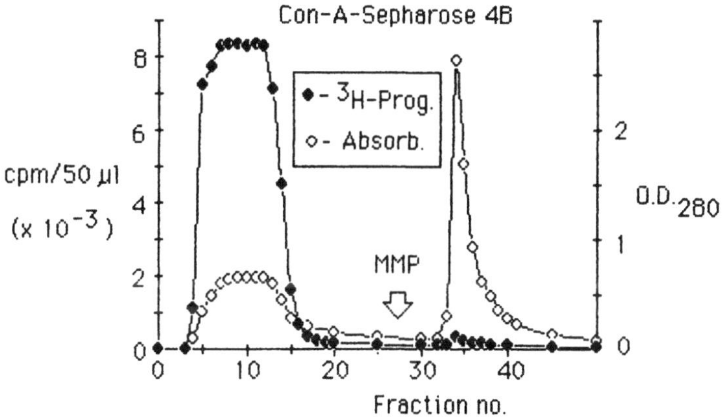

Affinity chromatography of pooled PBG-containing fractions on Concanavalin A- Sepharose 4B. Adsorbed protein was eluted using a 2 % methyl-mannopyranoside solution (MMP) as indicated by the arrow.Column dimensions: 2.5x10 cm, temperature : 4 °C. Each fraction equals 7.0 ml.

Gel filtration on Sephadex G-200 (Pharmacia, Sweden) was run in 2.6 x 40 cm columns packed according to the manufacturers instructions.. Up to 4 ml samples were applied to the column and eluted at a flow rate of 30 ml/h.

Protein from column effluents were monitored by their absorbtion at 280 nm using an Isco UA-5 on line UV monitor. Recovery of PBG was followed by its binding of ^{3}H-progesterone (81 Ci/mmol, Amersham, U.K.) which was added to serum prior to isolation.

A radioimmunoassay for PBG based on a double antibody-solid phase (DASP) design was developed using as trace ^{125}I-PBG labelled to a specific activity of ≈ 50 mCi/mg by the chloramine-T method.

Progesterone-binding cyst protein was isolated from cyst fluid using a combination of the methods listed above and also including an ion-exchange chromatography on DEAE-Sepharose 6 CL details of which will be described elsewhere (9).

Figure 3

Sephadex G-200

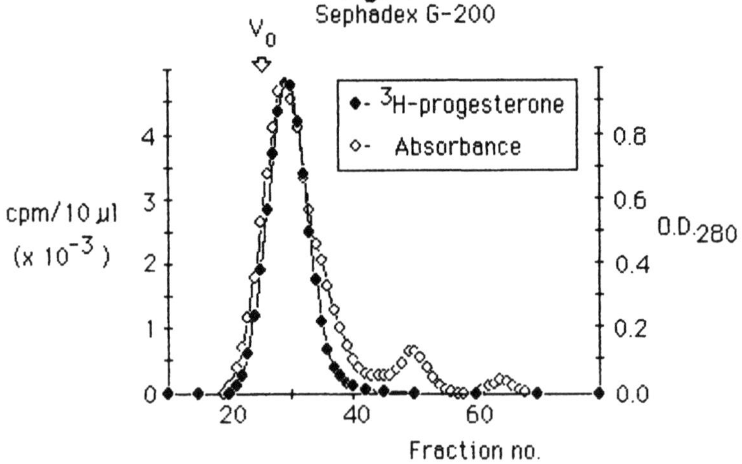

Gel filtration of purified PBG on Sephadex G-200. Pooled and concentrated material from the previous fractionations was applied in 4 ml buffer. Column dimensions: 2.6x35 cm, temperature: 4 °C

Immunisation of rabbits with PBG or PBCP were performed according to the protocol of Vaitukaties (10).

Results

An example of the first purification step, chromatography on hydroxyl-apatite, is shown in figure 1. PBG runs through the column unadsorbed and accompanied by only 5-8 % of total protein. Purification is about 15-20 fold and the yield is almost quantitative (table 1). The radioactivity associated with the second peak is tritiated progesterone and labelled degradation products bound to albumin and is devoid of PBG.

The next purification step makes use of the fact that PBG unlike most glycoproteins, does not interact with Concanavalin A. Affinity- chromatography of PBG on Concanavalin A- Sepharose 4 B is shown in figure 2. PBG again runs through the column unadsorbed (yield 95-98 %) accompanied by 21-26 % of applied protein (purification factor 4-5). This step also separates PBG quantitatively from the corticosteroid-binding globulin of the guinea pig (6). Analysis by polyacrylamide-gel electrophoresis reveals that PBG is essentially pure at this point and is suitable for most analytical purposes .This two-step procedure can be further simplified by running the two columns in series thus making PBG purification essentially a one step procedure.

If the PBG preparation is going to be used for immunization a final cleaning up step by gelfiltration on Sephadex G-200 is advisable. Figure 3 shows an example of such a step , PBG being eluted close to the void volume and well separated from some minor impurities.

Table 1

Method	Protein (Lowry)	Cumm. Recovery	Purif. factor
Start	2490 mg	100 %	-
Hydroxylapatite	166 mg	92.0 %	15
Con. A-Sepharose 4B	35 mg	89.7 %	4.7
Sephadex G-200	33 mg	84.7 %	1.1

The whole precedure can be completed in a few working days and gives a stable and highly pure product in yields exceeding 85 % and an overall purification factor of 60-80. The final product contains all molecular forms of PBG disrespective of charge/size conformation.

Rabbits immunized with PBG preparations purified as above responded albeit slowly, producing low titered antibodies to PBG. These antisera was judged to be monospecific by double immunodiffusion and by crossed immunoelectrophoresis.

Discussion

The first purification step employs Hydroxyl-apatite . PBG is not adsorbed by hydroxyl-apatite even at low concentration of phosphate buffer, i.e. at 10 mM or below. This is a property PBG has in common with all high affinity steroid-binding serum proteins hithereto described and may indeed be the only physicochemical property shared by this type of proteins.

Specific antisera to PBG has been used to look for PBG-like molecules in other animals. So far PBG immunoreactive proteins have not been detected in any domestic animal using conventional methods such as immunodiffusion and immuno electrophoresis. Using a more sensitive immunoenzymatic assay Perrot and Milgrom (11) were able to detect PBG in serum from non-pregnant guinea pigs and in the foetal circulation. Results obtained in this laboratory using a radioimmunoassay for PBG found levels of only 0.1 % of that in pregnancy to be present in foetal serum and serum from virgin animals. Similar levels were however seen in all animals studied (rat, mouse, goat,cow,horse,dog,cat, human) and is considered to be a matrix effect due to a non-specific interference with the binding assay. Regardless whether PBG is present in normal guinea pig serum or not, its concentration is too low to contribute significantly to the binding of endogenous steroids. Antiserum to PBG does not react with several pregnant hystricomorph rodent sera (11) despite a reported presence of pregnancy-induced progesterone- binding proteins in some new world species (Coypu, Viscacha, Degu) (12) Allthough PBG may look like a biochemical and immunological pecularity of the guinea pig it is not unprecedented. Guinea pig insulin for example, is very different from that of other species and is not recognized by antisera to bovine and porcine insulin.(13). Even though PBG does not seem to have any immunologially crossreactive counterpart in the animal kingdom there are proteins with steroid-hormone binding capability and some of them may be classified as pregnancy proteins. Examples are corticosteroid-binding globulin ,sex-hormone binding globulins in primates, alpha-foetoproteins of rat and mouse and uteroglobin in the rabbit.

Table 2

COMPARISON OF PBG WITH PBCP

Property	PBG	PBCP
Affinity for hydrophobic steroids	yes	yes
Charge heterogeneity at neutral pH	yes	yes
Affinity for hydroxylapatite	low	low
Isoelectric points	2.7-3.5	2.8-4.2
Carbohydrate content	40-70 %	40-50 %
Molecular weight	83 000	81 000
Hydrated radius	49 Å	47 Å
Frictional ratio	1.74	1.62
Electrophoretic mobility	α_1	α_1
Ass.constant for progesterone (4°O)	$1.4 \cdot 10^9$	$1.6 \cdot 10^6$
Affinity for concanavalin A	no	yes
Immunological crossreactivity	none	none
Number of subunits	1	3

Progesterone-binding proteins of lower affinity has been described in rat tissues such as the 9 S binding protein found in several reproductive tissues (14) and Prostatein secreted exclusively by the ventral prostate (15). In the human PBG resembles alpha-1-acid-glycoprotein,AAG, in several respects for instance in terms of acidity and carbohydrate composition. The steroid-binding affinity of AAG, however, is much lower (10^6 at 4 °C) and it has only half the molecular size.of PBG .There is one protein which is present in huge quantities (10-20 mg/ml) in cyst fluid found in gross cystic disease of the breast and which are remarkably similar to PBG both with respect to molecular size and chemical composition(14,15). The protein named PBCP, i.e. progesterone-binding cyst protein,is also present in blood in low levels (50 -100 micrograms/ml) dependent on age but not on sex. PBG is not elevated during pregnancy or changed during the menstrual cycle (16). The physicochemical properties are compared with those of PBG in table 2. No immunologic crossreactivity can be detected using specific antisera towards both proteins.

Acknowledgement

This work was supported by The Norwegian Society for Fighting Cancer (Norsk Forening til Kreftens Bekjempelse.).

References

1. Lea,O.A., A. Bessesen and K.F. Støa. 1976. Progesterone-binding globulin and testosterone-binding activity in guinea pig serum during pregnancy. Relationship to progesterone and oestrogens. Acta Endocr.(Kbh.) 81,367-378
2. Støa,K.F. and A. Bessesen .1975 .Identification and quantitation of 5 alpha-pregnanedione in human pregnancy serum. J. Steroid.Biochem. 6,21-25
3. Lea, O.A. 1973. On the steroid specificity of the pregnant guinea pig progesterone- binding globulin. Biochim.Biophys.Acta 322,68-74
4 Blanford, A.T.,W. Wittman,S. Stroupe and U. Westphal.1978. Steroid-protein interactions XXVIII. Influence of steroid structure on affinity to the progesterone-binding globulin. J. Steroid. Biochem.9,187-201
5. Metz,B.C., C. Micouin and E.M. Chambaz. 1977. Origine placentaire de la protein de liason de la progesterone chez le cobaye gravid; Etude immunohistologique. C.R.Acad.Sci.(Paris) 285,937-940
6. Lea, O.A. 1979. Interaction of steroid-binding serum proteins with concanavalin A-Sepharose 4B. Hormone Res. 11,240-253
7. Lea, O.A. 1973. Isolation and characterization of a progesterone- and testosterone-binding globulin from pregnant guinea pig serum. Biochim.Biophys. Acta 317,351-363

8. Burton, R.M., G.B. Harding, W.R. Aboul-Hosn, D.T. MacLaughlin and U. Westphal. 1974. Progesterone-binding globulin from the serum of pregnant guinea pigs, a polydisperse glycoprotein. Biochemistry 13,3554-3561
9. Lea, O.A. 1985. Properties of a progesterone-binding glycoprotein isolated from human breast cyst fluid. Manuscript in preparation.
10. Vaitukaitis, J.,J.B. Robbins,E. Nieschlag and G.T. Ross .1971. A method for producing specific antisera with small doses of immunogen.J.Clin.Endocr. 33,988-991
11. Perrot, M. and E. Milgrom. 1978. Immunochemical studies of guinea-pig progesterone-binding plasma protein. Endocrinology 103, 1678-1685
12. Heap, R.B., N.Ackland and B.J.Weir.1981. Progesterone-binding proteins in plasma of guinea-pigs and other hystricomorph rodents. J.Reprod.Fert. 63,477-489
13. Wriston, J.C. 1981. Biochemical pecularities of the guinea pig and some possible examples of convergent evolution. J.Mol.Evol. 17,1-9
14. Wilson,E.M., O.A.Lea and F.S.French.1977. 9S binding protein for androgens and progesterone. Proc.Natl.Acad.Sci.USA. 74,1960-1964
15. Lea,O.A., P.Petrusz and F.S.French.1979.Prostatein. A major secretory protein of the rat ventral prostate. J.Biol.Chem. 254,6196-6202
16. Pearlman, W.H., J.L. Gueriguian and M.E.Sawyer.1973. A specific progesterone-binding component of human breast cyst fluid. J.Biol. Chem. 248,5736-5741
17. Lea, O.A. 1985. Evaluation of a specific radioimmunoassay for progesterone-binding cyst protein. Manuscript in preparation.

BIOLOGICAL IMPLICATIONS OF PROGESTERONE-BINDING GLOBULIN IN THE GUINEA PIG

Oscar A. Lea

Hormone Laboratory and Department of Pharmacology, University of Bergen, Norway

Introduction

During pregnancy the guinea pig produces large quantities of a unique binding protein, Progesterone-binding globulin ,PBG, which show high binding affinity for several biologically important steroid hormones (1,2) among them progesterone and testosterone. PBG does not bind cortisol or oestrogens (table 1) and hence show a specificity (3) which distinquishes it from corticosteroid-binding globulin,CBG and sex-hormone binding globulins present in other species and also from the oestrogenbinding foeto-proteins of the rat and mouse. Unlike other steroid-binding proteins PBG is not regulated or induced by other steroid hormones.Apart from the first 3 weeks following parturition, PBG cannot be detected outside of pregnancy (4). PBG is present only in trace amounts in the foetal circulation(5) and is most likely produced by the syncytial trophoblastic cells of the placenta (6) and secreted unidirectionally into the maternal circulation. Serum PBG concentration rises sharply at days 18-22 of gestation and stays elevated troughout pregnancy (fig 1). The concentration of PBG reaches levels of $10-40 \cdot 10^{-6}$ M which represents an excess of 20 times that needed for binding endogenous progesterone. About 2/3 of PBG binding sites are free or available for binding of steroids of lower affinity like for instance testosterone (4).

Effects on Metabolic Rates

Concomitant with the rise in serum PBG there is a paralell rise in the circulating levels of progesterone (4.7), testosterone and 5α-dihydrotestosterone (8). This rise

Pregnancy Proteins in Animals

in steroid level is accompanied by a sharp drop in steroid metabolic clearance rate (9) and is in accord with a widely accepted axiom stating that high affinity steroid -binding proteins influence hormone deactivation by decreasing their uptake in metabolically active tissues and thereby lowering their rate of metabolism. In accordance with this concept it has recently been shown in perfusion studies using heterologous sera that serum from pregnant guinea pigs lowers the unidirectional transport of progesterone and testosterone into liver (10) and brain (11). Strangely,

Figure 1

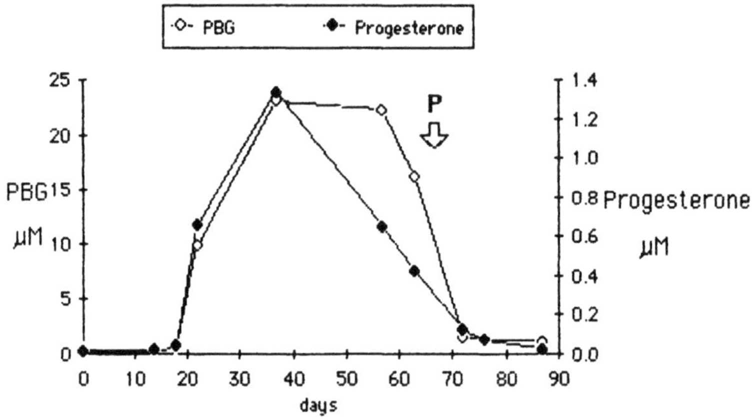

Concentration of PBG binding sites and progesterone during gestation in the guinea pig. Each point is the average value for a ten day period obtained from analysis of 7-11 animals. Arrow indicates parturition P.

this effect on progesterone is only observed with pregnant guinea pig serum and not with human serum despite of the fact that the latter contains corticosteroid-binding globulin ,CBG. which shows an affinity for progesterone matching that of PBG In most species incuding human, the metabolig clearance rate of progesterone is high (40-180 L/day/kg) compared to that of cortisol which is bound to CBG with an affinity equal to progesterone (12). In contrast to cortisol the metabolic clearance of progesterone does not decrease during pregnancy or following oestrogen administration even though there is a 2-3 fold increase in the level of CBG (12,13).

In order to find explanations for these controversies one have to look at the mechanism which regulates the uptake of steroid hormones into cells. and factors which are rate limiting in this respect. Despite of some evidence to the contrary (13), most published data seem to agree that steroid hormones are taken up by cells as the free hormone and that this transmembrane transport is not facilitated

Table 1

Steroid binding affinities of PBG at 4°C

Hormone	*Association constant* (10^7 M)
Gestagens.	
Progesterone	144.0
5α-dihydroprogesterone	140.0
Androgens.	
Testosterone	7.1
5α-dihydrotestosterone	21.5
Androstenedione	4.2
Oestrogens:	
Oestradiol	0.06
Oestrone	< 0.02
Glucocorticoids:	
Cortisol	0.29
Corticosterone	0.59

by any extracellular binding protein. The rate-limiting step in this mechanism could be either the diffusion of free hormone across the plasma membrane or the rate at which it dissociates from extracellular binding proteins. This rate is correlated to the association constant of the particular steroid-protein interaction and is very rapid ($T_{1/2}$ < milliseconds) for low affinity complexes as those of albumin and slow ($T_{1/2}$ in the order of seconds) for those of PBG and CBG (15). Progesterone dissociates much faster from human CBG than does cortisol and this may explain the controversy referred to above. The extent to which a steroid is bound to a high affinity binding protein will also influence rates of transport. In the pregnant guinea pig progesterone is almost quantitatively (>95%) bound to PBG. In the human less than 50 % appears to be bound to CBG and is mainly found associated with albumin and other proteins of high dissociation rate (16). Finally, the probability that a free steroid molecule at any instant will become reassociated with its high affinity carrier instead of diffusing through the cellular membrane will depend on the concentration of unoccupied binding sites of that carrier. In the human the concentration of CBG binding sites is of the same order as is total molar cortisol and progesterone concentration and hence the concentration of free binding sites considerably lower. In the pregnant guinea pig free PBG binding sites are available in a 20-fold molar excess.

The apparent deactivation caused by PBG occurs at a time when progesterone may no longer be needed for the maintenance of pregnancy. Ovariectomy after day 28 does not necessarily terminate pregnancy (17). The pregnant guinea pig also seem protected against the virilizing action of exogenous androgens with binding affinity towards PBG (18). Pharmacological doses of testosterone and dihydrotestosterone given to the mother will, however, masculinize the offspring (19).

Effects on Compartmentalization

The large increase in endogenous levels of progesterone, testosterone and dihydrotestosterone are to some extent paralelled by a rise in the foetal circulation. The foetal plasma progesterone concentration is about 1% of that found in the maternal circulation .Foetal concentration of androgens are lower than in the mother up to the period of gonadal differentiation (days 22-26 of gestation) when there is a dramatic increase in male but not in female levels (8,20) The placenta seem to act as a dialysis membrane allowing maternal free steroids to equilibrate between the maternal and foetal compartment in concentrations reflecting the distribution of PBG in these compartments. Similar conclusions can be drawn from studies using infusion of radiolabelled testosterone into the maternal circulation. The concentration of testosterone and metabolites with affinity for PBG (mainly androstenedione) was extremely low in the foetal tissues compared to that of maternal blood (21). So, while causing the concentrations of progesterone and potent androgens to rise in the maternal circulation due to a decreased clearance, PBG also prevents these steroids from entering the foetal compartment where they could interfere with gonadal differentiation and behavioral imprinting of the foetal brain (19).

Effects on Metabolic Pathways

Some indications that high-affinity steroid binding may influence the way a steroid is metabolised *in vitro* have been reported (22.23) . The guinea pig seem to be an ideal model for study of any such effect *in vivo* since PBG only

Figure 2

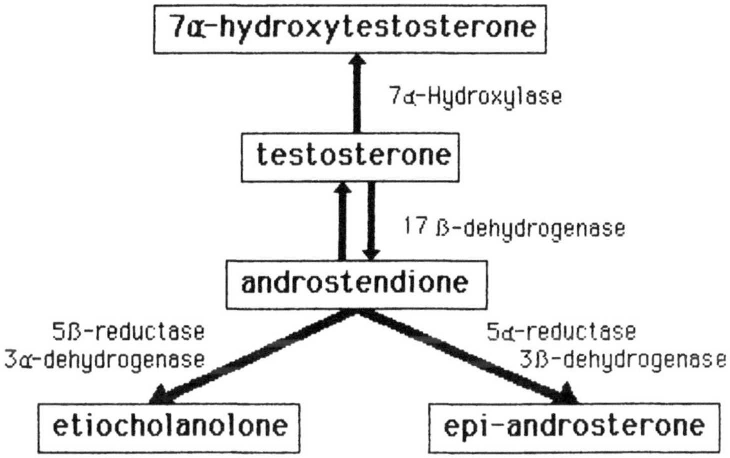

appears to be present in the pregnant animal. I have chosen to study the metabolism of radiolabelled testosterone given intravenously to nonpregnant and to pregnant animals in 49-56 days of gestation. Labelled urinary metabolites excreted during the first 48 hour period were isolated, identified and quantitated. Major metabolic pathways found to operate in the guinea pig are summarized in figure 2. The major urinary metabolite was 7α-hydroxytestosterone-glucuronide which comprised 21.3-35.8 % and 29.6-34.4 % of total radioactivity excreted in pregnant and nonpregnant animals,respectively (24). This metabolite appears to be rather unique to the guinea pig and probably is of hepatic origin. The guinea pig also produces more conventional metabolites ,epi-androsterone and etiocholanalone, which are derived from testosterone by reduction of the 5α or ß position. (see chart) and which are excreted both as glucuronides and sulfates.There was a striking difference between pregnant and nonpregnant animals in the relative amounts of epi-androsterone and etiocholanolone excreted, the 5α -reduced pathway (epiandrosterone) being preferred by pregnant animals (fig. 3). This shift in

Figure 3

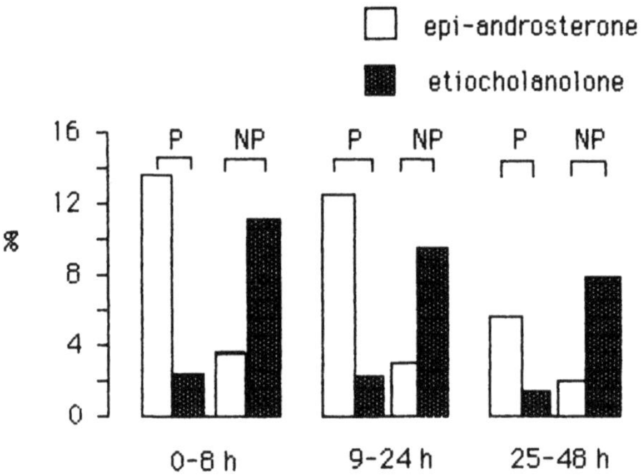

Excretion of radioactive 17-ketosteroids in urine at different time intervals following injection of labelled testosterone. Amounts are given in pere cent of total radioactivity excreted during each time interval. Measurements were obtained in 3-6 pregnant (P) or nonpregnant (NP) animals.

metabolic pathway is not readily explained by enzyme induction since hormonal changes during pregnancy are not likely to promote a change in 5 - reductase activity. Placental metabolism could theoretically contribute to formation of reduced metabolites since this tissue posess 5α reductase activity (25). The metabolite produced by this tissue *in vitro* is however androsterone and not epi-andreosterone which is found *in vivo*. Taken together with the evidence that endogenous hormones as well as trace amounts of labelled hormones apparently enter the foeto-placental compartment only in minor quantities, any contribution from these tissues on total testosterone metabolism can probably be neglected. An explanation left is that PBG influences the metabolic conversion of testosterone in extrahepatic tissues. by shifting the uptake of hormone between organs of different metabolic capability. It is conceivable that PBG shift the rate-limiting step in hormonal uptake from one controlled by diffusion to one limited by dissociation of the PBG-steroid complex during passage through the tissue. The net uptake of hormone in this case will be strongly dependant on the average capillary transit time in each anatomical region and for a given blood supply, uptake will be higher in tissues with thin and long capillaries. Apparently, tissues rich in 5 reductase activity belong to the last category. Allthough indirect, this study indicates strongly that high affinity serum binding may influence the qualitative aspects of steroid hormone metabolism. The pregnant guinea pig system offers a good model for further studies into the biological consequences of steroid-hormone interactions.

Acknowledgement

This work was supported by The Norwegian Society for Fighting Cancer (Norsk Forening til Kreftens Bekjempelse)

References

1 Lea, O.A. 1973. Isolation and characterization of a progesterone- and testosterone-binding globulin from pregnant guinea pig serum. Biochim. biophys.Acta . 317, 351-363

2. Milgrom,E., P.Allouch, M. Atger and E-E. Baullieu. 1973 Progesterone-binding plasma protein of pregnant guinea pig. J.Biol.Chem.248, 1106-1114

3. Lea, O.A. 1973. On the steroid specificity of the pregnant guinea-pig progesterone-binding globulin. Biochim.biophys.acta. 322, 68-74

4. Lea, O.A. ,A. Bessesen and K.F. Støa . 1976. Progesterone-binding globulin and testosterone-binding activity in guinea pig serum during pregnancy: Relationship to progesterone and oestrogens. Acta Endocr.(Kbh) 81,367-378

5. Perrot, M. and E. Milgrom. 1978. Immunochemical studies of guinea-pig progesterone-binding plasma protein. Endocrinology 103, 1678-1685

6. Metz, B.,C. Micouin and E.M. Chambaz.1977. Origine placentaire de la proteine de liason de la progesterone chez le cobaye gravide;etude immunohistolgique. C.R. Acad. Sci. Paris. 285, 937-940

7 Challis,J.R.G., R.B. Heap and D. Illingworth. 1971. Concentrations of oestrogen and progesterone in plasma of nonpregnant ,pregnant and lactating guinea pigs. J. Endocr. 51, 333-345

8. Rigaudiere, N.,P. Pradier and P. Delost. 1980. Free and bound androgens in the female guinea pig during gestation and after parturition and in the offsprings during the perinatal period. Acta Endocr.(Kbh.) 95, 101-109

9. Illingworth, D.V., R.B. Heap and J.S. Perry. 1970. Changes in the metabolic clearance rate of progesterone in the guinea-pig. J. Endocr. 48, 409-417

10. Pardridge, W.M. and L.J. Mietus. 1978. Transport of proteinbound hormones into liver in vivo. Am. J. Physiol. 237, E367-372

11. Pardridge, W.M. and L.J. Mietus. 1980. Transport of thyroid and steroid hormones through the blood-brain barrier of the newborn rabbit: Primary role of protein-bound hormone. Endocrinology ,107, 1705-1710

12. Little,B., R.B. Billiar, S.S. Rahman, W.A. Johnson, Y. Takaoka and R.J. White.1975. In vivo aspects of progesterone distribution and metabolism. Am. J. Obstet.Gynecol. 123, 527-534

13 Billiar, R.B., M. Jassani and B. Little. 1975. Estrogen and the metabolism of progesterone in vivo. Am. J. Obstet.Gynecol. 121, 877-883

14 Siiteri ,P.K., J.T. Murrai, G.L. Hammond, J.A. Nisker, W.J. Raymore and R.B. Kuhn. 1982. The serum transport of steroid hormones. Recent progress in hormone research. 38,457-510

15 Westphal U. 1978. Steroid-binding serum globulins: Recent results. In: Receptors and Hormone Action. Vol.2 (B.W. O'Malley and L. Birnbaumer, eds.). Academic Press, New York. pp.443-472

16 Lea, O.A. 1979. Interaction of steroid-binding serum proteins with Concanavalin A-Sepharose 4B. Hormone Res. 11, 240-253

17 Heap, R.B. and R. Deanasly. 1966. Progesterone in systemic blood and placentae of intact and ovariectomized pregnant guinea-pigs. J. Endocr. 34, 417-423

18 Diamond, M. and W.C. Young. 1963. Differential responsiveness of pregnant and nonpregnant guinea pigs to the masculinizing action of testosterone propionate. Endocrinology 72, 429-438

19. Goldfoot, D.A. and J.J. Van der Werff ten Bosch. 1975. Mounting behavior of female guinea pigs after prenatal administration of the propionates of testosterone, dihydrotestosterone and androstanediol. Hormones and Behavior.6, 139-148

20. Buhl, A.E., L.M. Paztor and J.A. Resko. 1979. Sex steroids in guinea pig fetuses after sexual differentiation of the gonads. Biol.Reprod. 21,905-908

21 Vreeburg, J.T.M., P.J.A. Woutersen, M.P. Ooms and J.J. van der Werff ten Bosch. 1981. Androgens in the fetal guinea-pig after maternal infusion of radioactive testosterone. J. Endocr. 88, 9-16

22. Mercier-Bodard, C., M. Marchut, M. Perrot, M-T. Picard, E-E. Baulieu and P. Robel. 1976. Influence of purified plasma proteins on testosterone uptake and metabolism by normal and hyperplastic human prostate in constant-flow organ culture. J. Clin. Endocrinol. Metab. 43, 374-386

23. Demisch, K., P. Birkelbach and W. Staib. 1971. Influence of albumin on testosterone metabolism in isolated perfused livers of guinea pigs. Z. Physiol. Chem. 252, 181-188

24. Lea, O.A. 1985. Testosterone metabolism in the pregnant and nonpregnant guinea pig. Acta Endocr. In press.

25. Grossman, S.B. and E. Bloch. 1973. Comparative placental steroid synthesis. II. C-19 steroid metabolism by guinea-pig placentas and fetal adrenals in vitro. Steroids. 21, 813-832

PREGNANCY INTERRUPTION BY A MONOCLONAL ANTIBODY THAT RECOGNIZES NON-T SUPPRESSOR CELLS IN MATERNAL LYMPHOID TISSUE

D.W. Hoskin, R.A. Murgita, S. Hamel

Department of Microbiology and Immunology, McGill University
Montreal, Canada H3A 2B4

K-O. Gronvik

Department of Immunology, The National Veterinary Institute,
Biomedical Center
Box 585, S-751 23 Uppsala, Sweden

Introduction

The fact that the mammalian fetus successfully develops in a latently hostile maternal immunological milieu has been explained by the existence of a set of well coordinated embryo-protective physiological and immunoregulatory processes (1). It would appear that several immunoregulatory mechanisms of both humoral and cellular origin act in conjunction to stabilize the delicately balanced maternal-fetal liason. Mixed lymphocyte reactivity (2), cytolytic T (3), and natural killer (NK) cell activity (4), mitogen stimulations (5), and antibody synthesis (6) can all be suppressed by pregnancy and/or newborn serum. Pregnancy-induced physiological changes in the lymphoid organs are accompanied by modulations in lymphocyte reactivity (7) which may be induced by changes in the levels of humoral factors such as alpha-fetoprotein (AFP) (6) and gestational hormones (8), but may also be due to increased numbers of naturally occurring immunoregulatory suppressor cells of T (9), macrophage (10) and non-T (11) origin. Such inhibitory cells have been shown to regulate a variety of *in vivo* and *in vitro* immune functions including antibody responses (12), autologous (13) and allogeneic (11) mixed lymphocyte reactions, and the generation of cytolytic T cells (14). Thus, elevated levels of pregnancy-associated immunoregulatory molecules and cells, as well as immunosuppressive trophoblast components (15) may each contribute to the overall maternal-fetal immunological homeostasis.

Pregnancy Proteins in Animals

Murine bone marrow (BM) contains at least two lymphocyte populations which proliferate in the presence of exogenous AFP (16). One type of AFP-reactive BM cell is agglutinated by the B cell-specific lectin soybean agglutinin (SBA). The second population of AFP-activated BM cells lack SBA receptors. In this paper we show that SBA^+ and SBA^- BM cells which have been pre-incubated with AFP display an enhanced capacity to suppress autologous mixed lymphocyte reactions (AMLR). We have previously demonstrated the presence of a non-T, B-like suppressor cell population in lymphoid tissue of primiparous isopregnant mice which can be positively selected for by agglutination with SBA (11). We have generated a hybridoma producing antibodies against a cell population enriched for the pregnancy-associated B suppressor cell (17). Cytotoxic pretreatment with these monoclonal antibodies largely abrogates the inhibitory activity of pregnancy-associated SBA^+ suppressor cells on AMLR. In addition, *in vivo* administration of such antibodies to pregnant mice of both inbred and outbred strains results in termination of a significant percentage of the pregnancies.

AFP Stimulated Bone Marrow Cells Display Enhanced Suppressor Activity

Several investigators have reported the presence of $Thy\text{-}1^-Ig^-$ inhibitory cells within murine BM (18,19). Phenotypically, these naturally occurring regulatory cells would seem to be very similar to the subset of SBA^+ BM cells which proliferate in the presence of AFP (16). SBA^- BM cells which are activated by AFP express a $Thy\text{-}1^+$ Lyt 1^+2^- phenotype (16). Neonatal animals, which are known to have elevated levels of serum AFP, harbour both non-T (11) and Lyt 1^+2^- T splenic suppressor cells (20). Moreover, physiologic concentrations of AFP have been shown to activate Lyt 1^+2^- inhibitory cells in cultures of normal adult murine spleen cells (20). This lead us to examine AFP-activated BM cells for possible *in vitro* suppressor function. BM cells were first separated into SBA^+ and SBA^- fractions and cultured for 72 to 120 hours in the presence or absence of 100 ug/ml AFP or normal mouse serum (NMS). The cells were then collected, washed and added to an AMLR consisting of neonatal CBA/J mouse thymocytes reacting against syngeneic adult spleen cells. As shown in Table 1, the addition of AFP-treated SBA^+ or SBA^- BM cells to assay cultures resulted in a significant suppression of AMLR when compared to NMS-treated controls.

Table 1. SBA^+ and SBA^- Bone Marrow Cells Stimulated with AFP Exhibit Enhanced Suppressor Function.

Cell Additions	Pre-incubated with	^{3}H-Thymidine Incorporation AMLR	Relative Response	% Suppression
		cpm ± SE		
SBA^+ BM	100 ug/ml NMS	34,638 ± 3,642	100	-
	100 ug/ml AFP	20,805 ± 3,135*	34	40
SBA^- BM	100 ug/ml NMS	137,510 ± 10,690	100	-
	100 ug/ml AFP	22,763 ± 2,386**	12	83

SBA^+ and SBA^- BM cells from normal adult CBA/J mice were cultured in the presence of 100 ug/ml NMS or AFP for 72 and 120 hours, respectively. 125,000 pre-cultured SBA^+ BM cells or 50,000 pre-cultured SBA^- BM cells were then added to AMLR consisting of 250,000 neonatal CBA/J thymocytes and an equal number of syngeneic adult spleen cells. Peak proliferation was measured at 96 to 120 hours of culture. Data are expressed as mean counts per minute of triplicate cultures plus or minus the standard error of the mean. SBA fractionation of BM cells was carried out according to the protocol of Reisner et al (21). Culture conditions and assay of proliferation are described elsewhere (22).

$$\text{Relative Response} = \frac{\text{cpm exp. mix} - \text{cpm exp. background}}{\text{cpm control mix} - \text{cpm control background}} \times 100$$

$$\%\ \text{Suppression} = \left(1 - \frac{\text{cpm exp.}}{\text{cpm control}}\right) \times 100$$

*$P < 0.05$;
**$P < 0.001$ compared to control and calculated according to Student's t-test.

The enhanced suppressor capacity of SBA^+ and SBA^- BM cells which have been cultured in the presence of AFP leads us to speculate that AFP-induced proliferation of BM-derived T and non-T cells *in vitro* may represent the initial phase of suppressor cell induction *in vivo*. Indeed, BM cells from pregnant animals with elevated levels of serum AFP are more potent suppressors of AMLR than virgin BM cells (data not shown).

Phenotypic Analysis of Pregnancy-Associated Non-T Suppressor Cells

Pregnant mice have been shown to harbour naturally occurring inhibitory cells of both T (9) and non-T type (11). We have recently found that a non-T suppressor cell is present in the spleen of primiparous isopregnant CBA/J mice (11). Table 2 summarizes the characteristics of this suppressor cell population.

Table 2. Characteristics of Pregnancy-Associated B-Like Suppressor Cells.

Density	1.056 - 1.067 g/ml
SBA Receptors	+
Surface Ig	+/-
Radiosensitive	+
Soluable Factor	+
T Cell Markers	-
Phagocytic	-
Adherent	-

Spleen cells from isopregnant CBA/J mice were subjected to the various treatments (11,13) and assayed for suppressor activity on AMLR.

Such cells can be positively selected for by agglutination with SBA and they localize at a density characteristic for B lymphocytes on discontinuous Percoll^R gradients (11). Further evidence for a unique pregnancy-associated B-like suppressor cell comes from data showing resistance to cytotoxic pretreatment with anti-T cell reagents and a lack of features associated with macrophages (11) or NK cells (data not shown). Rabbit anti-mouse Ig "panning" studies suggest that this suppressor population consists of both Ig^+ and Ig^- lymphocytes. Suppressor activity is radiosensitive and is mediated by a soluable factor (data not shown). The absence of conventional T cell markers coupled with the expression of SBA receptors suggests that these suppressor cells belong to an early B cell lineage or, alternatively, to a non-NK null cell population.

AFP may play a role in the induction of SBA$^+$ splenic suppressor cells during pregnancy. The kinetics of non-T suppressor activity parallels the natural rise and fall in serum AFP levels of gravid animals (data not shown). Moreover, these non-T inhibitory cells are phenotypically similar to the SBA$^+$ BM cells which are activated by AFP (16). AFP-activated SBA$^+$ BM cells exhibit enhanced suppressor function (see Table 1). Pregnancy-associated SBA$^+$ splenic inhibitory cells may therefore be derived from a subset of AFP-activated SBA$^+$ BM cells.

In Vitro and *In Vivo* Effects of Monoclonal Antibodies Against Pregnancy-Associated Suppressor Cells

Using a somatic cell hybridization technique (23) we have generated rat monoclonal antibodies of IgG class against unique cell surface structures on SBA$^+$ spleen cells obtained from gravid CBA/J mice. In Table 3 we show that cytotoxic pretreatment of SBA$^+$ pregnant spleen cells with antibodies produced by clone PSBA-1 efficiently abrogates the inhibitory activity exerted on AMLR. These monoclonal antibodies are cytotoxic for less than 1% of pregnant spleen cells as measured by viability counts in the trypan blue dye exclusion test.

Table 3. Deletion of Pregnancy-Associated Suppressor Cells by Monoclonal PSBA-1 Antibodies Plus Complement.

Cell Additions	Pretreatment	^{3}H-Thymidine Incorporation AMLR	Relative Response	% Suppression
		cpm ± SE		
-	-	196,512 ± 9,864	100	-
Pregnant SBA$^+$ spleen	-	87,265 ± 6,829*	32	56
Pregnant SBA$^+$ spleen	RC	84,948 ± 8,591*	28	57
Pregnant SBA$^+$ spleen	PSBA-1 + RC	164,286 ± 7,709	71	16

250,000 SBA$^+$ spleen cells from isopregnant CBA/J mice (minus those removed by antisera plus rabbit complement (RC) treatment) were added to AMLR consisting to 250,000 neonatal CBA/J thymocytes and an equal number of syngeneic adult spleen cells. Peak proliferation was measured at 96 hours of culture. Data are expressed as

mean counts per minute of quadruplicate cultures plus or minus the standard error of the mean. SBA fractionation of spleen cells was performed according to the protocol of Reisner et al (21). Negative selection protocol, culture conditions and assay of proliferation are described elsewhere (22). Hybridomas were generated by immunizing a 14 week old male Wistar rat with SBA^+ spleen cells from isopregnant CBA/J mice 3 times at 2 week intervals. Three days after the final injection the rat was sacrificed and 10^8 immune rat spleen and inguinal lymph node lymphocytes were hybridized with 2×10^7 thioguanine-resistant HGPRT-negative Balb/c-derived X-653 mouse myeloma cells. Hybridomas were cultured in microtitre plates and supernatants screened for the presence of cytotoxic antibodies by complement-mediated lysis of spleen cells from pregnant CBA/J mice. Supernatants from one culture designated PSBA-1 displayed cytotoxic activity which removed pregnancy-associated splenic suppressor activity.
*P < 0.001 compared to control by Student's t-test.

We were next concerned whether interaction by PSBA-1 antibodies with target antigens on pregnancy-associated immunoregulatory cells *in vivo* would destabilize immunological homeostasis during pregnancy. Antibodies were injected intravenously into inbred and outbred pregnant mice beginning on day 7 of gestation, the point at which non-T suppressor cells are first detected in our functional assays. Daily injections of PSBA-1 antibodies until day 14 of gestation resulted in total resorption of embryos in 43% of the pregnancies (see Table 4).

Table 4. Pregnancy is Disrupted by Monoclonal Anti-SBA^+ Antibodies.

Treatment	Complete Resorptions/Animals Treated	
	Syngeneic Pregnancy	Allogeneic Pregnancy
-	0/6	0/5
X-653	0/5	ND
CAB-1	ND	0/4
PSBA-1	3/7	3/7

Beginning on day 7 of gestation timed isopregnant CBA/J and allopregnant Swiss Webster female mice were injected intravenously with approximately 1 mg of ammonium sulphate precipitates derived from PSBA-1 or CAB-1 serum-free hybridoma culture supernatants, or with an equal volume of X-653 myeloma culture supernatant. Antibody injections were continued on a daily basis until day 14 of gestation. During this time interval all pregnant animals were closely observed and examined each day by palpitation for evidence of embryo resorption.

The uterine horns of animals in the process of resorbing their embryos were examined visually during the third week of gestation. They contained reduced numbers of poorly developed embryos but no macroscopic tissue damage could be seen. However, these mice revealed typical physiological changes for an ongoing pregnancy such as swollen uterine horns and decreased size and cellularity of their thymuses. Pregnant mice treated with supernatant from X-653 myeloma cultures or with monoclonal rat anti-CBA/J concanavalin A blast antibodies did not resorb their fetuses.

We can only speculate about the underlying mechanisms for the resorption of embryos following anti-suppressor cell antibody treatment but it is not unlikely that graft-versus-host-like reactions may have occurred within the fetal-placental unit. This may be due to PSBA-1-mediated interference with, or elimination of, immunoregulatory non-T suppressor cells in the mother. In addition, it is likely that PSBA-1 antibodies gain entry into the fetal circulation since immunoglobulins of the IgG class are known to cross the placenta. Cytotoxic pretreatment with PSBA-1 antibodies diminishes the inhibitory activity of neonatal spleen cells on AMLR (data not shown). Thus PSBA-1 monoclonal antibodies may also interfere with the immunoregulatory activity of suppressor cells found within the fetus.

It is evident from these studies that pregnancy-associated SBA^+ suppressor cells are of great importance in maintaining the stability of the murine maternal-fetal relationship. The elevated levels of inhibitory cells associated with pregnancy may have the essential function of maintaining the immune response to fetal antigens at an acceptable level which does not cause immunopathological damage.

Acknowledgement

This research was supported by grants from the National Cancer Institute of Canada and the Medical Research Council of Canada (MT-6470). D.W. Hoskin is the recipient of a fellowship from the Medical Research Council of Canada.

References

1. Gill, T.J. III, C.F. Repetti. 1979. Am. J. Path. 95, 465.

2. Pavia, C.S., D.P. Stites. 1979. J. Immunol. 123, 2194.

3. Smith, G. 1978. Transplantation 26, 278.

4. Barret, D.S., L.S. Rayfield, L. Brent. 1982. Clin. Exp. Immunol. 47, 742.

5. Weppner, W.A., J. H. Coggin, Jr. 1980. Cell. Immunol. 54, 193.

6. Murgita,R.A. 1976. Scand. J. Immunol. 5, 1005.

7. Anderson, D.J. 1978. Cell. Immunol. 41, 150.

8. Fuchs, T., L. Hammarstrom, C.I.E. Smith, J. Brundin. 1981. J. Reprod. Immunol. 3, 75.

9. Chaouat, G., G.A. Voison. 1979. J. Immunol. 122, 1383.

10. Hunt, J.S., L.S. Manning, G.M. Wood. 1984. Cell. Immunol. 85, 499.

11. Hoskin, D.W., D.C. Hooper, R.A. Murgita. 1983. Am. J. Reprod. Immunol. 3, 72.

12. Suzuki, R., T.B. Tomasi. 1979. J. Exp. Med. 150, 898.

13. Hoskin, D.W., R.A. Murgita, D.C. Hooper. 1983. In: Alpha-Fetoprotein and Congenital Disorders (G. Mizejewski and I. Porter, eds.). Academic Press (in press).

14. Chaouat, G., P. Monnot, M. Hoffman, G.A. Voison. 1982. Cell. Immunol. 68, 322.

15. McIntyre, J.A., W.P. Faulk. 1979. Cell. Immunol. 58, 202.

16. Hoskin, D.W., S. Hamel, D.C. Hooper, R.A. Murgita. 1985. Cell. Immunol. (in press).

17. Hoskin, D.W., R.A. Murgita, K-O. Gronvik. 1984. Scand. J. Immunol. 20, 173.

18. Duwe, A.K., S.K. Singhal. 1979. Cell. Immunol. 43, 372.

19. Dorshkind, R., C. Rosse. 1982. Am. J. Anat. 164, 1.

20. Murgita, R.A., D.C. Hooper, M. Stegagno, T.L. Delovitch, H. Wigzell. 1981. Eur. J. Immunol. 11, 957.

21. Reisner, Y., A. Ravid, N. Sharon. 1976. Biochem. Biophys. Res. Comm. 72, 1585.

22. Hooper, D.C., R.A. Murgita. 1981. Cell. Immunol. 63, 417.

23. Kohler, G., C. Milstein. 1975. Nature 256, 495.

TISSUE-SPECIFIC REGULATION OF GAMMA FETAL ANTIGEN EXPRESSION DURING MURINE DEVELOPMENT

P.J. Higgins*

Unit of Cell Biology and Genetics, Memorial Sloan-Kettering Cancer Center, New York, N.Y. 10021

Introduction

Progressive quantitative reductions in the tissue content of the serologically-defined murine gamma fetal antigen (GFA) is associated with postnatal development of the mouse (1). Recently, sites of embryonic hematopoiesis (fetal liver and spleen) were found to yield positive immunocytochemical reactions for the presence of GFA confirming previous suggestions, based on quantitative estimates of the GFA content of newborn mouse tissues (1), that areas of active hematopoiesis in the developing mouse constitute one source of GFA-enriched tissue. These data prompted quantitative analysis of a spectrum of hematopoietic adult tissues for GFA. This paper details the changing levels of GFA expression during late gestation and early postnatal murine development, the relatively high content of this antigen in adult bone marrow, and its modulation during in vitro erythroid differentiation. Evidence is presented supporting the tentative classification of GFA as a murine differentiation antigen.

* Present address: Dr. Paul Higgins, Chief, Laboratory of Cell and Molecular Biology, Veterans Administration Medical Center, Research Service 151
113 Holland Avenue, Albany, N.Y. 12208, U.S.A.

Pregnancy Proteins in Animals

Materials and Methods

Antibodies to GFA.- Antiserum to GFA was prepared as described (2). Briefly, New Zealand white rabbits were immunized with Freund's adjuvant-supplemented saline extracts of a 3-methylcholanthrene-induced C57BL/6 mouse fibrosarcoma. The gamma globulin fraction of anti-fibrosarcoma serum was partially absorbed with adult mouse serum and lyophilized extracts of adult internal organs (2). Immunoelectrophoretic analysis with this absorbed fraction of immune serum disclosed a single precipitin arc in the gamma migrating region in extracts of mouse fetal tissue (3). This antigen has been designated gamma fetal antigen (GFA).

Tissue preparation and quantitation of GFA content.- Extracts of adult or fetal/postnatal mouse tissues were prepared by homogenization in saline or 10^{-3}M tris buffers and clarified at 10,000 xg. Extract GFA content was determined by radial immunodiffusion assay (4); the radial immunodiffusion (RID) unit utilized for GFA quantitation is defined as:

$$\frac{\text{Diameter}^2\ \text{(precipitin disk)} - \text{Diameter}^2\ \text{(well)}}{\text{micrograms of protein assayed}}$$

Erythroleukemia cell culture and induced differentiation.- Friend murine erythroleukemia (FL) cells (5) were maintained in suspension culture in Dulbecco's modified Eagle's medium containing $2.5\text{x}10^{-2}$ M HEPES buffer (pH 7.3) and 1,000 mg glucose/ml, supplemented with 15% fetal bovine serum and 100 U/ml each of penicillin and streptomycin; L-glutamine was added freshly to each 500 ml of stock culture medium. For induction of differentiation, dimethylsulfoxide (DMSO) was added directly to cultures containing 1-2 x 10^5 cells/ml to a final concentration of 2% (280 mM). The extent of differentiation was determined 6 days later using the benzidine reaction to score for hemoglobin containing erythroid elements (6).

Flow cytometric measurement of cellular GFA content.- Equal numbers of control and DMSO-treated FL cells were collected by centrifugation at 900 xg. After washing 2x in Hank's balanced salts solution (HBSS), cell pellets were resuspended in 1 ml HBSS and forcefully ejected into 9 ml of ice cold acetone-ethanol (1:1) for fixation at -20°C for 24 hr. Fixed cells were collected by centrifugation at 900 xg, rehydrated by washing 2x in HBSS and treated with ribonuclease A (bovine pancreas, 3002 U/mg, Worthington Diagnostic Systems, NJ)(final concentration 200 U/ml HBSS) for 30 min at 37°C. Following ribonuclease digestion, FL cells were collected by centrifugation and resuspended in 1 ml of rabbit anti-GFA (1:50) or normal rabbit serum (1:50) for a 1 hr incubation on ice. The cells were then pelleted at 900 xg, washed 2x in HBSS and incubated for 1 hr with fluorescein isothiocyanate (FITC) conjugated IgG fraction of goat anti-rabbit IgG (1:100). After a final series of washes, cells were stained with propidium iodide (PI; 20 micrograms/ml HBSS) for 30 min at room temperature prior to analysis using an Ortho System 50H flow cytometer equipped with a 488 nm argon ion laser. Relative cellular DNA content was obtained on a cell-by-cell basis as the integrated red PI fluorescence signal. The pulse width of the PI signal was measured separately and utilized to gate out unwanted cell doublets (7). A third parameter, integrated green fluorescence, was measured and correlated with the red fluorescence (DNA content) of each cell. The integrated green (FITC) fluorescence was proportional to GFA content and independent of cell size (8). The mean fluorescence values for the total cell population and selected cell cycle compartments (G_1, S, G_2+M) based on DNA content were recorded and analyzed with an Ortho 2150 minicomputer. A minimum of 5×10^3 cells were measured for each sample.

Results

Quantitative estimations of total body GFA content of the develop- mouse (late gestation/early postnatal period) appear in Figure 1.

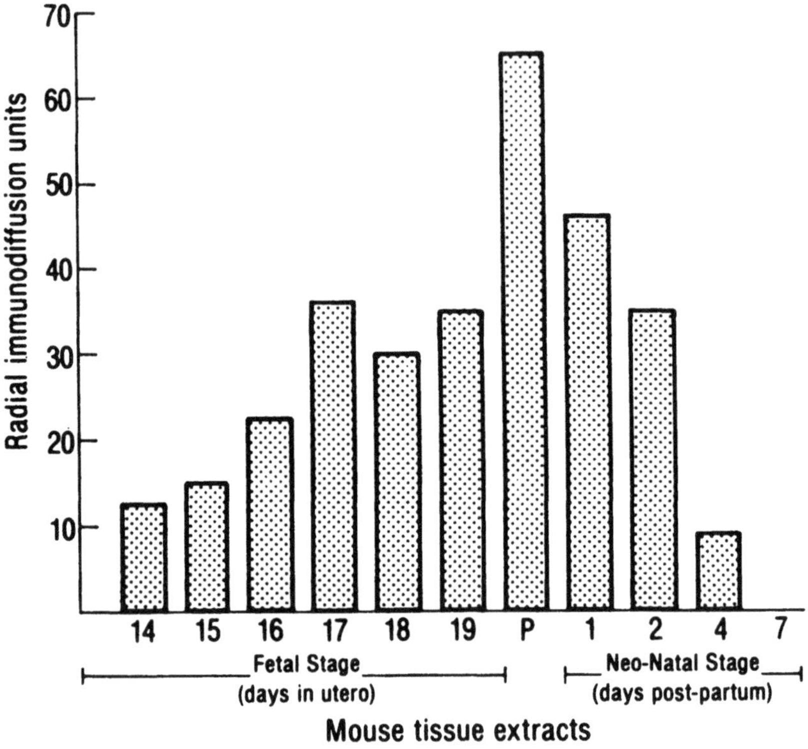

Figure 1. Concentration profile of tissue GFA content as a function of total body saline-extractable protein in the late gestation and early postnatal mouse as determined by radial immunodiffusion.

GFA was first detected in total fetal extracts after day 13 of gestation and persisted until 1 week postpartum. Since the total body GFA level was significantly greater at parturition, several individual tissues of the newborn mouse were analyzed for GFA content. High levels of GFA occurred in the splenic tissue of the newborn while extracts of brain, intestine, and kidney were GFA negative (Table 1).

Table 1. Quantitation of GFA in saline soluble extracts of mouse tissues by single radial immunodiffusion assay.

Tissue extracted[1]	Radial immunodiffusion units[2]
spleen	191.8 ± 23.8
thymus	24.6 ± 2.8
liver	10.0 ± 1.4
brain	0
intestine	0
kidney	0

[1] Organs removed from newborn animals immediately after parturition.

[2] Data represent the mean ± standard deviation of 8-10 assays. Antigen preparations which failed to yield precipitin disks were further assayed at both higher and lower antiserum dilutions and at various extract protein concentrations to eliminate the possibility of false negatives.

The significance of this unequal tissue distribution of GFA content is not readily apparent. Although splenic GFA levels decline considerably with age, from 198 ± 16 RID units for the newborn to 0.73 ± 0.01 RID units for the adult mouse (1), these data suggested that the fetal/neonatal spleen (and perhaps other hematopoietic tissues) may be involved in GFA biosynthesis (2) or at least be a principal site of its localization. To further examine these possibilities, a spectrum of normal adult mouse tissues was assayed for GFA content. It was immediately evident that hematopoietic tissues in general, and bone marrow in particular, contained high levels (relative to non-hematopoietic organs) of GFA (Table 2).

Table 2. Relative GFA content of adult mouse hematopoietic tissues.

Tissue extracted[1]	Radial immunodiffusion units[2]
bone marrow	31.8 ± 2.3
spleen	0.8 ± 0.1
thymus	0.3 ± 0.1
non-hematopoietic[3]	0.04 ± 0.01

[1] Tissues were obtained from 8-10 week old mice, homogenized in 10^{-3} M tris buffer, and the 10,000 xg soluble fraction assayed for GFA by radial immunodiffusion.

[2] Data represent mean ± standard deviation of 6-10 assays/tissue.

[3] Consensus results of individual assays on heart, liver, kidney, brain, ovary, seminal vesicle, lung, diaphragm, colon, stomach and small intestine.

Preliminary immunocytochemical experiments indicated that certain erythroid and non-erythroid blast elements in the adult mouse bone marrow were reactive with anti-GFA serum (Pizzi et al., submitted). Since FL cells provide a convenient system to examine specific aspects of erythroid differentiation under controlled in vitro conditions, a flow cytometric approach was utilized to quantitate, on a cell-by-cell basis, the GFA content of FL cells as a function of growth state and cell cycle compartment. Treatment of FL cells with DMSO for 6 days results in partial completion of the erythroid differentiation program; the percentage of cells containing hemoglobin rose from 0.5% to 82%. Paralleling the accumulation of differentiated cells in the C_{1D} compartment [equivalent in DNA content to G_0/G_1 cells (Figure 2)], DMSO-treated cells manifested considerably lower GFA-associated immunofluorescence (Figure 2). The shift to lower GFA content can be quantitated following subtraction of nonspecific fluorescence (as determined by substitution of normal rabbit serum for the primary anti-GFA serum). Thus, G_{1D} cells were calculated to have 34% less GFA on average than G_o/G_1 control cells. The remaining S and G_2+M phase cells in the DMSO-treated cultures demonstrated an even greater loss of GFA (37% and 47%, respectively) such that the DMSO-treated population taken as a whole contains 46% less GFA (Figure 2).

Discussion

Total body content of GFA gradually increases in the fetal mouse during late gestation, peaks at parturition, and then rapidly declines thereafter. Comparisons between newborn and adult indicated that after birth progressive quantitative reductions in the GFA content of particular tissues occur during postnatal development (1). In the adult mouse, however, hematopoietic tissues (particularly bone marrow) expressed high levels of GFA relative to non-hematopoietic tissues. A tentative assignment of GFA as a normal differentiation antigen, thus, appears to be supported by several lines of evidence including: the progressive restriction in GFA tissue distribution during postnatal development of the mouse, its quantitative varation in several tissues of the adult mouse, and by direct measurements of differences in the cellular GFA content between exponentially growing (blast stage) and differentiated populations of Friend erythroleukemia cells. The temporal relationship between GFA reduction and FL cell differentiation remains to be defined. Current work is focused on elucidation of whether GFA down-regulation is linked to withdrawal of FL cells from the cell division cycle (and therefore represents an early event in differentiation) or is an aspect of late stage erythroid differentiation.

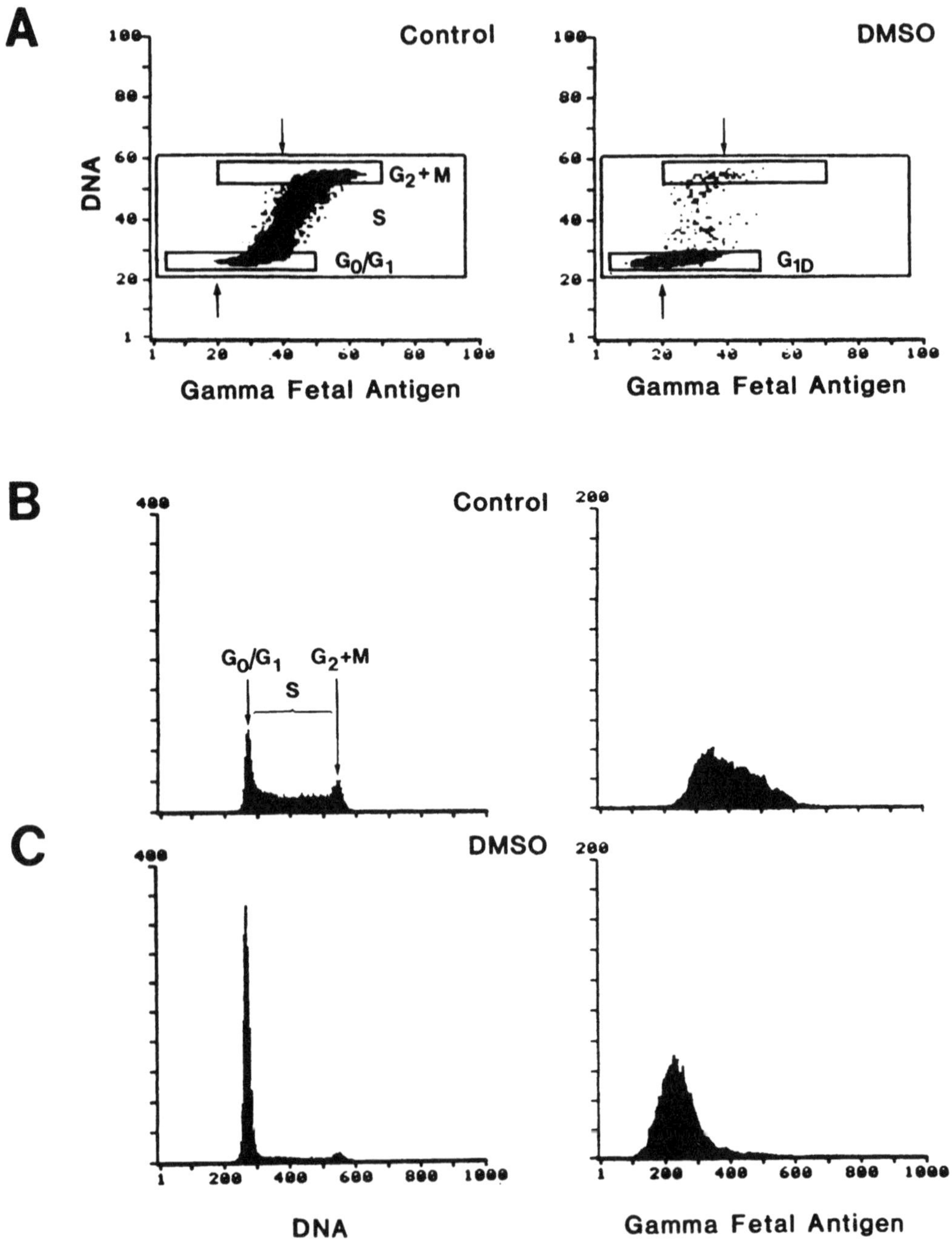
A
Control
DMSO
DNA
G2+M
S
G0/G1
G1D
Gamma Fetal Antigen
Gamma Fetal Antigen
B
Control
G0/G1
S
G2+M
C
DMSO
DNA
Gamma Fetal Antigen

Figure 2. GFA expression as a function of DNA content in control and DMSO-differentiated FL cells. (A) Scattergrams of control and DMSO-treated FL cells. Each point represents the relative red (PI) fluorescence specific for DNA content and green (FITC) fluorescence which is proportional to GFA content of individual asynchronously-growing FL cells (control) or FL cells differentiated by treatment with 2% DMSO for 6 days (DMSO). The rectangular boxes denote the position of postmitotic G_0/G_1 or G_{1D} cells in control and DMSO-treated cultures, respectively, as well as the premitotic G_2+M compartment. The arrows indicate minimum GFA levels for G_0/G_1 and G_2+M control cells. Note that a portion of the G_{1D} and most of the G_2+M populations in DMSO-treated cells have GFA levels lower than the corresponding controls. (B) Single parameter DNA (left panel) and GFA (right panel) content histograms for control cells. All values are in arbitrary units. Note substantial S phase fraction in the control cell population indicating a high level of proliferation. (C) As in B, the DNA (left panel) and GFA (right panel) distributions for DMSO-treated cells are indicated. All histograms are normalized. Evident is the accumulation of DMSO-treated cells in G_{1D} (left) and the shift in GFA content (right) to lower values.

Acknowledgement

This research was supported by NIH grant no. CA-25285 awarded by the National Cancer Institute and American Cancer Society Special Institutional Grant no. 7.

References

1. Higgins, P.J., C. Tong, E. Borenfreund, R.S. Okin, A. Bendich. 1979. Br. J. Cancer 40, 253.

2. Tong, C., E.H. Stonehill, P.J. Higgins, A. Bendich. 1978. Eur. J. Cancer 14, 147.

3. Higgins, P.J., C. Tong, E. Borenfreund. 1981. Oncology 30, 340.

4. Higgins, P.J., E. Borenfreund. 1980. Biochim. Biophys. Acta 610, 174.

5. Traganos, F., P.J. Higgins, C. Bueti, Z. Darzynkiewicz, M.R. Melamed. 1984. J. Natl. Cancer Inst. 73, 205.

6. Orkin, S.H., F.I. Harosi, P. Leder. 1975. Proc. Natl. Acad. Sci. 72, 98.

7. Sharpless, T., F. Traganos, Z. Darzynkiewicz, M.R. Melamed. 1975. Acta Cytol. 19, 577.

8. Sharpless, T.K. 1979. In: Flow Cytometry and Sorting (M.R. Melamed, P.F. Mullaney and M.L. Mendelsohn, eds.). Wiley, p.359.

PREGNANCY ASSOCIATED α_2-GLYCOPROTEIN ANALOGUES IN MONKEY AND DOG

D. Szabó
Institute of Biochemistry, University Medical School Pécs, Hungary

P. Gőcze, G. Than
Department of Obstetrics and Gynecology, University Medical School Pécs, Hungary

Introduction

The isolation of human pregnancy associated α_2-glycoprotein (α_2PAG), the preparation of monospecific antiserum, and normal and pathological serum levels of the protein have been published earlier by us simultaneously with others (1). To make further progress it was necessary to find analogous animal serum proteins. Monospecific anti-human α_2PAG serum, developed in rabbit cross reacted only with sera of monkies, while other mammalian sera did not react unequivocally positively, a finding which is in accordance with that of Bohn and Lin (2,3). Lin's immunological studies showed, that except for the chimpanzee, α_2PAG-s of all monkey species investigated are only partially identical to that of human and therefore they cannot be measured using anti-human α_2PAG serum. However, the isolation of α_2PAG analogue from serum of pregnant rhesus monkey made the quantitative determination possible and has been reported by us in 1977 (4). A wide-ranging study of α_2PAG analogues in different monkey species was performed by Lin and gave the possibility of further comparative investigations. As experiments with monkies are very expensive it was necessary to find other analogous laboratory animals.

Hau constructed a murine model for the study of all types of pregnancy proteins (5). von Schoultz and Stigbrand demonstrated a little cross reaction between human α_2PAG antiserum and serum of pregnant beagle dog, and prepared anti-dog α_2PAG serum at first (6). The aim of the present work was to isolate dog α_2PAG analogue and to develop a method for its measurements and to compare the molecular structure of these analogues.

Pregnancy Proteins in Animals

Materials and Methods

Blood was collected from the foreleg vein of german sheepdogs and stored at -20°C until use. Anti-dog α_2PAG rabbit serum was supplied by H. Bohn (Behringwerke, Marburg).

Gel filtration was carried out on Sephadex G-200 column (2.5 x 90 cm), equilibrated with 30 mM phosphate buffer pH 7.4. Protein content was recorded using ISCO UA-5 absorbance monitor. Fractions were tested immunologically by the Laurell method using anti-dog α_2PAG serum (20 ul/ml agarose). Ion exchange chromatography was performed on DEAE-Sepharose column (1.5 x 30 cm), equilibrated with the above buffer. Elution was carried out using a linear NaCl gradient 0.0-0.3 or 0.1-0.2 M.

Preparative isoelectric focusing: the sample was injected into a prefocused ISCO Model 212 electrofocusing column filled with sucrose gradient 400 g/l-50 g/l containing 1.5% Pharmalyte pH 4-6.5 . After 48-hours focusing (1000 V, 0.5 mA) the fractions were collected.

The protein content of pure samples was determined by the Lowry's method.

Criteria of homogenity: samples were checked by electrophoresis on polyacrylamide gel or on polyacrylamide gradient gel (PAA 2/16 Pharmacia), and by electrofocusing on analytical rods.

Molecular weight determinations were carried out by gel filtration on Sephadex G-200 or Sepharose CL-6B column using Pharmacia Calibration Kit. Subunit structure was studied by SDS electrophoresis on polyacrylamide rods. Sample preparation: 1% SDS, 1% mercaptoethanol, 100°C, and 5 min.

Results and Discussion

Isolation of dog α_2PAG analogue: At the gel filtration, fractions containing α_2PAG were localized in the first peak (see Fig. 1). The active fractions were then combined and subjected to ion exchange chromatography. The fractions shaded in Fig. 2 were collected and dialyzed against 10 mM phosphate buffer

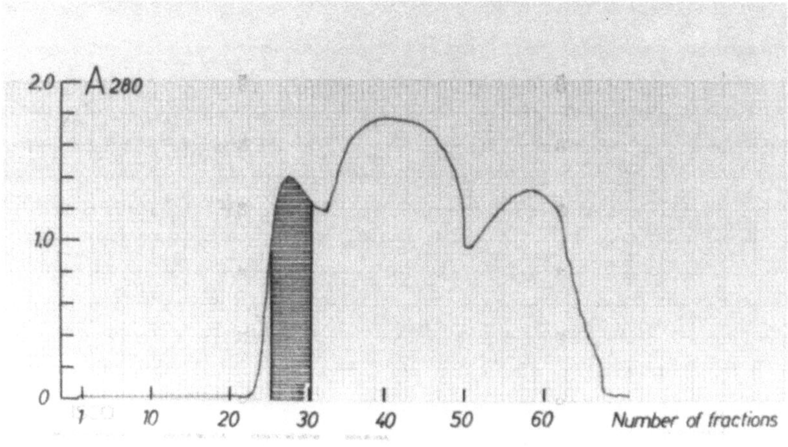

Fig. 1 : Gel filtration on Sephadex G-200 column. Fractions marked with lines contain α_2PAG

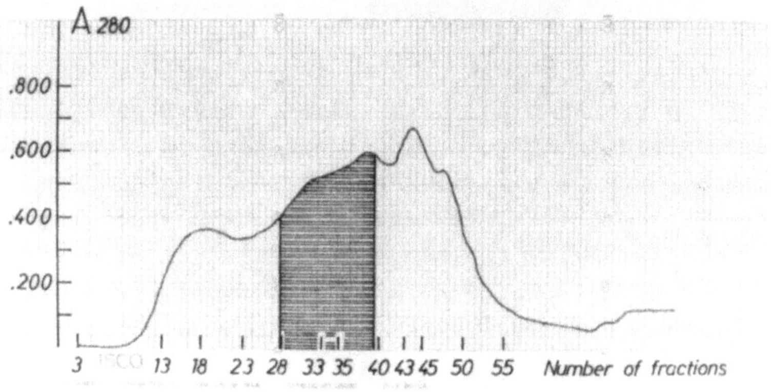

Fig. 2 : Ion exchange chromatography on DEAE-Sepharose column

pH 7.4 and the ion exchange chromatography step was repeated using a less steep gradient. Most active fractions were combined, concentrated and dialyzed as above, and the sample injected into the electrofocusing column. α_2PAG was focused to the band corresponding a pH range of 4.9 to 5.4 , as can be seen in Fig. 3 .

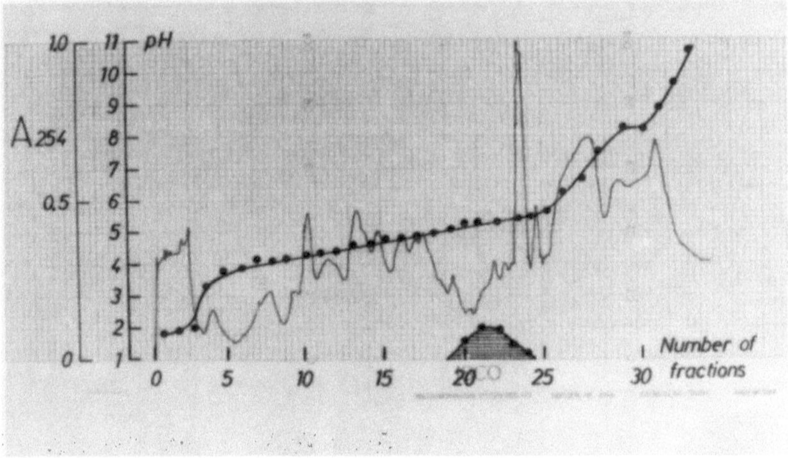

Fig. 3 : Isoelectric focusing on ISCO Model 212

The protein content of the dialyzed and concentrated sample was determined by Lowry's method and used as standard in the electro-immunoassay. On the basis of the above detailed purification scheme a similar molecular character may be supposed as for those isolated from human and monkey.

Purity control: The protein seemed to be homogenous as only a single band was seen on polyacrylamide electrophoresis on 7.5% gel and on the analytical electrofocusing. The isoelectric point was found to be pH 5.11 . α_2PAG seemed to be homogenous also on immunoelectrophoresis. However, applying a special gradient gel (PAA 2/16 Pharmacia) two bands, localized closely to each other could be detected (see Fig. 4). It successed to separate the two bands on micropreparative scale and they could be examined individually (Fig. 5). The two variation of α_2PAG did not show any difference in their immuno-reactivity and in their isoelectric point. Due to the known molecular sieving effect of the gradient gels, only small difference in their molecular weights can be assumed.

Determination of molecular weight and subunit structure: the molecular weight of dog α_2PAG was found to be 420,000 daltons using the gel filtration method and Pharmacia standards. Following effective reduction, only one band could be observed on

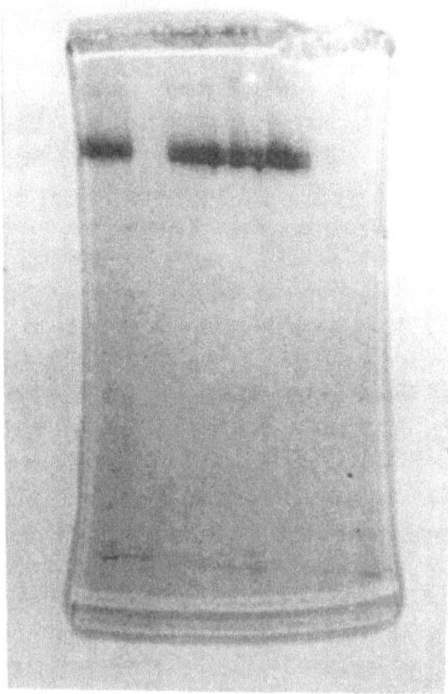

Fig. 4 : Polyacrylamide gradient gel electrophoresis on Pharmacia PAA 2/16 plate

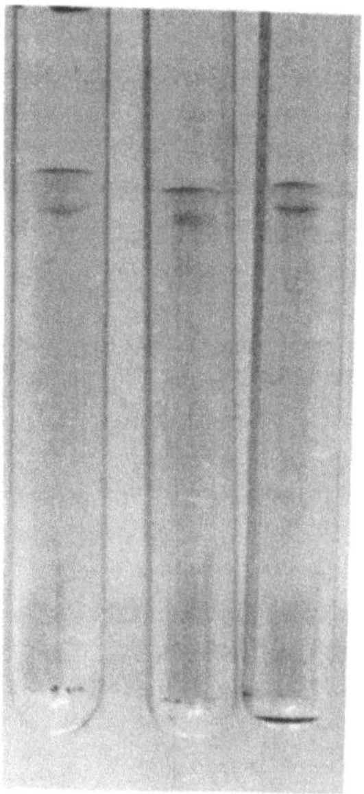

Fig. 5 : Variation of dog α_2PAG. From left to right: 1. Isolated variation I. 2. The two variations in mixed state. 3. Isolated variation II.

the SDS electrophoresis corresponding a molecular weight of 195,000. Both of the above mentioned variations of the α_2PAG gave similar results, thus the minimal difference in molecular weight was not detectable with this technique.

Stimson made a comprehensive study concerning to the molecular weight of human α_2PAG and established a range of 365,000 to 389,000 depending on the type of ultracentrifugal determination (7). He described further a "deviation" from standard behaviour on gel filtration, namely the protein showed higher values measured by this technique. Human α_2PAG is composed of two identical subunits as in the case of dog analogue showed above. Similar data were found investigating the molecular weight and subunit composition of rhesus monkey α_2PAG in our earlier work. These data indicate interesting physico-chemical similarities between human, monkey and dog α_2PAG and rise the possibility to consider them homologous proteins. These similarities are supported by similar electrophoretic mobility and immunological cross reactions. In contrast to this, the murine analogue seems to be homogenous having a molecular weight of 150,000 , which does not differ substantially from the subunit molecular weight of the human, monkey and dog analogues.

Quantitative determinations of α_2PAG analogue in dog sera: Using isolated α_2PAG as standard, a method has been developed with Laurell's electrophoresis in antiserum containing gel. The range of the quantitation lies between 13.6-218 mg/l , and the sensitivity is about 5 mg/l.

The α_2PAG analogue has been measured in the sera of nonpregnant animals and found to be present at relatively high levels, 218 $\pm$107 mg/l in males and 305 $\pm$185 mg/l in females, respectively. These values are essentially higher than the normal levels of men, especially in the case of male animals. Similar results were found in rhesus monkey but the number of animals was too low to make exact evaluation.

To study the α_2PAG level in maternal serum during pregnancy, blood samples were taken from 5 dogs before and during pregnancy every third day. The results of this experiment was summarized in Fig. 6. The values, increasing gradually showed a maximum at 6th week (1460 $\pm$653 mg/l), then decreased significantly. This tendency continoued after delivery and reached the non-pregnant level at about 90th day. It is worth to mention that α_2PAG concentration was always higher during the period of rut than before it. The maximum value of α_2PAG concentration is observed at the end of the second trimester which shows some parallelism with data obtained on mice as measured by Hau (5). In contrast, in human the concentration reaches a maximum at the third trimester. A further difference is that the ratio of pregnant/nonpregnant α_2PAG levels is also much higher in human, in other words the α_2PAG level rises more significantly. Our few observation showed also elevation in rhesus monkey, in contrast to Lin's data (3).

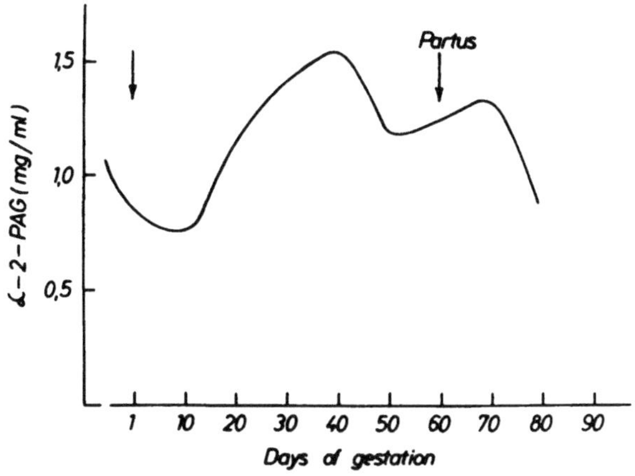

Fig. 6 : Serum level of dog α_2PAG during pregnancy

Conclusion:

1.) α_2PAG analogues have been isolated from sera of pregnant rhesus monkey and dog.

2.) Proteins were characterized physico-chemically and established that their molecular weight, subunit structure, isoelectric point and electrophoretic mobility were similar to the human α_2PAG.

3.) Quantitative assays were developed, and serum levels of α_2PAG analogue in control and pregnant animals were also presented. Dog seemed to be an useful model to study the physiological role of α_2PAG.

References

1. Than, G., I. Csaba, D. Szabó. 1974. Lancet ii, 1578.

2. Bohn, H., H. Ronneberger. 1973. Arch. Gynecol. 215, 277.

3. Lin, T.M., S.P. Halbert, R. Plasencia. 1976. Clin.exp.Immun. 26, 609.

4. Szabó, D., G. Than, I. Csaba. 1977. IRCS Med.Sci. 5, 476.

5. Hau, J. 1982. In: Pregnancy Proteins (J.G. Grudzinskas, B. Teisner, M. Seppala, eds.). Academic Press Sydney, p. 435.

6. von Schoultz, B., T. Stigbrand. 1976. Acta Endocrin. 81, 379.

7. Stimson, W.H., D.M. Farquharson. 1978. Int.J.Biochem. 9, 839.

TISSUE AND CELLULAR SITES OF PREGNANCY ASSOCIATED ALPHA$_2$ GLYCO-PROTEIN (α_2PAG) SYNTHESIS IN PREGNANT RATS

E.J. Sarcione, D. Delluomo, M. Zloty and W. Biddle

Department of Clinical Pharmacology and Therapeutics, Roswell Park Memorial Institute, New York State Department of Health and the Physiology Department, Roswell Park Graduate Division, State University of New York at Buffalo, New York, 14263, U.S.A.

Introduction

It is well documented that elevated serum α_2PAG levels are found in pregnant and estrogen treated women and in patients with diverse types of neoplasms including breast cancer (1,2). Although less is known regarding the organ and cellular origin of α_2PAG during pregnancy and in breast cancer patients, several reported synthetic sites include a) peripheral blood leucocytes and monocytes from pregnant women, and normal liver (3), b) cultured malignant monocytoid and lymphoid cell lines (4) and c) primary breast cancer tissue and cultured cell lines (5). In a previous report, a rat α_2PAG analogous to its human counterpart was demonstrated in pregnant, breast cancer-bearing and estrogen treated animals (6). Immunological cross-reaction and antigenic identity between α_2PAG analogues detected independently in rats (6,7) and in mice (8,9) has been demonstrated (10). Furthermore, immunologic cross-reaction and partial antigenic identity between murine and human α_2PAG's was reported (11).

In the present study, this rat model system was used to investigate: a) organ and cellular sites of α_2PAG synthesis in pregnant and fetal animals, and b) stimulatory mechanisms involved in cellular α_2PAG synthesis.

Pregnancy Proteins in Animals

Results

Figure 1 shows that all maternal and fetal tissues and cells studied incorporated appreciable amounts of ^{14}C-leucine radioactivity into immunochemically isolated α_2PAG and acid precipitated total cell protein (open) when compared to puromycin inhibited controls (solid) *in vitro*. A comparison based on weight of tissue incubated indicated that maternal and fetal liver were the most active producers of α_2PAG. On the other hand, a comparison based on α_2PAG as a percent of total cell protein radioactivity revealed that maternal peripheral blood leucocytes and fetal liver synthesized more of this pregnancy protein. These results suggest either that a) all maternal and fetal tissues and cells studied synthesized α_2PAG or more likely that b) α_2PAG synthesis observed in all tissues

Fig. 1.

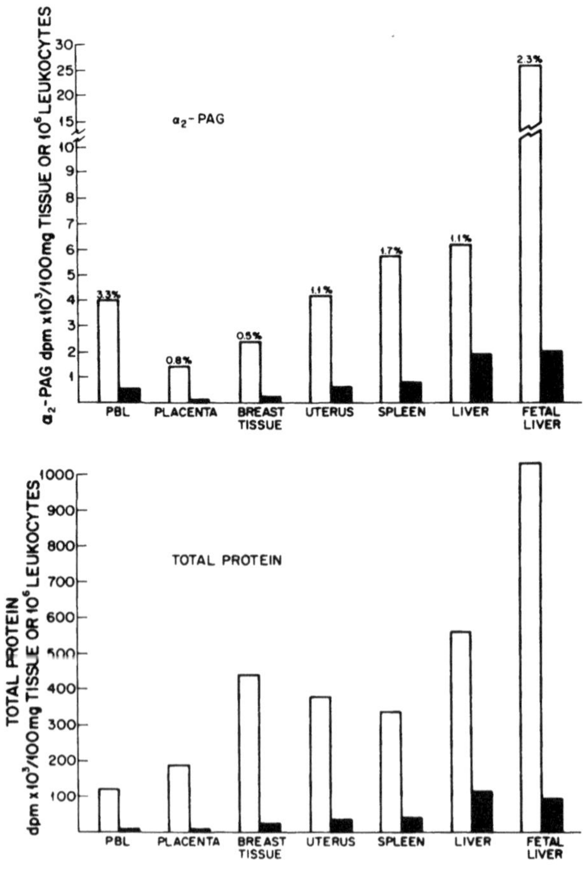

was due to peripheral blood leucocytes which exit the circulation and infiltrate these tissues. Taken together the previously reported and current evidence that both breast cancer cells and fetal liver synthesize α_2PAG suggest the intriguing possibility that this pregnancy protein should also be classified as an "oncofetal" protein.

To answer the question, is cellular α_2PAG synthesis specifically stimulated during pregnancy?, we examined protein synthesis by peripheral blood mononuclear cells obtained by Ficoll gradient centrifugation of heparinized whole blood. Figure 2 shows that peripheral blood mononuclear leucocytes obtained from pregnant rats synthesized five times more α_2PAG per cell during 24 hours _in vitro_ than did these same mononuclear cells obtained from normal adult females. During the same time, however, mononuclear leucocytes from both pregnant and normal animals synthesize approximately equal amounts of total cell protein. This indicates that α_2PAG synthesis by peripheral blood mononuclear leucocytes was specifically stimulated during pregnancy, and that the results obtained were not a reflection of increased total cell protein synthesis.

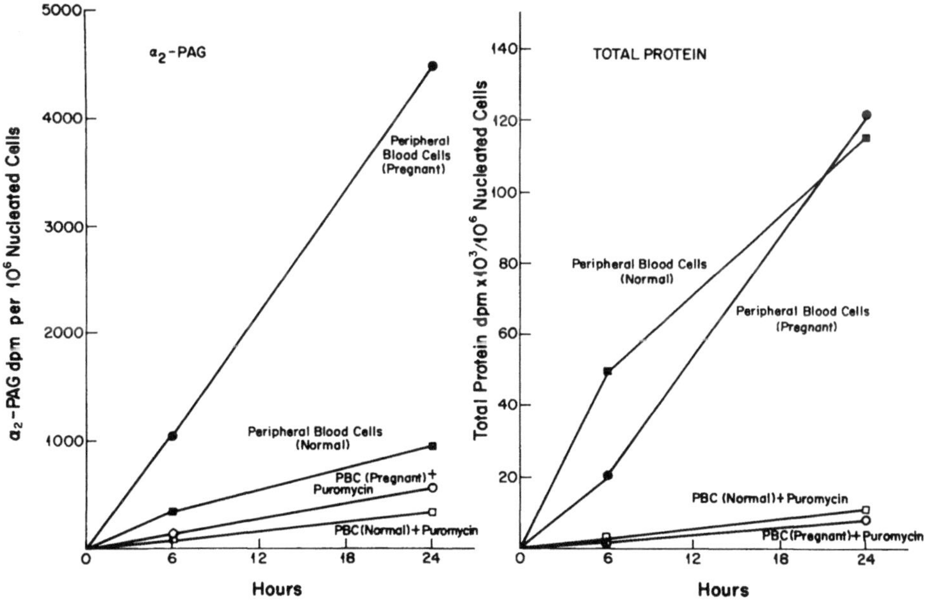

Fig. 2.

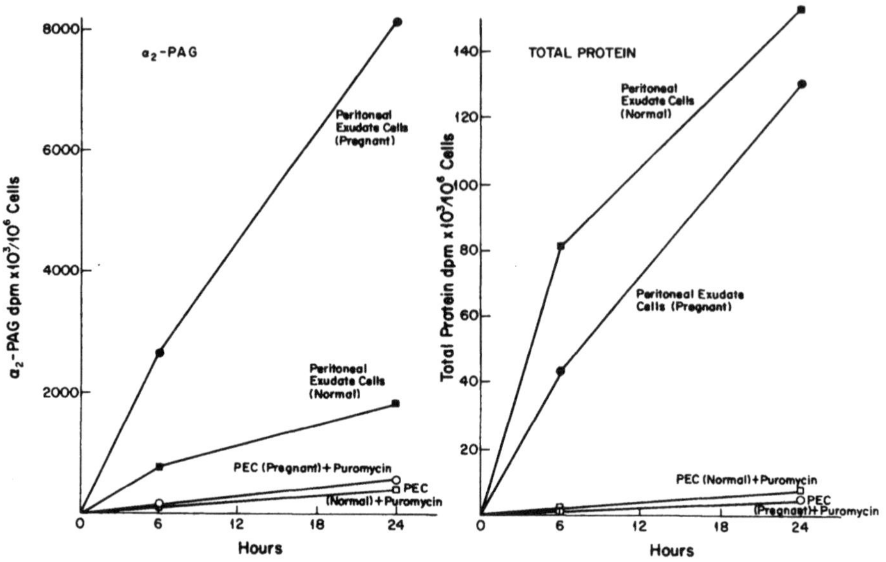

Fig. 3.

To investigate the involvement of tissue macrophages in α_2PAG production, peritoneal exudate cells were harvested 3 days after intraperitoneal administration of paraffin oil, then these cells were further separated into plastic adherent macrophages and non-adherent cells. A direct comparison in Figure 3 shows that peritoneal exudate macrophages from pregnant rats synthesized five times more α_2PAG during 24 hours than did these same cells obtained from normal adult females. When compared based on percent of total cell protein radioactivity, tissue macrophages from pregnant rats produced four times more α_2PAG than did macrophages from normal animals. These results indicate that α_2PAG synthesis by tissue macrophages was also specifically stimulated during pregnancy.

Figure 4 shows that double immunodiffusion of radiolabeled protein extracts, containing added carrier rat pregnancy sera, against rabbit anti-rat α_2PAG sera resulted in a single continuous immuno precipitin line shown on the left, and produced a positive superimposable autoradiograph shown on the right for maternal liver (L), breast tissue (BT), spleen (SPL), uterus (UT), placenta (PL), and fetal liver (FL) (top two), and also for peripheral blood mononuclear leucocytes (PBL) and peritoneal exudate macrophages (PEC)

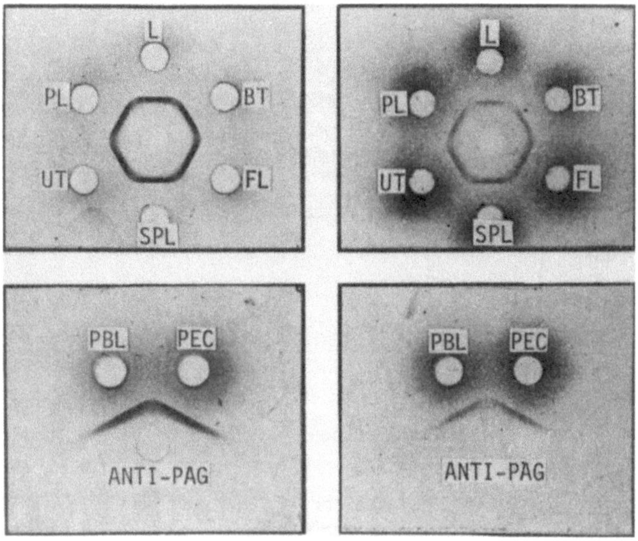

Fig. 4.

obtained from pregnant rats (lower two). These observations confirm the monospecificity of the α_2PAG antisera used throughout these studies, and demonstrate α_2PAG synthesis by these maternal and fetal tissues and cells by a method independent of the immunochemical isolation procedure.

To establish which types of peripheral blood mononuclear leucocytes synthesize α_2PAG during pregnancy, the mononuclear cell fraction obtained from whole blood by Ficoll gradient centrifugation was again further separated into plastic adherent "monocytes" and non-adherent "lymphocytes". Although these are operational definitions of these cells, microscopic examination revealed that greater than 99% of the adherent cells were morphologically similar to monocytes and greater than 97% of the non-adherent cells were similar to lymphocytes. Figure 5 shows that a) both maternal and fetal peripheral blood "monocytes" and "lymphocytes" synthesized α_2PAG and b) a per cell comparison indicates that "monocytes" synthesized 3-5 times more α_2PAG than "lymphocytes". It should be emphasized, however, that since whole blood contains approximately ten times more lymphocytes than monocytes, it's likely that the total lymphocyte population in peripheral blood produces more α_2PAG than do the total number of monocytes. Furthermore, the evidence that both fetal monocytes and lymphocytes produce more α_2PAG per cell than

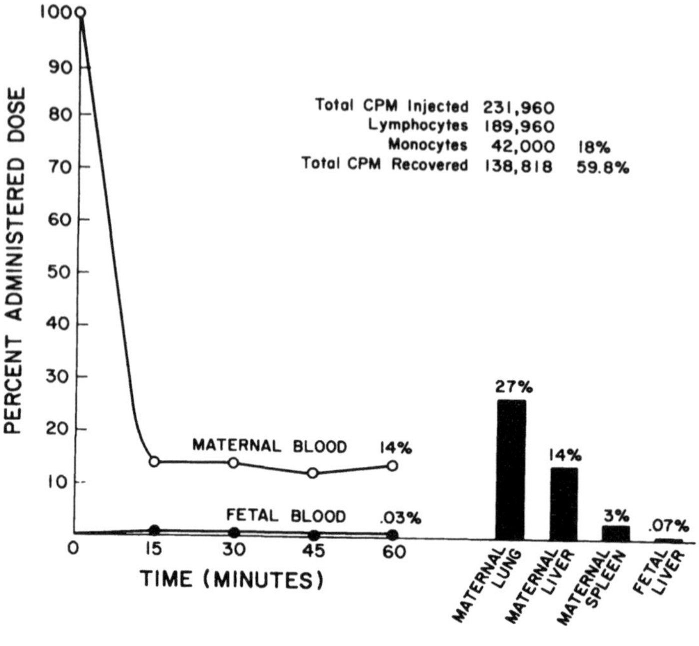

Fig. 5.

their maternal counterparts, supports the conclusion that fetal mononuclear leucocytes per se rather than translocated maternal cells are responsible for the α_2PAG synthesis observed in fetal blood.

Since some maternal rat (12) and human (13,14) lymphocytes have been reported to enter the fetal circulation, we decided to evaluate the magnitude of this transplacental passage of maternal leucocytes to the fetus. This was done by radiolabeling maternal peripheral blood mononuclear leucocytes with ^{14}C-leucine incorporation into cellular protein in vitro, then infusing these washed cells intravenously into pregnant near-term rats. Figure 6 shows that, as expected, these radiolabeled leucocytes disappeared rapidly from the maternal circulation, and a large proportion of the total administered radioactivity accumulated in lung, liver and spleen. During this same time, however, only 0.03% of the total administered maternal leucocyte radioactivity appeared in the fetal circulation and did not increase over time, and only 0.07% of this cellular radioactivity accumulated in fetal liver. Similar results

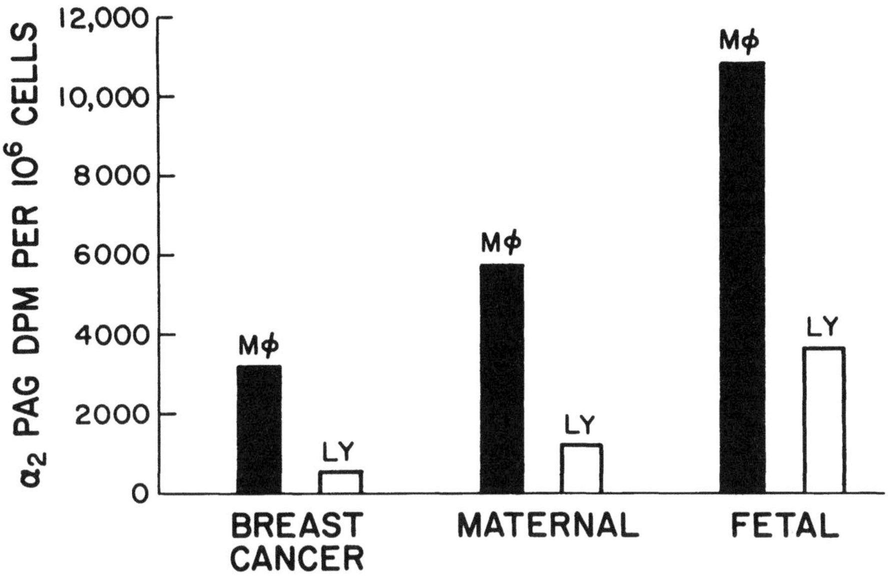

Fig. 6.

(not shown) were obtained when this study was extended to 24 hours. We conclude that relatively few maternal monocytes and lymphocytes normally appear in the fetal circulation and therefore, it's likely that fetal monocytes and lymphocytes *per se* synthesized α_2PAG.

Conclusions

A) Rat maternal peripheral blood monocytes and lymphocytes and tissue macrophages are cellular sites of α_2PAG synthesis during pregnancy.

B) α_2PAG synthesis by maternal peripheral blood monocytes and lymphocytes and tissue macrophages is specifically stimulated above normal cell levels during pregnancy.

C) The humoral-endocrine mechanisms responsible for specific stimulation of α_2PAG synthesis in maternal mononuclear cells *in vivo* are currently under investigation.

D) It is likely that most of the α_2PAG synthesis observed in non-circulating maternal tissues can be attributed to stimulated monocytes and lymphocytes which infiltrate all tissues.

E) Rat fetal peripheral blood monocytes and lymphocytes per se apparently also synthesize α_2PAG.

F) Specific stimulation of α_2PAG synthesis observed in these fetal mononuclear cells is assumed to be effected by the same humoral-endocrine mechanisms responsible for stimulating α_2PAG synthesis in maternal mononuclear cells.

G) The evidence obtained in this study indicates that rat α_2PAG is an appropriate animal model system to investigate the biologic function of this ubiquitous pregnancy protein.

References

1. Thomson, A.W. and C.H.W. Horne. 1980. Biological and clinical significance of pregnancy-associated α_2-glycoprotein (α_2-PAG)- A review. Invest. Cell Pathol. 3. 295-309.

2. Horne, C.H.W., A.W. Thomson and S.S. Armstrong. 1982. In: Immunology of Human Placental Proteins. Placenta (Suppl. 4). Praeger Publishers. pp. 33-50.

3. Stimson, W.H., D.M. Farquharson, A. Shepherd and J.M. Anderson. 1979. Studies on the synthesis of pregnancy associated α_2-glycoprotein by the liver, placenta and peripheral blood leucocyte populations. J. Clin. Lab. Immunol. 2. 235-238.

4. Lundgren, E., M.G. Damber, G. Roos, B. von Schoultz, T. Stigbrand, K. Nilsson, and J.J. Alexander. 1979. Cell lines with spontaneous secretion of pregnancy-associated α_2-globulin. Int. J. Cancer 24. 45-48.

5. Sarcione, E.J., D. Delluomo and M. Zloty. 1983. Pregnancy-associated alpha$_2$ glycoprotein (α_2PAG) synthesis by human breast cancer tissue and cultured cell lines. Int. J. Cancer 31. 143-147.

6. Sarcione, E.J. and D.S. Delluomo. 1981. Demonstration of pregnancy-associated α_2-glycoprotein (α_2-PAG) in pregnant, estrogen-treated and breast cancer-bearing rats. Oncodevelopmental Biol. and Med. 2. 345-354.

7. Porstmann, T. and H. Schmechta. 1980. Isolation of a pregnancy-associated globulin (PAG) from the rat and preparation of a specific antiserum (anti-PAG). Acta Biologica Medica Germanica 39. 449-487.

8. Hau, J., P.E. Svendsen, B. Teisner and S.E. Svehag. 1978. Studies of pregnancy-associated murine serum proteins. J. Reprod. and Fertility 54. 239-243.

9. Waites, G.T. and S.C. Bell. 1984. Identification of a murine pregnancy-associated serum protein (α_1PAP) as a female-specific acute-phase reactant. J. Reprod. and Fertility 70. paper 797.

10. Hau, J. and T. Porstmann. 1984. Characterization of the analogues to human pregnancy-associated α_2-glycoprotein (α_2PAG, PZP) isolated in the mouse and rat. Lab. Animals 18. 344-348.

11. Hau, J., J.G. Westergaard, P. Svendsen, A. Bach, and B. Teisner. 1981. Comparison of pregnancy-associated murine protein-1 and human pregnancy zone protein. J. Reproductive Immunol. 3. 341-349.

12. Beer, A.E., R.E. Billingham and S.L. Yang. 1972. Maternally induced transplantation immunity, tolerance and runt disease in rats. J. Experimental Med. 135. 808-826.

13. Schroder, J. 1974. Passage of leukocytes from mother to fetus. Scand. J. Immunol. 3. 369-373.

14. Field, E.J. and E.A. Caspary. 1971. Is maternal lymphocyte sensitisation passed to the child? Lancet ii. 337-342.

ISOLATION, QUANTITATION AND SOME PHYSICO-CHEMICAL AND BIOLOGICAL PROPERTIES OF A PREGNANCY-ASSOCIATED GLYCOPROTEIN (PAG) IN THE RAT

T. Porstmann, A.Lukowsky, G.M. Müller, H.D. Volk
Institute of Medical Immunology,

B. Porstmann, E. Nugel
Institute of Pathological and Clinical Biochemistry,

R. Schade
Institute of Pharmacology and Toxicology, Humboldt-University
Medical School (Charité), 1040 Berlin-GDR, Schumannstr. 20/21

H. Schmechta, B. Kopetz
Institute of Forensic Medicine, Military Medical Academy,
1242 Bad Saarow, GDR.

Introduction

Numerous factors are considered to be responsible for fetal maintenance throughout gestation despite the expression of paternal histocompatibility antigens on fetal cells (Rev. 1). Research of the biological function, particularly the investigation of possible immunosuppressive properties of pregnancy-associated and specific proteins play a great part to detect an altered immune response of the mother during pregnancy. One of these proteins in humans, the pregnancy-associated α_2-glycoprotein, discovered by Olivelli and Rugieri in 1958 (2) is now well characterized with respect to physico-chemical data but contradictory results exist concerning the kind of effect and also the efficiency of the immunosuppression (Rev.3). With the exeption of a heterotopic heart transplant study in the mouse (4) all results of an immunosuppression by human α_2-PAG have been received from in-vitro experiments. Research into in-vivo effects are restricted to animal models, however the strong immunogenity of α_2-PAG requires to search for its analogue in the respective species.

Pregnancy Proteins in Animals

1. Isolation of Rat PAG

For the isolation pooled serum from pregnant Wistar rats and male PAG-negative animals picked by double-dimension immunodiffusion technique was subjected to fractionated precipitation with ammonium sulphate (final conc. 1.4 and 2.2 mol/l). The precipitate of the latter was equilibrated against PBS, gel filtrated on Sephadex G-200 (Pharmacia, Uppsala) and the rising part of the second protein peak of the male animal's pool serum was used for the immunization of rabbits, while the corresponding part of the pool serum of the pregnant rats was further purified by affinity chromatography on Con A-Sepharose (Pharmacia, Uppsala) with a linear gradient of α-D methyl-mannoside (0.01 to 0.12 mol/l). The PAG-containing fraction, eluted from 0.04 to 0.1 mol/l α-D methyl-mannoside, was subjected to ion exchange chromatography on DEAE-Sephacel (Pharmacia, Uppsala). Applying a linear ionic strenghth gradient of 0.015 to 0.15 mol/l NaCl in 0.01 mol/l phosphate buffer, pH 6.8 PAG was eluted from 0.05 to 0.12 mol/l NaCl and shows a purity greater than 75 % (5).

Final purification of PAG was performed by negative immunosorbent chromatography on the insolubilized IgG fraction of the immunized rabbits. In the procedure, as summarized in fig. 1, the yield of purified PAG was 5 to 6 mg per 100 ml serum.

For reasons of immuno-chemical quantitation of PAG rabbits and guinea pigs were immunized with the purified protein and the developed antisera, purified by negative affinity chromatography on serum of male rats insolubilized to Sepharose.

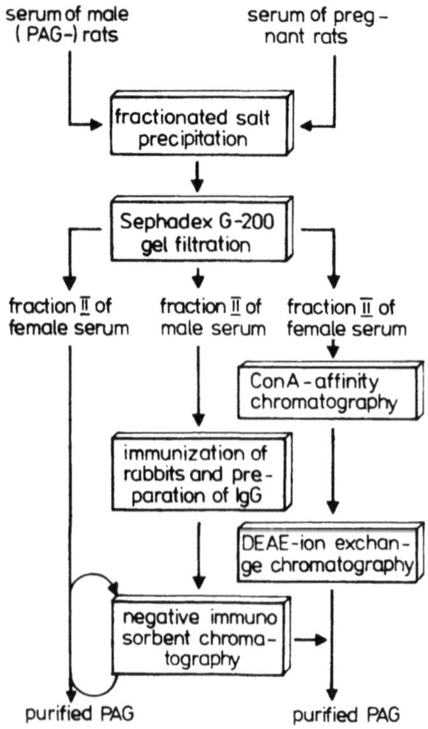

Figure 1

2. Physico-Chemical Properties of Rat PAG

As α_1-globulin PAG migrates in agarose gel electrophoresis (Fig.2) but shows identical electrophoretic mobility like IgG in the polyacrylamide gel electrophoresis and is covered by the IgG band if untreated serum is separated.

Purified PAG shows microheterogenity concerning its isoelectric point with a major band at pH 4.4. Variations in the carbohydrate components and in the content of sialic acid may be responsible for this fact because allelic variants of the protein as a possible reason for this heterogenity were excluded by immunochemical analysis.

The molecular weight varies between 147 and 175 kD depending on the method used. The reason for overestimating the molecular weight by 15-20 % by analytical gel filtration (Tab. 1), may be due to the presence of sialic acid (6).

Heating the PAG molecule up splits it into 2 subunits in the presence of mercaptoethanol (7).

PAG is a glycoprotein reacting with different lectins, which is indicative of the presence of terminal mannose, glucose and galactosamine (Tab. 1).

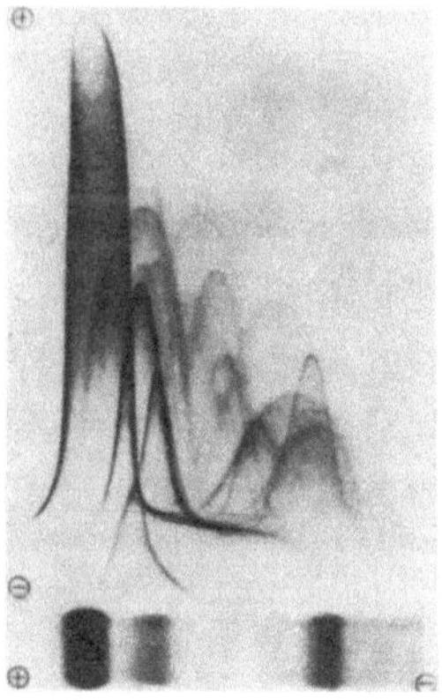

Figure 2
Crossed-immunoelectrophoresis carried out as intermediate gel technique with monospecific anti-PAG from the rabbit

Electrophoretic mobility		
Agarose gel elph.	α_1-region	
Elph. migr. vel.	5.16×10^{-5} $cm^2 v^{-1} sec^{-1}$	
PAAG elph.	α_2-region	
Rel. mob.	R_f 0.34 (5% PAAG)	
Isoelectric point		
Prep. IEF gradient 3.5-10	pH 4.4	
Anal. IEF gradient 4-6.5	pH 4.4-4.9	
Molecular weight		
Gel filtration	175 000	
Ultracentrifugation	147 000	$S^0_{20,w} = 6.74 \pm 0.03$ S
SDS PAAG elph	150 000	
SDS PAGE with mercapt red.	80 000	
Lectin binding		
Lentil Lectin	++ Mannose, Glucose	
Concanavalin A	++ Mannose	
Helix pomatia Lectin	++ Galactosamin	
Extinction coefficient	$E^{1\%}_{280} = 10.48 \pm 0.5$	

Table 1

3. Quantification of Rat PAG

Due to a lack of sensitivity using immunodiffusion techniques for the quantitation of rat PAG a highly sensitive indirect two-site binding enzyme immunoassay (EIA) was developed and used together with an EIA for human α_2-PAG to study cross-reactivity between the both proteins (Fig.3).

- Solid phase coating with 20 mg/l IgG in carbonate buffer at 20°C for 24 h
- Sample incubation at 37° C for 2 h
- Incubation of 20 mg/l sandwich antibody at 37° C for 2 h
- Incubation of 1 mg/l conjugate at 37°C for 2 h
- Enzyme reaction with H_2O_2 and ABTS[R] at 20°C for 30 min
- BF-separation: 3 times washing with PBS-Tween
- Diluent: PBS 0,05% Tween 5 % sheep or horse serum
- Reaction volume: 200 µl

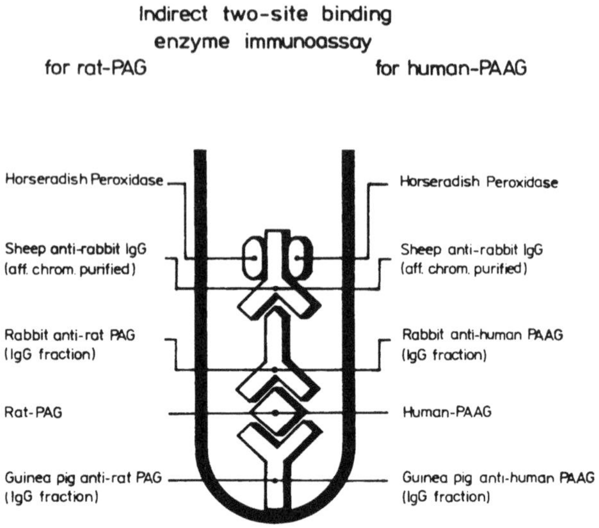

Principle of the two-site binding EIA for human α_2-PAG and rat PAG

The detection limit for rat PAG and human α_2-PAG respectively was 2 µg/l, the working range of the assays was 5 - 200 µg/l.
The identical slope of the dose-response curves received with purified rat PAG and with native sera varying in their PAG content exclude a denaturation of the protein by the preparation procedure and enable its use as standard material.

By means of the EIA PAG was detected in all female and male rats of various strains. Male rats with PAG serum concentrations in the µg-range per litre gave a level a thousand times lower than females (8).

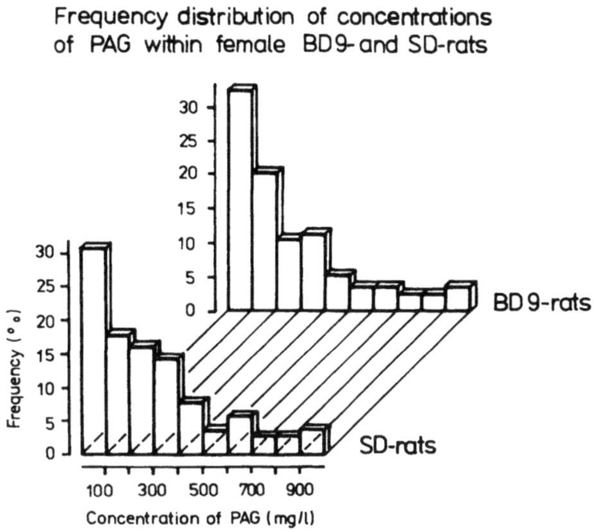

Figure 4

The examination of 200 male SD and BD IX rats, 18-20 weeks old, revealed a strain-dependent distribution of the PAG concentration. The geometric mean value of PAG concentrations was higher by 2.7 fold in BD IX rats with 164 µg/l as compared to SD rats having a geometric mean value of 60 µg/l. Investigating the same number of female animals, these differences were not found between the two strains (Fig.4 and 5).

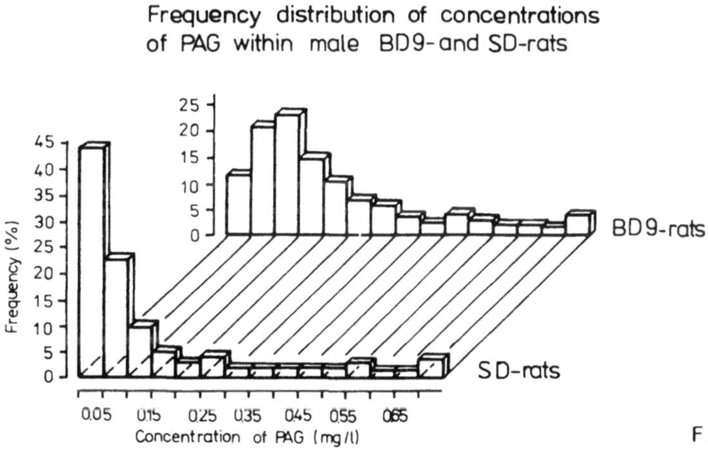

Figure 5

PAG serum levels are approximately log.-normally distributed. Inbred and outbred strains show the same extent of variation with respect to their PAG concentrations.

4. Regulation of Rat PAG

PAG in its serum level is subject to control by sexual hormones. The application of oestrogen significantly raises the PAG serum level both with female and male animals depending on the dose. In contrast to this, in both sexes testosterone leads to a considerable drop in the PAG serum concentration also depending on the dose (Fig.6).

Moreover castration leads to an increase in PAG with males and a PAG decrease with females. The substitution of the castrates with androgen and oestrogen respectively restores the PAG level just as before castration (9). Administration of testosterone to 1-day old female Wistar rats and removal of gonads in neonatal male rats resulted in a significantly diminished increase of PAG in adult female and an enhanced increase in adult male rats after oestrogen treatment when compared to animals untreated during the neonatal life. This effect was no longer demonstrable if the treatment was performed on the 7th day after birth (10).

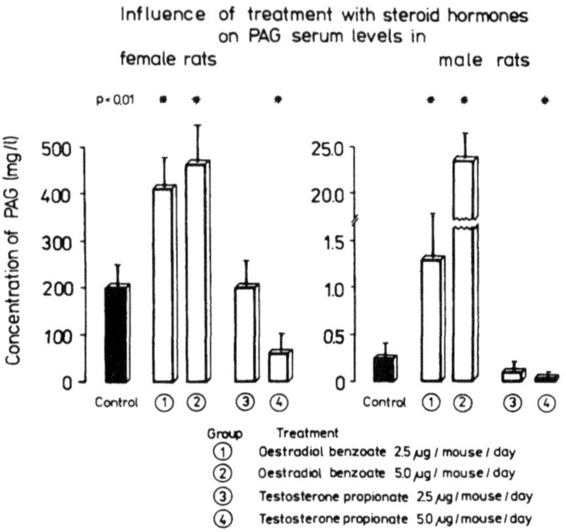

Figure 6
Hormones were administered subcutaneously over a period of 3 weeks

Owing to the sex hormone dependent regulation the PAG serum levels of two weeks old male and female Lewis 1U rats did not differ from each other. With the onset of sexual maturity male rats exhibit a drop in and female rats an increase in PAG serum levels (Fig. 7).

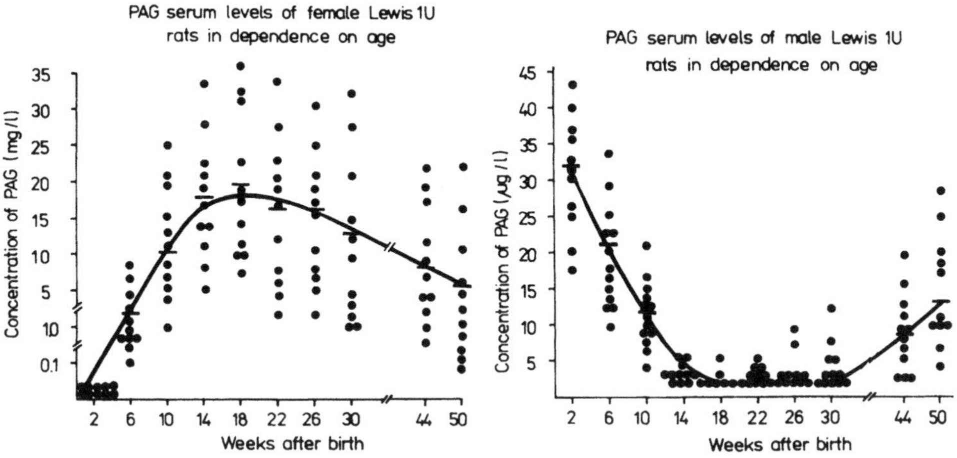

Figure 7: Note the starting decrease and increase of the PAG serum level in female and male rats at the end of their first year (for comparison see chapter 6).

A more complex regulation of rat PAG is evident from experiments, in which animals were treated with catecholamines and glucocorticoids. Single application of adrenalin and dexamethasone decreases the PAG serum level significantly depending on the dose, which is still more pronounced after repeated injections (Table 2).

PAG serum levels of female Wistar rats
after treatment with adrenalin and dexamethasone

Adrenalin s.c. (mg/kg)	PAG (mg/l)	Dexamethasone* (mg/kg) s.c.	PAG (mg/l)	Dexamethasone** (mg/kg) s.c.	PAG (mg/l)
0.9 % saline	388.1	0.9 % saline	388.1	0.9 % saline	295.0
0.2	305.4	0.07	352.5	0.04	252.4
0.4	231.4	0.20	227.5	0.20	160.4
0.8	221.3	0.60	235.0	0.80	107.0
1.6	169.3	1.80	221.8	3.20	61.1

geometric mean values of n=10 24 h after *single injection
**repeated injections (4×)

Table 2

Summarizing all the results it is assumed that the PAG serum level is directly under control of gonadal hormones and indirectly regulated by adrenal hormones. Furthermore the hypothalamo-pituitary-liver axis which modulates the density of receptors for oestrogen on hepatocytes during neonatal life (11) influences the response of the PAG levels to oestrogen in adult rats in this way. The preliminary conception of the regulation of rat PAG is given in figure 8.

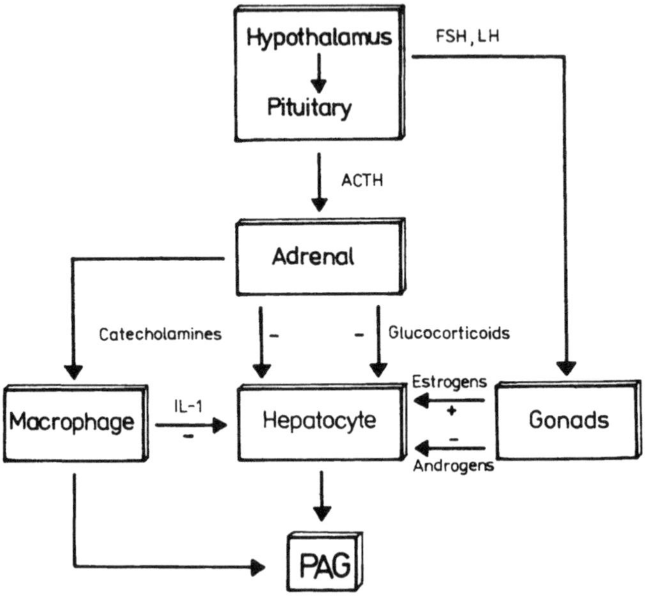

Through indirect fluorescence antibody technique using rabbit anti-PAG IgG and FITC labeled sheep-IgG directed against rabbit IgG, PAG was detected in the cytoplasma of hepatocytes from female rats (Fig. 9).

Figure 9a: Negative control with "non-immune" rabbit IgG

Figure 9b: Positive control with anti - PAG rabbit IgG

Figure 9a

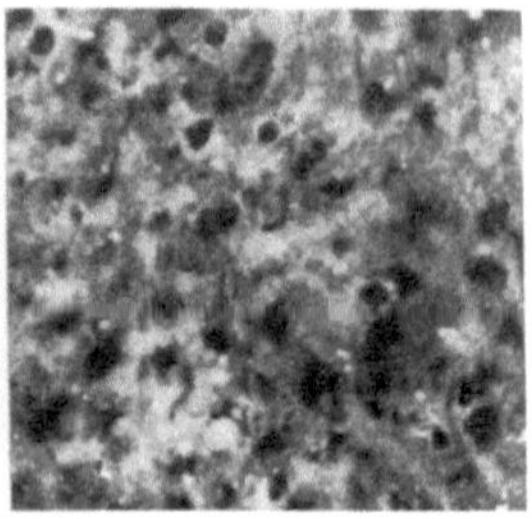

Figure 9b

5. Influence of Pregnancy, Tumour Growth and Inflammation on the Rat PAG Serum Level

An increase in the PAG serum concentration does not occur before the eighth day of pregnancy. During the second trimester of pregnancy the PAG serum level rises rapidly with maximum values being reached around the 15th day (11). The PAG level rise varies individually. However, it generally comes to factors 3 to 5. During the last trimester of pregnancy till the day of delivery the PAG serum level drops dramatically to its initial concentration before conception and even falls below this value during the period of lactation.

No differences were found apart from a significantly lower geometric mean value of PAG in syngenically mated rats as compared to allogenically mated once, at a very early stage of pregnancy still before the official increase in PAG and after delivery the reason of which remains obscure (Fig.10).

PAG serum levels during allogenic (♀R/Bln × ♂R/Bln) and syngenic (♀R/Bln × ♂BD IX/Bln) pregnancy

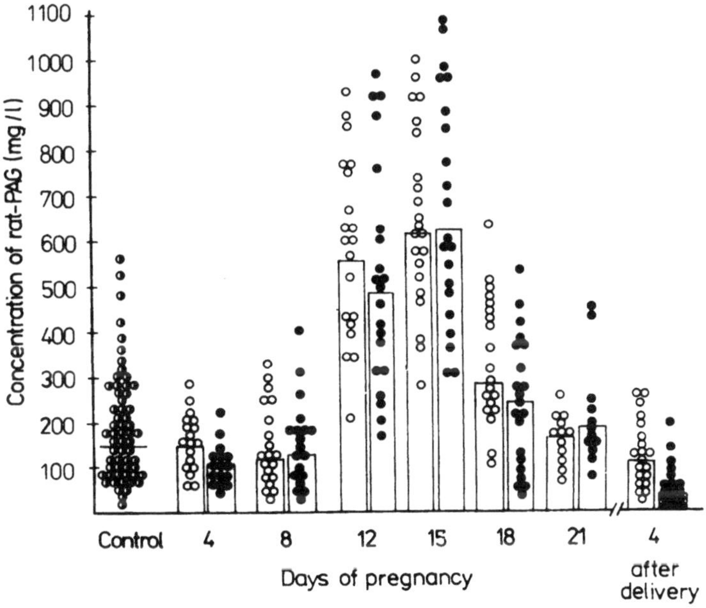

This experiment indicates that a stimulation of the immune system by alloantigens does not stimulate the PAG synthesis.

In the group of allogenically mated rats even a drop in the PAG concentration was noticed once without any consequences for the further course of pregnancy and the offsprings (Fig. 11). In general there was no correlation between maximum serum levels of PAG, litter size and birth weight.

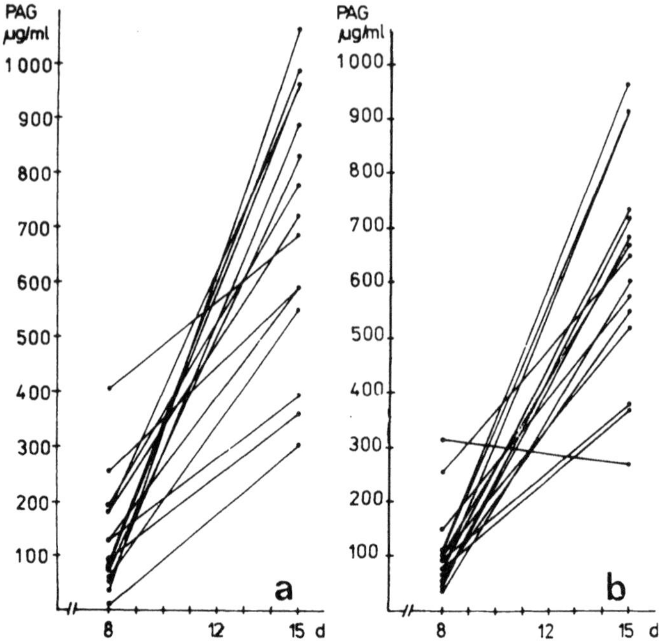

Figure 11a : PAG serum level rise in syngenic pregnancy
Figure 11b : PAG serum level rise in allogenic pregnancy

PAG was followed up in BD IX rats grafted with a slowly growing methylcholanthrene-induced sarcom and SD rats grafted with a progressively growing Yoshida sarcom.
In both strains the PAG levels of male rats remained uninfluenced, while the PAG serum concentrations decreased in female rats as tumours were growing. There was a clear connection between tumour size and time and extent of the serum level drop (table 3). With the Yoshida sarcom the PAG serum level had fallen significantly as compared to the initial values already 6 days after implanting the tumour, while in animals with methylcholanthrene-induced tumours a drop was noticed only 18 days after tumour inoculation. However the PAG concentrations of the females in no case dropped down to the range of PAG concentrations of males. The low- key PAG

basal synthesis in male rats may possibly be subject to a different regulation than the PAG synthesis in females (13).

PAG serum levels in relation to the growth of the Yoshida-sarcom

Day after tumour transplantation	Diameter of tumour mm ($\bar{x} \pm s$)	female SD rats n	female SD rats mg/l	male SD rats n	male SD rats µg/l
-21 (Control)		48	170	40	104
6	14.2 + 1.0	8	34	8	233
11	27.9 + 2.7	8	21	8	135
16	35.7 + 3.4	8	4	7	198

Table 3

In inflammation processes the PAG serum levels differ according to sexes. While, after intramuscular injection of 0.5 ml of complete Freund Adjuvance (Difco Laboratories, Detroit) the PAG level in female SD rats fell significantly, it rose considerably in male rats (Fig.12). Still more pronounced was the PAG drop in female rats after intramuscular injection of streptococci from a geometric mean value of 605 µg/l prior to the treatment to values between 0.26 and 17 µg/l 6 days after the injection (14.). Thus rat PAG is to be ranged in the group of negative acute phase proteins. (15)

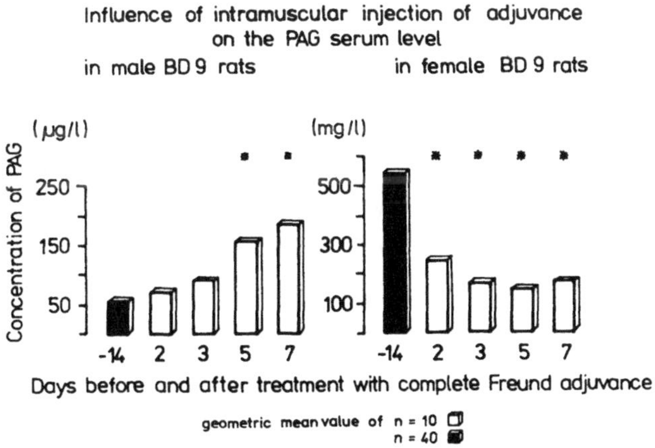

Figure 12

6. Analogues to Rat PAG

Using our affinity-chromatographically purified antisera in the double-dimension immunodiffusion technique, a strong cross-reactivity was noticed only with a serum protein of pregnant mice (Fig. 13).

1,4 pooled sera from pregnant rats
2 pooled sera from pregnant mice
3 pooled sera from pregnant women
5 anti-human α_2-PAG
6 anti-rat PAG

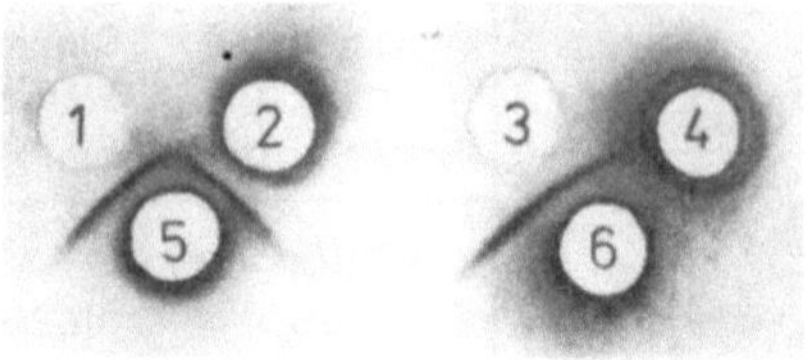

The immunochemical identity of the following proteins was now demonstrated by countercurrant line-immunoelectrophoresis comparing the different antisera (7).
Pregnancy-associated murine protein (PAMP-1) Hau et al.,
Pregnancy-associated serum protein (α_1-PAP), Waites and Bell,
Pregnancy-associated serum protein (α_2-PAG), Sarcione and Delluomo,
Pregnancy-associated glycoprotein (PAG), Porstmann and Schmechta.

All proteins were influenced by gonadal hormones in the same way. For PAMP-1 furthermore an effect of pituitary hormones on its serum level was demostrated by hypophysectomy and replacement therapy with hormones (16). However, an increase of α_2-PAG serum levels in breast cancer-bearing SD rats (17) and elevated levels of α_1-PAP during an acute phase response in female C57Bl/10 mice (18) were reported, which is contrary to our findings.

Due to a lack of sensitivity necessary to demonstrate a cross-reactivity between human α_2-PAG and rat-PAG by immunodiffusion techniques this issue was checked again by EIA, a much more sensitive technique of proof. Both the use of rat sera with different concentrations of PAG in the EIA for human α_2-PAG and vice versa the use of human sera with PAG concentration differences about factor 750 in the rat-PAG EIA resulted in absorbances in both assays which did not exceed the threefold of the standard deviation of the blank values. The cross-reactivity of purified human α_2-PAG and rat-PAG with antisera produced in mammels was

less than 0.05 per cent in the EIA for rat PAG.
However, similarities of rat PAG and human α_2-PAG with respect to hormonal regulation (19), serum level changes during pregnancy (20) and growing age (21,Fig.14) and the site of production, liver cells (22) and macrophages (23) brought us on to assume, that both proteins in the respective species are identical.

Figure 14: Increased serum levels of human α_2-PAG with growing age are caused by a decreased testosterone production, together with an increased serum level of oestrogen, due to insufficient elimination (24).

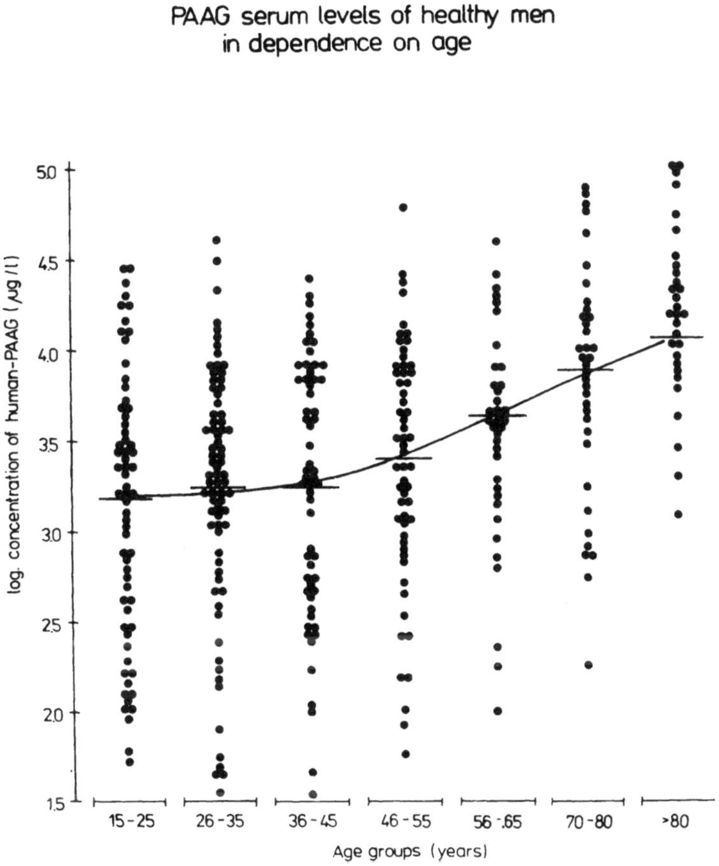

Immunological cross-reaction between human α_2-PAG and PAMP-1, the rat PAG homologue in the mouse, was demonstrated by countercurrant line immunoelectrophoresis using antibodies raised in hens against the proteins (25).

7. Immunosuppression by Rat PAG

There is evidence from in-vitro experiments that human α_2-PAG can exert an immunosuppressive effect. Inhibition of lymphocyte transformation and mixed lymphocyte reactions are triggered by an impairment of T cell function by human α_2-PAG. However, the presence of macrophages was necessary to improve the suppressive effect by human α_2-PAG indicating an interference with the process of antigen recognition and/or presentation (26). The human α_2-PAG blocked BCG induced inhibition of leucocyte migration (27) and a diminished answer of lymphocytes sensitized in the presence of human α_2-PAG in the macrophage electrophoretic mobility (MEM) test (28,29) allow though different explanations for the way possible for α_2-PAG action. For this reason we investigated in the MEM-test the influence of lymphokine liberation and lymphokine target cell interaction by rat PAG.

The principle of the MEM-test is that sensitized T lymphocytes (T_{DH}cells) secrete after repeated contact with the antigen macrophage slowing factor (MSF). After binding to a receptor with terminal fucose in the membrane of peritoneal exudate macrophages this factor leads to a decrease in the transmembrane potential which results in a slowing migration of macrophages in the electric field (30).

To exclude an altered electrophoretic mobility (EM) of macrophages by charge shift following PAG binding, peritoneal cells were incubated in medium, containing either 50 mg/l rat PAG or rat albumin (Fig.15). EM was measured by an Opton-Cytopherometer (Zeiss, Oberkochen) and automatic time counting. All tests were designed as blind studies.
The calculation of migration inhibition in the main experiments can be seen from figure 16. Each column represents the mean value and its standard deviation (S.D.) of the EM of 50 cells, each measured twice. The broken lines represent the twofold of the S.D. of either the positive or negative control (hatched area) and were taken as significance level.

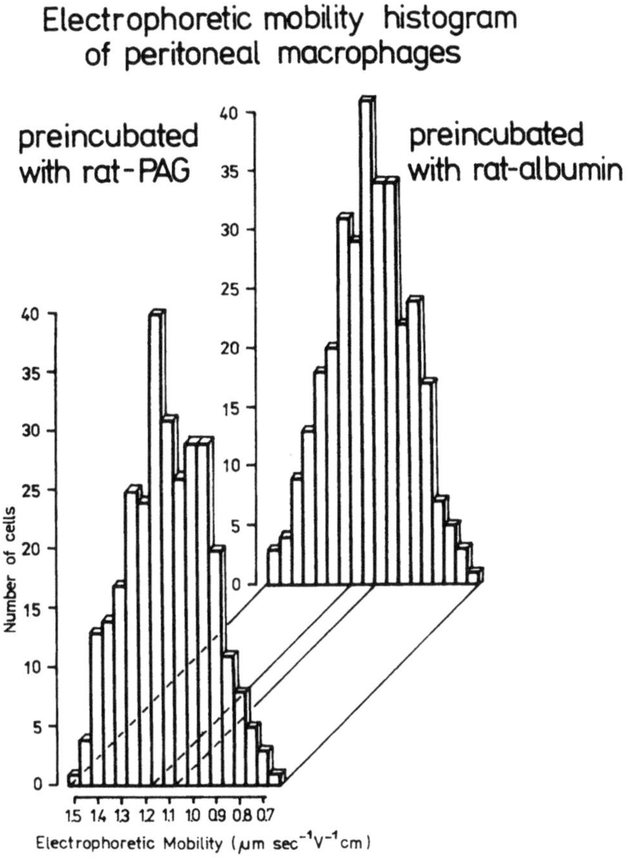

Figure 15

For the experiments spleen cells of BD IX rats which had developed a cell mediated immunity against purified protein derivate (PPD) were incubated with their recall antigen in the presence of various concentrations of purified rat PAG or rat albumin for 45 minutes. Afterwards the cell free supernatant was added to guinea pig peritoneal macrophages and incubation was performed for 30 minutes. For reason of control macrophages were incubated with MSF positive and MSF negative lymphocyte supernatants before measurement of EM.

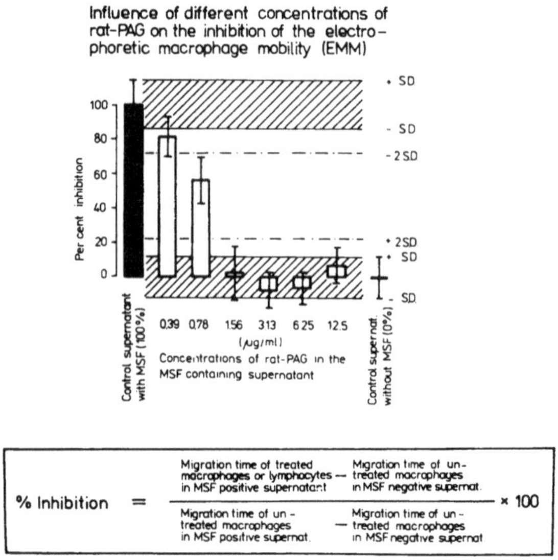

Figure 16

The addition of albumin to the spleen cells did neither affect the antigen recognition and lymphokine liberation nor prevented the action of MSF on macrophages.

The application of PAG to the spleen cells leads to a diminished effect of the MSF depending on the dose. If the PAG concentration exceeds 1 µg/ml in the incubation medium the cell supernatant, after antigen exposure of the spleen cells, will not cause any slowing down of the EM of the macrophages in comparison with MSF negative spleen cell supernatant (Fig. 16).

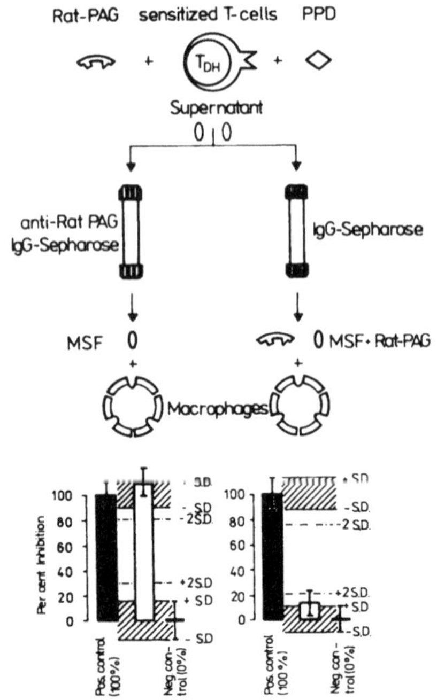

This result permits several explanations for the effect of rat PAG:

- PAG binds to T_{DH}cells and prevents antigen recognition. This would entail a missing MSF liberation,
- PAG binds to MSF and prevents the latter's binding to and effect on macrophages,
- PAG competes with MSF for binding to the receptor on macrophages,
- Despite MSF binding to the macrophages PAG prevents the subsequent membrane depolarization.

Figure 17

To clear up these various possibilities from one part of the supernatant containing recall antigen, MSF and PAG the latter was removed by immunosorbent chromatography at anti-PAG IgG-Sepharose. The other part was subjected to chromatography at insolubilized "non-immune" IgG of the same species. While the PAG-free supernatant causes a slowing down of the EM of the macrophages which does not differ from MSF positive control supernatant in its effect, the supernatant still containing PAG does not slow down the EM. Here the velocity of the macrophages is not different from cells which were incubated with MSF negative control supernatant (Fig.17).

These results clearly show that rat PAG neither does prevent antigen recognittion by sensitized T lymphocytes and MSF liberation nor does bind the MSF. PAG only acts on the efferent limb (macrophages) in this system which runs counter to the interpretation of the action of human α_2-PAG derived from the MEM-test (28,29).

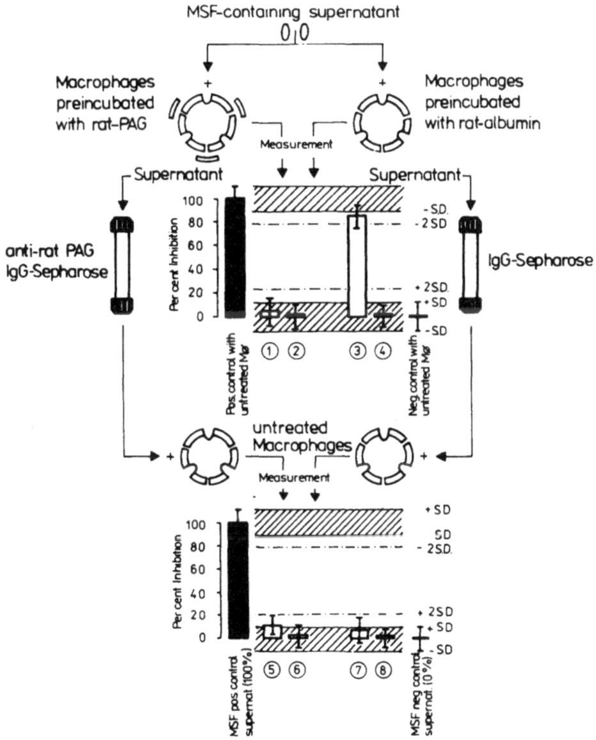

To find out a possible blockade of the MSF receptor on macrophages by PAG the cells were preincubated with rat PAG or albumin for 30 min. and then washed 3 times with medium. After application of MSF containing supernatant EM of macrophages was measured.

Macrophages preincubated with PAG did not show any reaction on MSF comparing their EM rate with untreated or albumin pretreated cells. (Fig. 18 upper part ① vs. ③

Figure 18

If the effect of PAG is nothing but a receptor blockade for MSF the latter should be still present in the supernatant, and the transfer of the supernatant to new untreated macrophages should trigger a slowing down of their EM. For this reason supernatant from the pretreated macrophages was subjected to chromatography at anti-PAG IgG-Sepharose in order to prevent a possible transfer of PAG and applied to new untreated macrophages. However the supernatant was just as ineffectiv now as was the supernatant from the macrophages pretreated with albumin (Fig. 18 lower part (5) vs. (7)). Of the latter one must assume that it is MSF negative since the lymphokine had bound to the albumin pretreated macrophages already during the first incubation period (compare (3) and (7) , the albumin system with (1) and (5), the PAG system of fig. 18).

These results indicate that MSF has bound to macrophages no matter whether pretreated with PAG or albumin which rules out a receptor blockade by PAG. However PAG prevents the membrane depolarization of macrophages after MSF binding to the receptor.

Since a large number of mononuclear cells, among them also peritoneal macrophages have membran-bound "lectin-like" receptors for mannose (31), galactosamine (32) and glucosamine (33), and effector cell target cell interactions but also cell glycoprotein interactions can be suppressed by monosaccharides (34,35, 36), the influence of simple sugars on rat PAG effectiveness was tested. Mannose, glucose, galactose and their amino derivates were used, because these sugars were detected on PAG by lectin binding.
If PAG is incubated with the macrophages in the presence of either mannose or galactose and their amino compounds, respectively, or if a subsequent incubation of the macrophages with these sugars succeeds a PAG preincubation, the PAG effect will be sublated. In spite of the presence of PAG the macrophages react on MSF positive supernatant again by slowing the EM. Both sugars clearly exhibit a dose-dependent dimunition of the PAG effect. The minimum concentration that still considerably hampers the effect of PAG was 15 mmol/l for mannose and 20 mmol/l for galactosamine, while 40 mmol/l of both sugars already entirely cancel out the PAG effect (Fig. 19).

Dose-dependent influence of galactosamine on inhibited MSF action upon macrophages by PAG

Dose-dependent influence of mannose on inhibited MSF action upon macrophages by PAG

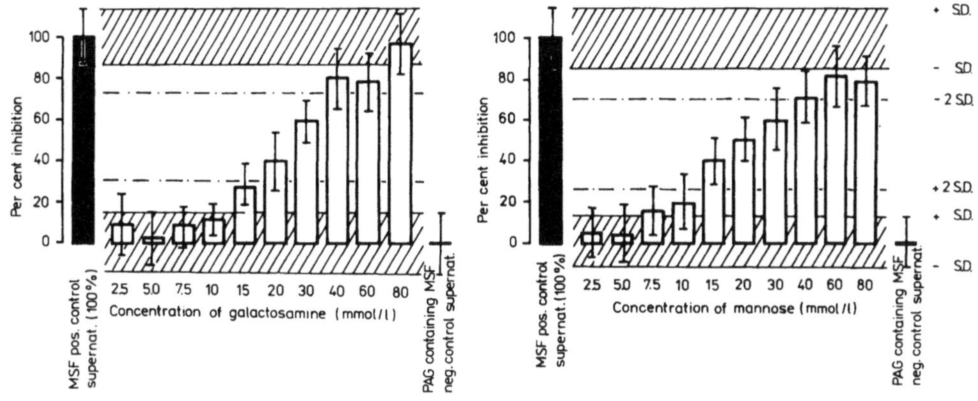

Figure 19

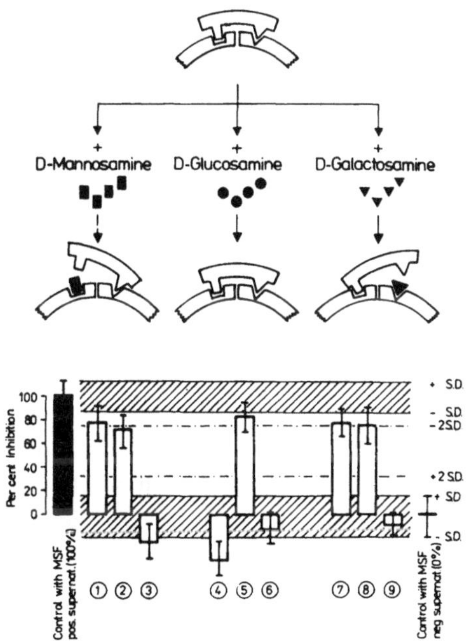

The application of even maximum concentrations of 80 mmol/l of glucosamine did not yield any effect on the PAG action (Fig. 20).

Fig 20

Man. pretreated macrophages with MSF containing supernatant in the presence (1) and absence (2) of PAG

Gluc. pretreated macrophages with MSF containing supernatant in the presence (4) and absence (5) of PAG,

Gal. pretreated macrophages with MSF containing supernatant in the presence (7) and absence (8) of PAG.

From these results there is evidence that PAG binds to "lectin like" receptors with galactosamine and mannose specificity on the macrophage surface through its terminal sugars. This may result in a cross-linking of membrane proteins with disturbance of membrane fluidity and also of membrane potential changes after binding of MSF, because for sublating the PAG effect only one of the two sugars is necessary (Fig. 20).

The influence of PAG on cell mediated immune reactions in-vivo was checked by footpad swelling under PAG treatment.
Mice received an intravenous injection of 750 µg purified PAG before i.v. application of SRBC. A comparative group was treated instead of PAG with an identical volume of PBS before SRBC application. Five days later cellular immune reaction was challenged by subcutaneous injection of SRBC into the left hind footpad while PBS was injected into the right hind footpad. Over a period of 48 h footpad swelling was determined every 12 h and compared with the footpad volume prior to the reaction by the liquid displacement technique.
The footpad swelling of the animals pretreated with PAG comes to only 45 % of that of untreated animals 24 h after the reaction was challenged. While the latter it almost remained constant until 48 h after the onset of reaction, it had at that time almost returned to the volume of the contralateral PBS-treated footpad in animals pretreated with PAG (Fig. 21 a).
The cell mediated immune reaction was nearly abrogated by PAG, likely due to an influence on antigen uptake and/or antigen presentation by spleen macrophages.
Furthermore the effect of a local PAG treatment on the challenge of the reaction was tested in sensitized mice. For this purpose the sensitized mice were subcutaneously inoculated with 75 µg of purified PAG together with SRBC into the left hind footpad and with cells only into the right hind footpad.
No differences between the left and right hind footpads were noticed at 24 and 48 h after challenge which argues against a local effect of PAG (Fig. 21 b). But a reduction of PAG to ineffective concentrations by diffusion before the mononuclear infiltration set in cannot be excluded as reason for its ineffectiveness.

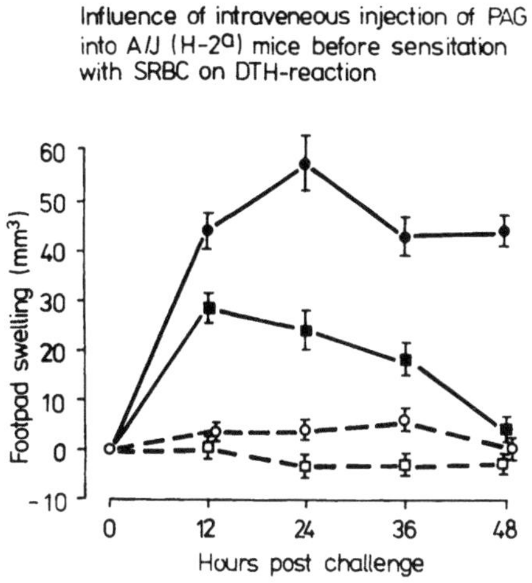

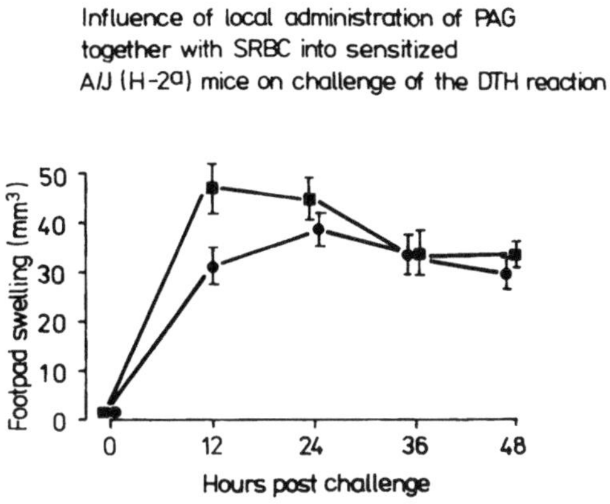

Figure 21 a Figure 21 b

Footpad volume prior challenge with sheep red blood cells (SRBC) was adjusted to zero. Each point represents the mean value ± standard deviation of the volume of 10 footpads.

●——● PBS pretreated mice injected with SRBC for challenge (left footpad)

○- -○ Contralateral footpad after challenge with PBS

■——■ PAG pretreated mice injected with SRBC for challenge (left footpad)

□- -□ Contralateral footpad challenged with PBS

■——■ Reaction challenge with SRBC in the presence of PAG

●——● Reaction challenge with SRBC in the absence of PAG

Conclusion

From the presented data it is obviously that rat PAG is the analogue to human α_2-PAG. It is under control of gonadal and adrenal hormones and reacts as negative acute phase protein. It acts immunosuppressive by interaction with "lectin like" membrane receptors on macrophages in-vitro and in-vivo in concentrations which occur during pregnancy.

References

1. Jacoby, D.R., L.B. Olding, and M.B. Oldstone. 1984. Immunologic Regulation of fetal-maternal balance. Adv. Immunol. 35, 157-208.

2. Olivelli, F. and P. Ruggieri. 1958. Sulla presenza e sul significato degli antigeni placentari nel siero di donna gravida. Minerva Ginecol. 10, 953-958.

3. Svendsen, P., T. Stigbrand, B. Teisner, J. Folkersen, M-G.Damber, B. von Schoultz, E. Kemp and S-E. Svehag. 1978. Immunosuppressive effect of human pregnacy zone protein on H-2 incompatible mouse heart allografts. Acta Pathol. Microbiol. Scand. 86, 199-201.

4. Porstmann, T. 1983. Schwangerschafts-assoziiertes α_2-Glykoprotein(PAAG). In: Eiweisse (P.Linneke, ed.). Akademie-Verlag, Berlin. pp. 215-224.

5. Porstmann, T. and H. Schmechta. 1980. Isolation of a pregnancy-associated globulin (PAG) from the rat and preparation of a specific antiserum (anti-PAG). Acta biol. med. germ. 39, 479-487.

6. Alhadeff, J.A. 1978. Gel filtration of sialoglycoproteins. Biochem. J. 173, 315-319.

7. Hau, J. and T. Porstmann. 1984. Characterization of the analogues to human pregnancy-associated α_2-Glykoprotein (α_2-PAG, PZP) isolated in the mouse and rat. Lab. Animals 18, 344-348.

8. Porstmann, T., H. Schmechta, E. Nugel, B. Porstmann, B. Kopetz and A. Lukowsky. 1982. Quantitation of a pregnancy associated globulin in the rat by two-site binding enzyme immunoassay. Acta biol. med. germ. 41, 953-960.

9. Schade, R., F. Götz, W. Bürger and G. Dörner. 1981. Effects of testosterone and estradiol on a sex-specific rat serum protein in intact and castrated female and male rats. Endokrinologie 78, 139-148.

10. Schade, R., R. Tönjes and G. Dörner. 1984. Sex-specific concentrations of plasma proteins in rats determined by sex hormones in perinatal life. Exper. Clin. Endocrinol. 81, 257-261.

11. Gustafsson, G-A., A. Mode and G. Norsted. 1983. Central control of prolactin and estrogen receptors in rat liver - expression of a novel endocrine system, the hypothalamo-pituitary-liver axis. Pharmacol. and Toxicol. 23, 259-278.

12. Schmechta, H., T. Porstmann, E. Nugel, B. Kopetz and B. Porstmann. 1982. Verhalten des trächtigkeitsassoziierten Alpha-1-Glykoproteins (PAG) der Ratte während der Trächtigkeit. Z. Versuchstierk. 24, 294-298.

13. Schmechta, H., T. Porstmann, B. Kopetz, E. Nugel, B. Porstmann and D-J. Rolle. 1983. Verhalten des trächtigkeitsassoziierten Alpha-1-Glykoproteins (PAG) der Ratte beim Wachstum experimenteller Transplantationstumoren (Methylcholanthren-induziertes und Yoshida-Sarcom). Z. Versuchstierk. 25, 63-68.

14. Kopetz, B. 1982. Nachweis eines immunsuppressiven trächtigkeitsassoziierten α_2-Glykoproteins der Ratte und enzymimmunologische Ermittlung der Serumkonzentration bei unterschiedlich behandelten tumortragenden und tumorfreien Tieren. Doctoral Thesis, Bad Saarow.

15. Schade, R., R. Petzold, K. Göhler, H. Leder, M. Leder, F. Götz and A. Friedrich. 1985. Possible modulation of the acute-phase response in rats - dexamethasone- and stress induced changes of acute-phase protein blood levels. Biogenic Amines 2, 1-12.

16. Hau, J., Gidley-Baird, A.A., B. Teisner, P. Svendsen and J. G. Westergaard. 1983. The influence of pituitary and gonadal hormones on serum levels of pregnancy-associated murine protein-1. Acta Endocrinol. 104, 23-26.

17. Sarcione, E.J. and D.S. Delluomo. 1981. Demonstration of pregnancy-assoziated α_2-Glykoprotein (α_2-PAG) in pregnant, estrogen-treated and breast cancer-bearing rats. Oncodevelop. Biol. Med. 2, 345-354.

18. Waites, G.T. and S.C. Bell. 1984. Identification of a murine pregnancy-associated serum protein (α_1-PAP) as a female-specific acute-phase reactant. J. Reprod. Fert. 70, 581-589.

19. Robinson, J.C., W.T. London and J.E. Pierce. 1966. Observations on the origin of pregnancy-associated plasma proteins. Am. J. Obstet. Gynecol. 96, 226-230.

20. Stimson, W.H. 1975. Variations in the serum concentration of a human pregnancy-associated α-macroglobulin during pregnancy and after delivery. J. Reprod. Fert. 43 579-582.

21. Folkersen, J., B.Teisner, N. Grunnet, J. G. Grudzinskas, J. G. Westergaard and P. Hindersson. 1981. Circulating levels of pregnancy zone protein: normal range and the influence of age and gender. Clin. Chim. Acta. 110, 139-145.

22. Bohn, H. 1972. Charakterisierung der schwangerschaftsassoziierten Glykoproteine als akute Phase Proteine. Ihr Nachweis im Serum von Patienten mit Tumoren und anderen Erkrankungen. Arch. Gynäkol. 213, 54-72.

23. Stimson, W.H. and J. C. Blackstock, 1975. Synthesis of a pregnancy-associated α-macroglobulin by human leucoytes. Experientia 31, 371-373.

24. Porstmann, T., H. Schmechta, A. Lukowsky, B. Porstmann and H. Wietschke, 1983. Enzymimmunoassay für das schwangerschaftsassoziierte alpha$_2$-Glykoprotein (PAAG) - Referenzbereiche für Männer. Dt. Gesundh.-Wesen 38, 123-126.

25. Hau, J., J. G. Westergaard, P. Svendsen, A. Bach and B. Teisner, 1981. Comparsion of pregnancy-associated murine protein -1 and human pregnancy zone protein. J. Reprod. Immunol. 3, 341-349.

26. Stimson, W.H. 1975. Pregnancy-associated α-macroglobulin (PAMP) and non-specific immunosuppression in human pregnancy. Behring Institute Mitteilungen 57, 42-49.

27. Birkeland, S.A., B. Teisner, W. Schilling, E. Kemp, G. Thomsen Pedersen, S.-E. Svehag. 1979. Effect of pregnancy zone protein on leucocyte migration inhibition, lymphocyte transformation and rosette formation by lymphocytes. Acta pathol. et microbiol. scand. 87, 235-240.

28. Straube, W., R. Hofmann, B. Klausch, J. Günther, H.-L. Jenssen and H. Köhler. 1975. The biological function of the pregnancy zone protein. In: Protides of the biological fluids. Proceedings of the 24th Colloquium Brugge. (H.Peeters, ed.) Pergamon Press, Oxford, pp. 197-201.

29. Bauer, H.W., H. Bohn and W. Ax. 1978. Immunologische Potenz des schwangerschaftsassoziierten Alpha$_2$-Glykoproteins (Alpha$_2$-PAG). Onkologie 1, 119-123.

30. Field, E.J. and E.A. Caspary. 1970. Lymphocyte sensitization: An in vitro test for cancer?. Lancet II, 1337-1341.

31. Straus, W. 1983. Mannose-specific binding sites for horseradish peroxidase in various cells of the rat. J. Histochem. Cytochem. 31, 78-84.

32. Nagamura, Y. and H. Kolb. 1981. Rat peritoneal macrophages express two different lectin-like receptors on the cell-surface. In: Lectins. Biology, Biochemistry, Clinical Biochemistry, Vol. 1 (T.C. Bog-Hansen ed.). Walter de Gruyter, Berlin . New York. pp 33-42.

33. Monsigny, M., C. Kieda and A-C. Roche. 1979. Membrane Lectins Biol. Cellulaire. 36, 289-300.

34. Werkmeister, J., J.C. Roder, C. Curry and H. F. Pross. 1983. The effect of unphosphorylated and phosphorylated sugar moieties on human and mouse natural killer cell activity : Is there selective inhibition at the level of target Recognition and lytic acceptor site. Cell. Immunol. 80, 172-186.

35. Weir, D.M., L.M. Grahame and H.M. Ögmundsdottir. 1979. Binding of mouse peritoneal macrophages to tumour cells by a "lectin-like" macrophage receptor. J. Clin. Lab. Immunol. 2, 51-54.

36. Fukazawa, Y., K. Kagaya and H. Miura. 1981. Biological and biochemical characterization of macrophage activating factor (MAF) in murine lymphocytes: Role of mannopyranosyl residue of the MAF molecule in macrophage activation. Microbiol. Immunol. 25, 1163-1172.

STUDIES ON MURINE α_1-PREGNANCY-ASSOCIATED PROTEIN (α_1-PAP) AND ITS RAT ANALOGUE AND THEIR RELATIONSHIP TO HUMAN α_2-PAG

*Gillian T. Waites and S.C. Bell

Department of Obstetrics & Gynaecology and Biochemistry, University of Leicester, University Road, Leicester LE1 7RH, U.K.

*Present address: AFRC Institute of Animal Physiology, Babraham, Cambridge CB2 4AT, U.K.

Introduction

In the search for laboratory animal models to aid the study of human pregnancy-associated serum proteins, several rodent proteins have been described (1 - 5). In a recent study (6), after immunization of rabbits with pooled pregnancy sera from mice, antibodies were detected against two pregnancy-associated proteins, α_1- and α_2-PAP. Murine α_1-PAP was considered to be pregnancy-specific, due to its absence from sera of virgin female and male mice of the inbred C57BL/10 strain, until serum concentrations in females in the acute-phase of an inflammatory response were found to be elevated by at least 36-fold (6). The sex-restriction of this response was found to be a reflection of regulation of α_1-PAP acute-phase synthesis by gonadal steroids (7). Substantial levels of α_1-PAP were also detectable in the sera of a number of other inbred females during pregnancy and since serum levels in all equivalent non-pregnant inbred strains were below the sensitivity of the assay, this represented at least a 200-fold increase during pregnancy. It was thus paradoxical to find that complete immunological cross-reactivity could be demonstrated (cited 7, 8) between α_1-PAP and pregnancy-associated murine protein - 1 (PAMP-1) described by Hau et al. (9) which was readily detectable in the sera of non-pregnant females of the outbred NMRI:Bom strain used and which only increased 5 - 6-fold during gestation. The physicochemical characteristics of PAMP-1 and α_1-PAP were found to be very similar; both proteins exhibiting

Pregnancy Proteins in Animals

α_1-electrophoretic mobility, native molecular weights of 140 K and 150 K, respectively, and reaching peak serum concentrations at mid-pregnancy, falling prior to parturition, which supported the suggestion of their analogy.

With the prime objective of these investigations being to establish animal models for the study of human proteins, it was of importance that immunological cross-reactivity could be demonstrated between PAMP-1 and the well-studied (for review see 10) human pregnancy-associated α_2-glycoprotein (human α_2-PAG) using antisera raised in the phylogenetically distant hen (11). Furthermore, cross-reactivity has also been demonstrated (8, 12) between PAMP-1, α_1-PAP and two proteins detected in the sera of pregnant rats, pregnancy-associated globulin (rat PAG; 3) and pregnancy-associated α_2-glycoprotein (rat α_2-PAG; 5), independently put forward as analogues of human α_2-PAG. Rat PAG, like PAMP-1, was readily detectable in the sera of non-pregnant females and increased during pregnancy while α_2-PAG, like murine α_1-PAP, was detectable only in the sera of pregnant females and not in normal females or males. Thus, although the demonstration of immunological cross-reactivity implied a close relationship between these rodent proteins, there are some anomalies which require explanations.

In this study, the relationship between α_1-PAP and PAMP-1 has been further explored by investigating the serum levels of murine α_1-PAP in a number of inbred and outbred strains, and using antiserum to murine α_1-PAP analogues have been searched for in a number of species, including the rat. Immunohistochemical localization of α_1-PAP, also described, is of relevance in understanding the relationship between the rodent pregnancy proteins and human α_2-PAG and their possible function in pregnancy.

Materials and Methods

Animals

The inbred strains of mice used were C57BL/10 ScSn ($H-2^b$), C57BL ($H-2^b$), B10.BR ($H-2^k$), CBA/Ca ($H-2^k$), C3H ($H-2^k$), Balb/C ($H-2^d$) and A ($H-2^{k/d}$). A random mating colony of BM mice and the outbred

MF/1 strain were also employed. Pregnant animals of the appropriate strain combination were obtained by overnight pairing of females with males of proven fertility and the morning on which a vaginal plug was detected was designated day 0 post coitum (p.c.) of pregnancy. Sera were collected and stored as previously described (6). Pooled sera from non-pregnant females of the outbred NMRI:Bom strain were kindly provided by Professor J. Hau. The rats used were of the inbred hooded, Wistar Leicester, Wab and outbred Sprague-Dawley strains from which blood for sera was similarly obtained via the tail vein or cardiac puncture. Following overnight pairing of females and males of the same strain, the day of detection of spermatozoa in vaginal smears was designated day 0 of pregnancy.

Pregnant females were bled by cardiac puncture. Rats subject to acute-phase stimulation were pre-bled via the tail vein and one week late received 0.5 ml turpentine in arachis oil (1.1; v/v) subcutaneously at 2 sites and were cardiac bled 48 hours later. Sera from non-pregnant and pregnant Salamanders were kindly provided by Dr M.T. Badet and from non-pregnant female and male golden hamsters and pregnant women in the third trimester.

Antisera

Monospecific rabbit anti-mouse α_1-PAP was prepared as previously described (6). For immunohistochemical studies an IgG fraction was prepared using sodium sulphate precipitation followed by ion exchange chromatography (12). Rabbit anti-mouse serum amyloid P-component (SAP) was prepared by Dr A.M. Bell as described previously (13) and goat anti-PAMP-1 by Professor J. Hau (2).

Immunoelectrophoretic techniques

Crossed and tandem immunoelectrophoresis were performed in a 1% (w/v) agarose gel using a barbital buffer pH 8.6 (Sigma) with a voltage gradient of 25 V/cm for 1½ hours in the first dimension and 10 V/cm overnight in the second dimension. Concentrations of α_1-PAP in sera were determined using monospecific rabbit anti-murine α_1-PAP in rocket immunoelectrophoresis employing the same gel-buffer system with wells receiving 5 μl and subject to electrophoresis overnight at 10 V/cm. Levels of α_1-PAP were expressed as arbitrary units per ml (AU/ml), where 1 AU was defined as the peak height obtained with a 1:30 dilution of a standard preparation of pooled murine pregnancy serum. The lower limit of sensitivity of the assay was 0.03 AU/ml and it exhibited a precision of $\pm$ 3%.

Molecular weight estimation

Rocket immunoelectrophoresis was used to determine the elution value of α_1-PAP from pooled female Wistar Leicester rat serum (3 ml) fractionated on a Sephacryl S300 (Pharmacia) column (41.5 x 2.5 cm) equilibrated and run in 0.1 M Tris-HCl containing 0.5 M NaCl (pH 8.0) and the molecular weight estimated by interpolation from a standard curve of Kav values calculated from a number of proteins of known molecular weight. Pooled murine pregnancy serum was subject to fractionation under identical conditions.

Indirect immunoperoxidase (PAP) staining for α_1-PAP

The following tissues obtained from non-pregnant female mice of the outbred MF/1 and inbred C57BL/10 strains, and from non-pregnant Sprague-Dawley rats, were fixed in acetic formol saline, embedded in paraffin and sections cut at 5 µm: small and large bowel, liver, Peyer's patch and salivary gland. Sections of each tissue were stained for α_1-PAP using an IgG cut of rabbit anti-serum against murine α_1-PAP by the indirect PAP technique outlined elsewhere in these proceedings (Udagawa, Waites, Armstrong, _et al_.). The specificity of the rabbit anti-murine α_1-PAP antiserum was verified by absorption with partially purified mouse α_1-PAP (12).

Results

Strain-dependent variations in the serum concentration of murine α_1-PAP

Using a rocket immunoelectrophoretic assay with a lower limit of sensitivity of 0.03 AU/ml, α_1-PAP was undetectable, or present at very low concentrations only, in all the inbred strains investigated and also in the random mating BM strain (Figure 1). This

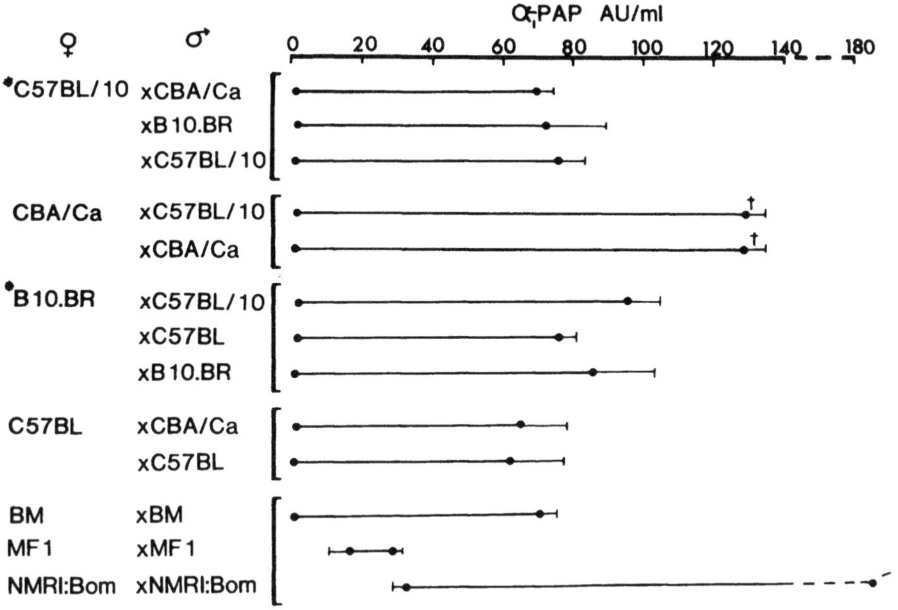

Figure 1. Increase in the serum concentrations of α_1-PAP in a variety of mating combinations on day 12 p.c. of pregnancy. Values are mean ± s.e.m. (6 < N < 30). *Indicates congenic mice. †Values are significantly higher than in all other pregnancies (0.01 < 0.001). ´converted to AU α_1-PAP/ml from Hau et al. (9).

was in contrast to the substantial levels of the immunochemically cross-reactive protein PAMP-1, reported by Hau to be present in the sera of females of the outbred NMRI: Bom strain (9). On screening sera from this outbred strain in the α_1-PAP assay, levels detected (31.9 ± 8.2 AU/ml; n = 10) were at least 80-fold higher than in any inbred strain. Furthermore, in a second outbred (MF/1) strain examined serum concentrations of α_1-PAP were also substantially higher than in the inbred strains (17.7 ± 7.0 AU/ml; n = 5). During pregnancy, elevated serum levels of α_1-PAP were detected in all strains and in all mating combinations examined (Figure 1) although whereas in the inbred strains, this represented at least a 200-fold increase over non-pregnancy levels, in the MF/1 females only a 2-fold increase was observed, comparable to the 5 - 6-fold elevation in NMRI:Bom pregnancy (9). The absolute levels were also variable between strains with CBA/Ca and NMRI:Bom females being notably high, while MF/1 females were low (Figure 1).

Identification and characterization of a protein in female rat sera exhibiting immunochemical cross-reactivity with murine α_1-PAP

Employing antiserum to murine α_1-PAP, analogues were searched for in other species. The hamster female protein (14) although sex-restricted in a similar way to α_1-PAP did not exhibit immunochemical cross-reactivity, as assessed by immunodiffusion, with α_1-PAP, but did show reactivity with murine serum amyloid P-component (SAP), its purported analogue (15). No cross-reactivity could be demonstrated with Salamander protein P described by Badet (16) nor with human third trimester pregnancy serum. However, complete cross-reactivity was obtained between murine α_1-PAP and a protein present in the sera of female rats.

The rat protein was shown by crossed-immunoelectrophoresis to have an equivalent electrophoretic mobility to murine α_1-PAP and in tandem immunoelectrophoresis rat and mouse serum run against anti-murine α_1-PAP produced single arcs which were confluent. The molecular weight of rat α_1-PAP was estimated by gel filtration chromatography to be 150 K, identical to murine α_1-PAP under the same conditions. Using the anti-murine α_1-PAP serum, the levels of rat α_1-PAP in a variety of samples were determined by rocket immunoelectrophoresis. Substantial levels of rat α_1-PAP of the same order as α_1-PAP in outbred mice, were found in non-pregnant females of each strain examined, but never in males. A kinetic profile of α_1-PAP during rat pregnancy (Figure 2) was determined using the inbred Wistar Leicester strain where serum concentrations increased from day 10 (41.2 ± 9.7 AU/ml; n = 5) reaching maximum observed levels (79.3 ± 12 AU/ml; n = 4) representing almost a 3-fold increase on day 13 of pregnancy; values then falling to a plateau prior to parturition. As shown in Figure 2, this was comparable to the profile of α_1-PAP during murine pregnancy. A similar increase was also seen in females of the Sprague-Dawley and Wab strains when assessed on day 13. Of further interest was the finding, as with the murine protein, of altered serum concentrations of rat α_1-PAP following acute-phase stimulation; however, in contrast to the elevation in inbred mice

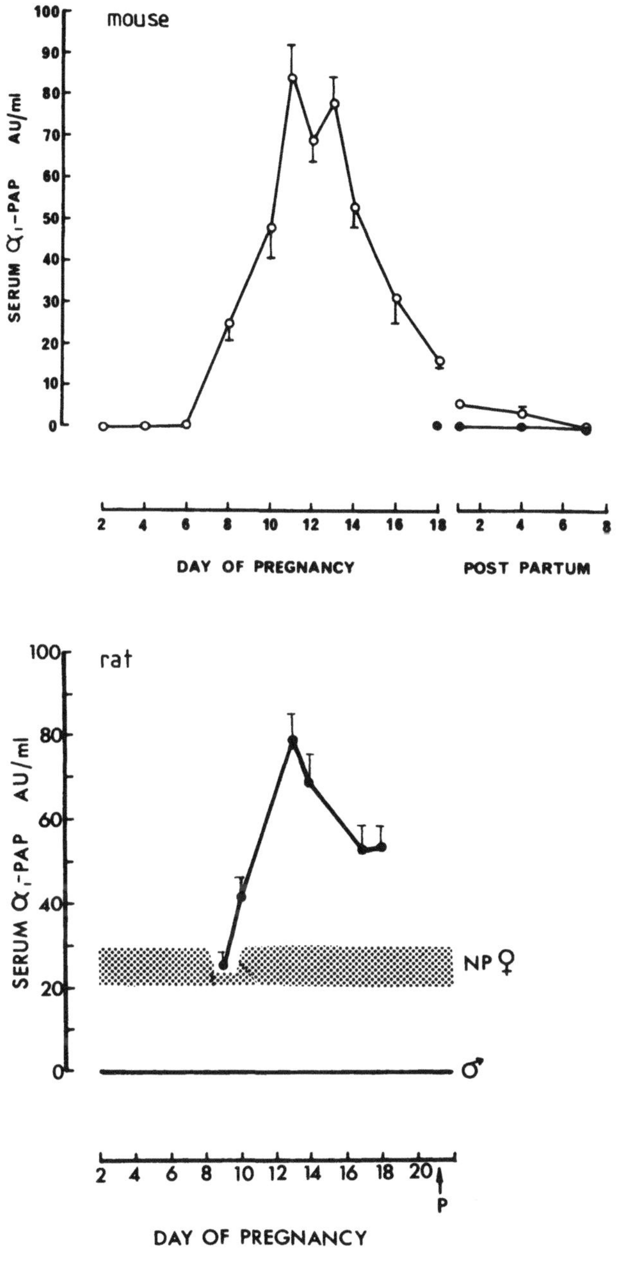

Figure 2. Serum concentrations of α_1-PAP during pregnancy in the mouse (C57BL/10 x CBA/Ca; ○ maternal ● fetal) and in the rat (Wistar Leicester). Values are mean $\pm$ s.e.m. ($3 < N < 14$). Levels were undetectable in normal mice and male rats; hatched

area represents mean ± s.e.m. in non-pregnant female rats. Levels were assessed using rabbit anti-mouse α_1-PAP in rocket immunoelectrophoresis where 1 AU was defined as the peak height obtained with a 1:30 dilution of a standard of pooled murine pregnancy serum. P = parturition.

(7), in the rat a mean individual decrease of 24% was seen (group mean ± SD : normal, 29 ± 5.5 AU/ml, n = 5; acute-phase, 18.8 ± 3.9 AU/ml, n = 4; 0.01 > p > 0.001).

Immunohistochemical localization of α_1-PAP in mice and rats

As shown in Table 1, in non-pregnant mice of the MF/1 strain, prominent α_1-PAP positive cells were evident within gut-associated lymphoid tissue and in the lamina propria of small and large bowel mucosae. Positively staining hepatocytes were also observed, usually surrounding the blood vessels and portal tracts, although staining was rare in salivary glands. A similar pattern of α_1-PAP

Table 1. Immunoperoxidase staining for α_1-PAP in rodent tissues and α_2-PAG in the human for comparison.

Tissue	Mouse		Rat	Human[1]
	MF/1	C57BL/10	Sprague-Dawley	
Small bowel mucosae	+	-	±	+
Large bowel mucosae	+	-	+	+
Peyer's Patch	+	-	±	+
Salivary gland	±	-	±	+
Liver	+	-	±	-[2]

\+ Positive staining

± Equivocal positive staining

\- Very litte or no positive staining

[1] Taken from Horne et al. (17)

[2] A.W. Thomson, Y. Udagawa, personal communication.

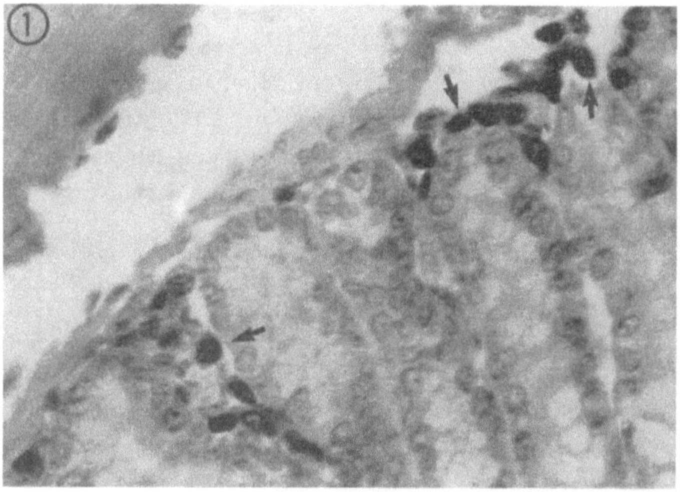

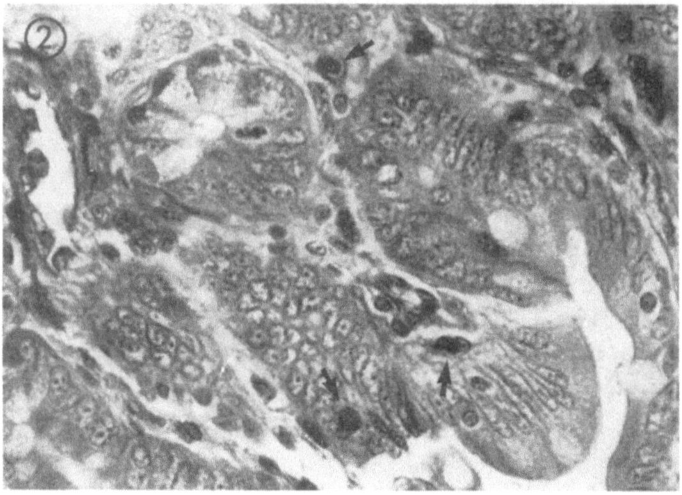

Plate 1

Tissues stained for α_1-PAP using rabbit anti-murine α_1-PAP serum in an indirect immunoperoxidase (PAP) technique, counterstained with haematoxylin.

Figure 1. Non-pregnant MF/1 mouse colonic mucosa showing positive α_1-PAP cells within the lamina propria. Absorption of the antisera with partially purified α_1-PAP greatly reduced staining. (x 600).

Figure 2. Non-pregnant Sprague-Dawley rat colonic mucosa. Despite the increased background, positively stained α_1-PAP cells can be seen to have a similar distribution to the MF/1 mouse colon. (x 600).

positive staining was detected in non-pregnant Sprague-Dawley rats, although the background staining was more intense. In contrast, no significant staining for α_1-PAP was observed in any of the tissues examined from non-pregnant C57BL/10 females. Plate 1 compares the staining pattern seen in these mouse and rat tissues. Absorption and NRS controls were all negative.

Discussion

Murine α_1-pregnancy-associated protein (α_1-PAP) has now been detected in the sera of non-pregnant females of the outbred strains investigated in contrast to the low or undetectable levels in all inbred strains. This extreme inter-strain variability of α_1-PAP levels enables the placement of animals into two categories: (i) low endogenous α_1-PAP producers with very low or undetectable levels of the protein in sera of non-pregnant females, increasing several hundred-fold during pregnancy and (ii) high endogenous α_1-PAP producers with serum concentrations at least 40-fold higher in non-pregnancy than the above strains and exhibiting as minimal as a 2-fold increase during pregnancy. The finding of substantial non-pregnancy levels of α_1-PAP in some strains of mice, similar to PAMP-1 (9), explains the previous paradoxical demonstration of immunological cross-reactivity between these two proteins (12, 8), which in all other respects appeared to be analogous. The difference in the endogenous production of α_1-PAP in these groups was too great to be explained by an acute-phase elevation of α_1-PAP in high endogenous producer strains as previously suggested (6) and was unrelated to age. Although it seemed that high endogenous levels were related to out-breeding populations, this was not the case as the random bred BM females had low endogenous levels of α_1-PAP whilst all the inbred Bom-derived strains employed by Hau were high endogenous producers (personal communication). No such differences were observed during pregnancy where all animals exhibited a substantial, albeit variable, pregnancy-induced elevation in serum α_1-PAP levels. It is suggested that at least two types of synthesis with differential control are operative; a pregnancy-independent source, constitutive in high endogenous strains and inducible in other strains, and a pregnancy-specific source present in all animals.

In the search for α_1-PAP analogues in other species, a protein was detected in the sera of female rats exhibiting complete immunological cross-reactivity with murine α_1-PAP. Basic physicochemical characterization showed that similar to the murine protein, rat α_1-PAP exhibited an α_1-electrophoretic mobility, a molecular weight of 150 K and was found to increase during pregnancy reaching peak levels (representing a 2 - 3-fold increase) at the same developmental stage as α_1-PAP in the mouse, i.e. mid-gestation. Due to the high endogenous levels of α_1-PAP observed in all strains of non-pregnant females examined and the increase during pregnancy, the rat would seem to fall into the same category as the outbred mouse strains. Rat α_1-PAP also behaved as an acute-phase reactant although, in contrast to the mouse, this was with a decrease in serum concentrations. Similar strain or sex-dependent differential responses to traumatization have been reported for the other acute-phase reactants SAP and hamster FP (18, 19) and do not therefore negate the analogy between murine α_1-PAP and the rat protein.

The demonstration of immunological cross-reactivity between PAMP-1 and human α_2-PAG (11) and later (8) between PAMP-1, α_1-PAP and two other rat proteins independently proposed as analogues of human α_2-PAG (PAG, 3; α_2-PAG, 5) suggested a group of analogous proteins which could be used to study the human protein. However, immunological cross-reactivity, especially if demonstrated using a phylogenetically removed species, as employed in the case of PAMP-1 and human α_2-PAG, does not provide proof of analogy between two proteins. In fact, the murine and human proteins are physicochemically dissimilar and their kinetic profiles during pregnancy do not correlate. However, although human α_2-PAG has previously been shown to be synthesized by peripheral blood monocytes and/or T and B cells, possibly the liver (see 10) and placenta (20) much of this remains controversial and the most definitive association between α_2-PAG and a specific cell type comes from the recent immunohistochemical findings of Horne _et al_. (17) showing concomitant α_2-PAG and IgA-producing plasma cell staining in the intestinal mucosae and gut-associated lymphoid tissue (GALT) in normal subjects. Employing a similar indirect immunoperoxidase (PAP) staining technique, evidence has been presented here showing that this unique distribution of α_2-PAG positive cells can be reproduced using anti-murine α_1-PAP antiserum on equivalent mouse and rat tissues. In non-pregnancy, the staining seemed to be reflective of peripheral serum levels, being found only in high endogenous α_1-PAP producer females. This was strengthened by the findings presented in a separate paper in these proceedings (Udagawa, Waites, Armstrong et al.) of enhanced staining for α_1-PAP in non-pregnancy-associated sites in MF/1 females during gestation and the appearance of these sites in pregnant C57BL/10 females, with the additional staining of cells in the mesometrial decidua and placental cytotrophoblast in both strains. Furthermore, there is some evidence of an association between murine α_1-PAP and IgA-producing plasma cells (21) also demonstrated using antiserum against rat α_2-PAG (22). If this localization reflects function, then it is suggested that α_1-PAP of the mouse and rat, and the other rodent proteins described, may represent the functional analogue of human α_2-PAG.

Acknowledgements

G.T.W. was in receipt of a Postgraduate Studentship Award from the Medical Research Council.

References

1. Lin, T.M., S.P. Halbert and D. Keifer. 1974. Pregnancy-associated serum antigens in the rat and mouse. Proc. Soc. Exp. Bio. Med. 145, 62-66.

2. Hau, J., J.G. Westergaard, P. Svendsen, B. Teisner, A. Bach and G. Thomsen-Pedersen. 1982. The development of a murine model for the study of human pregnancy zone protein (PZP, α_2-PAG) and pregnancy-specific β_1-glycoprotein (SP-1, PSβG). In: Immunology of Human Placental Proteins (A. Klopper, ed.). Placenta (Suppl. 4). Praeger. pp. 51-65.

3. Porstmann, T. and H. Schmechta. 1980. Isolation of a pregnancy-associated globulin (PAG) from the rat and preparation of a specific antiserum (anti-PAG). Acta Biol. Med. Germ. 39, 449-487.

4. Schade, R. and W. Burger. 1980. Quantitative examinations of some rat proteins by means of absorbed anti-pregnancy and anti-feto sera. Experientia 36, 125-126.

5. Sarcione, E.J. and D.S. Delluomo. 1981. Demonstration of pregnancy-associated α_2-glycoprotein (α_2-PAG) in pregnant, oestrogen-treated and breast cancer bearing rats. Onco. Dev. Biol. Med. 2, 345-354.

6. Waites, G.T. and S.C. Bell. 1984. Identification of a murine pregnancy-associated serum protein (α_1-PAP) as a female-specific acute-phase reactant. J. Reprod. Fert. 70, 581-589.

7. Waites, G.T. and S.C. Bell. 1984. Regulation of a murine α_1-pregnancy-associated protein by gonadal steroids during the acute-phase response. J. Endocr. 101, 315-318.

8. Hau, J. and T. Porstmann. 1984. Characterization of the analogue to human pregnancy-associated α_2-glycoprotein (α_2-PAG, PZP) isolated in the mouse and rat. Laboratory Animals 18, 344-348.

9. Hau, J., P.E. Svendsen, B. Teisner and S.E. Svehag. 1978. Studies of pregnancy-associated murine serum proteins. J. Reprod. Fert. 54, 239-243.

10. Thomson, A.W. and C.H.W. Horne. 1980. Biological and clinical significance of pregnancy-asociated α_2-glycoprotein (α_2-PAG): a review. Invest. Cell Pathol. 3, 295-309.

11. Hau, J., J.G. Westergaard, P. Svendsen, A. Bach and B. Teisner. 1981. Comparison of pregnancy-associated murine protein-1 and human pregnancy zone protein. J. Reprod. Immunol. 3, 341-349.

12. Waites, G.T. 1984. Studies on murine pregnancy-associated serum proteins. PhD thesis. University of Bristol.

13. Waites, G.T., A.M Bell and S.C. Bell. 1983. Acute-phase serum proteins in syngeneic and allogeneic mouse pregnancy. Clin. Exp. Immunol. 53, 225-232.

14. Coe, J.E. 1977. A sex-limited serum protein of Syrian hamsters: definition of female protein and regulation by testosterone. Natl. Acad. Sci. U.S.A 74, 730-733.

15. Coe, J.E., S.S. Margossian, H.S. Slayer and J.A. Sogn. 1981. Hamster female protein: a new pentraxin structurally and functionally similar to C-reactive protein and amyloid P-component. J. Exp. Med. 153, 977-991.

16. Badet, M.T. 1984. Specific and non-specific immunosuppressive factors in Salamander pregnancy serum. J. Reprod. Immunol. 6, 299-311.

17. Horne, C.H.W., S.S. Armstrong, A.W. Thomson and W.D. Thompson. 1983. Detection of pregnancy-asociated α_2-glycoprotein (α_2-PAG), an immunosuppressive agent, in IgA producing plasma cells and in body secretions. Clin. Exp. Immunol. 51, 631-638.

18. Mortensen, R.F., K. Beisel, N.J. Zeleznik and P.T. Le. 1983. Acute-phase reactants of mice. II. Strain-dependence of serum amyloid P-component (SAP) levels and response to inflammation. J. Immunol. 130, 885-889.

19. Coe, J.E. and M.J. Ross. 1983. Hamster female protein: a divergent acute-phase protein in male and female Syrian hamsters. J. Exp. Med. 157, 1421-1433.

20. Chemnitz, J., J. Hau, P.E. Svendsen and B.C. Christensen. 1982. Immunochemical localization of pregnancy-associated murine proteins in maternal and placental tissue. Biblthca. Anat. 22, 87-92.

21. Udagawa, Y., S.S. Armstrong, G.T. Waites, S.C. Bell, C.H.W. Horne and A.W. Thomson. 1985. Immunohistochemical localization of murine α_1-pregnancy-associated protein (α_1-PAP) in non-pregnant females - a comparative study with human pregnancy-associated α_2-glycoprotein (α_2-PAG). Clin. Exp. Immunol. (in press).

22. Udagawa, Y., S.S. Armstrong, B. Angus, A.W. Thomson, E.J. Sarcione and C.H.W. Horne. 1984. Immunohistochemical localization of pregnancy-associated alpha$_2$-glycoprotein (alpha$_2$-PAG) in the rat - a pointer to possible functional analogy with human alpha$_2$-PAG. Int. Symp. Materno-fetal Immunobiol. Cambridge, p. 21.

IMMUNOASSAY AND IMMUNOHISTOCHEMICAL LOCALIZATION OF MURINE PREGNANCY-ASSOCIATED PROTEIN (α_1-PAP) IN VIRGIN AND PREGNANT MICE: STUDIES ON 'LOW' AND 'HIGH' ENDOGENOUS STRAINS

Y. Udagawa, Sylvia S. Armstrong, A.W. Thomson

Immunopathology Laboratory, Department of Pathology,
University of Aberdeen, U.K.

Gillian T. Waites , S.C. Bell

Department of Obstetrics and Gynaecology,
University of Leicester, U.K.

Introduction

A pregnancy-associated serum protein of the mouse with α_1-electrophoretic mobility (α_1-PAP) and an estimated molecular weight of 150,000 has recently been detected in pregnant C57BL/10 mice and in females of the same strain after application of an inflammatory stimulus. Serum levels of α_1-PAP during pregnancy in female C57BL/10 mice mated with males of various strains have also been documented (1). The serum profile of this protein, which reaches maximum levels at mid-pregnancy, is similar to that of human pregnancy-associated α_2-glycoprotein (α_2-PAG) during pregnancy (1,2), although unlike α_2-PAG levels of α_1-PAP fall prior to parturition. Several observations have led to the recent suggestion that murine α_1-PAP might be analogous to human α_2-PAG (see Waites and Bell, this volume). Although undetectable in the serum of normal male and female C57BL/10 mice (1) α_1-PAP has recently been demonstrated in sera of normal female MF1 mice (3).

Using a combination of indirect immunoperoxidase (PAP) and indirect immunofluorescence staining procedures, we have demonstrated α_1-PAP within Ia-positive tissue leucocytes in normal female MF1 but not C57BL/10 mice (3). These cells included

Pregnancy Proteins in Animals

mononuclear cells within gut-associated lymphoid tissue (GALT) (lamina propria and Peyer's patches). Our results suggested that, like human α_2-PAG (4,5), murine α_1-PAP might also be synthesized by mononuclear leucocytes, such as B cells and monocytes. Some α_1-PAP positive hepatocytes were also observed in MF1 mice, which suggested that, as with other murine acute-phase proteins (6-8), the liver might be one of the sites of α_1-PAP synthesis.

The aims of the present study were to determine (i) whether α_1-PAP-positive cells could be demonstrated in MF1 and C57BL/10 mice during pregnancy; (ii) whether serum levels of α_1-PAP in MF1 and C57BL/10 strains might be related to the incidence of α_1-PAP-positive tissue leucocytes or hepatocytes and (iii) whether α_1-PAP could be demonstrated within the feto-placental unit.

Materials and Methods

Animals

Virgin female and syngeneically pregnant mice were obtained from colonies maintained at the University Animal Department, Foresterhill, Aberdeen (MF1) and at the University of Leicester (C57BL/10). Sprague-Dawley rats were bred at the University of Aberdeen.

Antisera

Monospecific rabbit antiserum against mouse α_1-PAP, prepared as described previously (1), was used in the indirect PAP staining technique. The swine anti-rabbit and PAP reagents used in the immunoperoxidase staining procedure were obtained from Dakopatts A/S, Denmark. Monoclonal rat anti-mouse Ia was purchased from Sera-Lab. Ltd. Human α_2-PAG antiserum, raised in rabbits, was from Behringwerke AG, West Germany.

Immunoassay of murine α_1-PAP

Serum concentrations of α_1-PAP in mouse sera were measured by rocket immunoelectrophoresis (9), using rabbit anti-mouse α_1-PAP as specified by Waites and Bell (1). Levels of α_1-PAP were expressed as arbitrary units per ml (AU/ml), where 1 AU was defined as the peak height obtained with a 1:30 dilution of a standard preparation of pooled pregnancy serum. The lower limit of sensitivity of the assay was 0.03 AU/ml and it exhibited a precision of $\pm$ 3%.

Tissues

The following tissues obtained from groups of four virgin or pregnant mice (8th, 12th or 18th day of gestation) and from non-pregnant Sprague-Dawley rats were fixed in acetic acid formol saline: small intestine, colon, liver, Peyer's patch and feto-placental units. Biopsies of the human tissues were fixed in neutral buffered formalin. These tissues were embedded in paraffin and sections cut at 5 μm.

Indirect PAP staining for mouse α_1-PAP

Sections of each tissue were stained by the indirect immunoperoxidase (PAP) technique of Gallon et al (11). The rabbit anti-mouse α_1-PAP (IgG fraction) was used at a dilution of 1:50; the swine anti-rabbit and the PAP complex, each at a dilution of 1:100. Peroxidase activity of the specifically bound PAP immune complex was demonstrated by the diaminobenzidine reaction.

Absorption controls

The specificity of the rabbit anti-mouse α_1-PAP antiserum used in the study was verified by absorption with partially purified mouse α_1-PAP (12), which markedly reduced subsequent staining for α_1-PAP using the immunoperoxidase procedure.

Enumeration of α_1-PAP positive cells

For each animal, the mean incidence of α_1-PAP positive cells in five arbitrarily selected fields per section of tissue was determined at a magnification of x 400, using an eyepiece grid which defined an area of $^1/_{64}$mm^2.

Immunofluorescence staining for mouse Ia

PAP stained sections were rinsed through phosphate-buffered saline (PBS), then incubated with rat anti-mouse Ia (1:150 dilution) for 30 min at room temperature. Sections were then rinsed twice in PBS and FITC rabbit anti-rat Ig (Sera Lab. Ltd.) applied undiluted for 30 min. After rinsing, the sections were mounted in buffered glycerine with paraphenylenediamine.

Statistics

The significance of differences between means was established using Student's 't' test.

Results

Serum levels of α_1-PAP in MF1 mice

The mean serum concentration of α_1-PAP in virgin MF1 females was 15.7 $\pm$ 5.6 AU/ml (n = 7). However, α_1-PAP levels increased significantly to day 12 (mid gestation), representing an approximate two-fold increase (27.3 $\pm$ 8.1 AU/ml; n = 6) ($0.01 < p < 0.001$). Values then fell during the remainder of gestation, until, at day 18, serum levels of α_1-PAP (4.4 $\pm$ 1.9 AU/ml; n = 6) were below those of virgin female mice ($p < 0.001$) (see Fig 5).

Serum levels of α_1-PAP in C57BL/10 mice

Although originally undetectable in the sera of virgin C57BL/10 females obtained from the colony maintained at the University of Bristol, in the group of C57BL/10 females employed in this study, very low levels of α_1-PAP could be detected in the sera of non-pregnant females (0.37 $\pm$ 0.31 AU/ml; n = 20) which increased almost 200-fold by day 12 of pregnancy (68.9 $\pm$ 19.0 AU/ml; n = 14) ($p < 0.001$) and subsequently fell until at day 18 the mean serum concentration was 16.5 $\pm$ 5.7 AU/ml (n = 10) (see Fig 5).

α_1-PAP positive cells in virgin female MF1 and C57BL/10 mice

Examination of small intestine, colon, Peyer's patches and liver obtained from normal, virgin MF2 females revealed α_1-PAP positive cells in each of these sites. Mononuclear cells with heavily staining cytoplasm could readily be observed in the lamina propria of gut mucosa (Fig 1); in Peyer's patches, these cells were more numerous (Fig 2). Within the liver however, α_1-PAP positive cells (hepatocytes) were relatively scarce (Fig 3). In contrast to MF1 females, virgin C57BL/10 mice showed a very low incidence of α_1-PAP staining cells in any of these tissues (Fig 4a).

Incidence of α_1-PAP positive cells during pregnancy

As with serum α_1-PAP levels, the incidence of α_1-PAP staining leucocytes in gut associated lymphoid tissue rose during gestation (Fig 4b). In MF1 mice, the increase from non-pregnant values to day 12 gestation was approximately two-fold for small bowel mucosa and Peyer's patches (Table 1). However, over the same period, no increase of equivalent significance in the numbers of staining cells within large bowel mucosa or liver was observed. Between days 12 and 18 of gestation, there was a marked fall in the mean incidence of α_1-PAP staining tissue leucocytes.

In C57BL/10 mice, a more dramatic increase in the incidence of staining mononuclear cells was found, the levels attained on day 12 of gestation being similar to those in the low endogenous strain (Table 2). As with MF1 females, α_1-PAP positive mononuclear leucocytes became less evident towards term (day 18) whilst staining hepatocytes reamained comparatively few in number throughout gestation. The relationships between serum α_1-PAP levels and the incidences of α_1-PAP positive cells in each mouse strain are summarized in Fig 5.

Identity of α_1-PAP positive mononuclear leucocytes

Further examination of α_1-PAP positive leucocytes in Peyer's patches using a combined indirect immunoperoxidase (PAP)/indirect

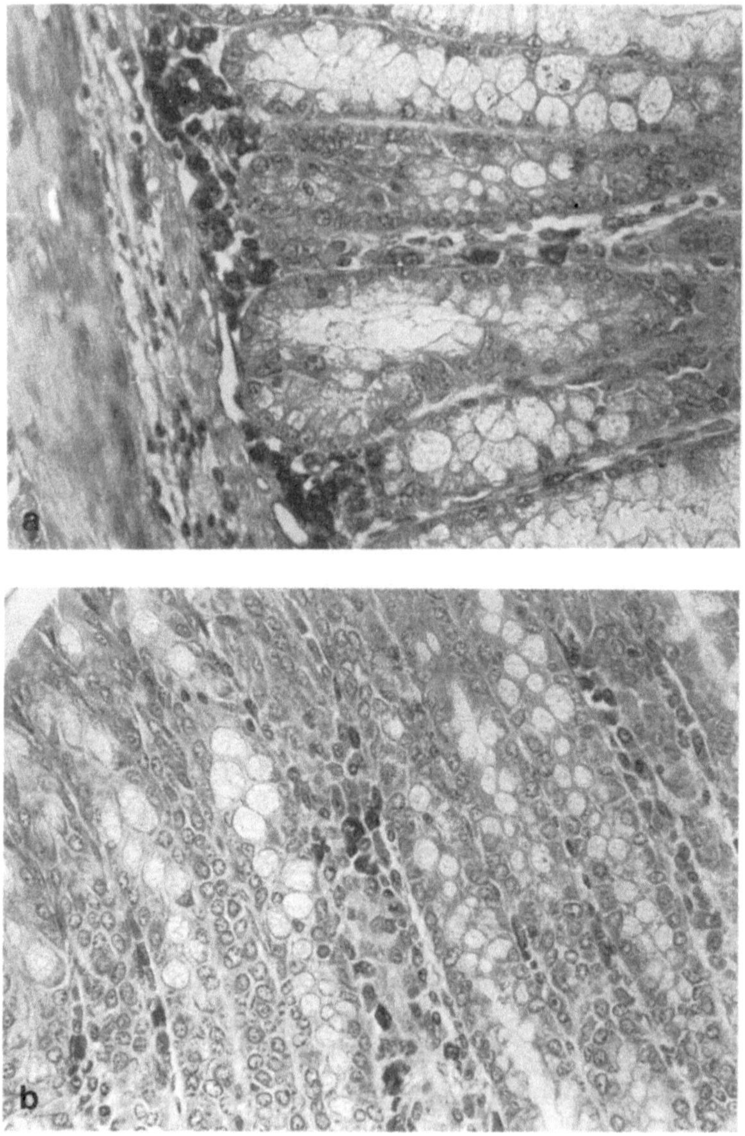

Fig 1 Non-pregnant MF1 mouse intestine stained for α_1-PAP using the indirect immunoperoxidase (PAP) procedure. (a), small bowel and (b) colon, showing α_1-PAP positive cells in the lamina propria (x 600).

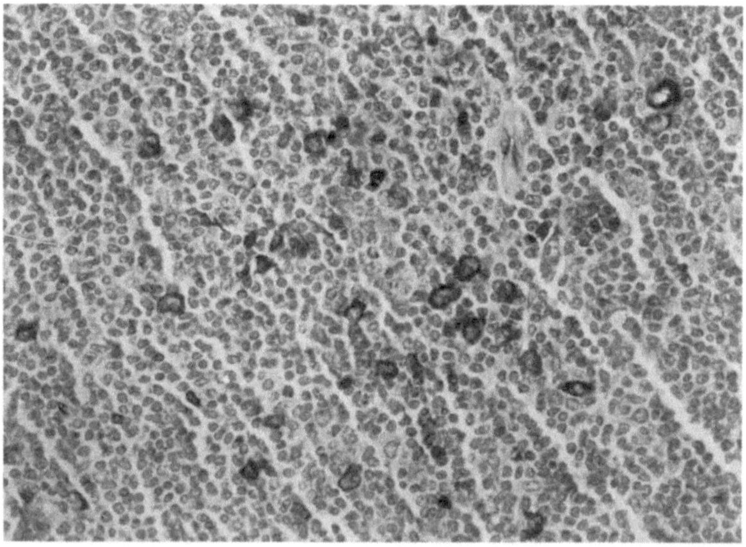

Fig 2 Non-pregnant MF1 mouse Peyer's patch. Cytoplasmic staining for α_1-PAP is clearly visible within numerous mononuclear leucocytes (macrophages) (x 600)

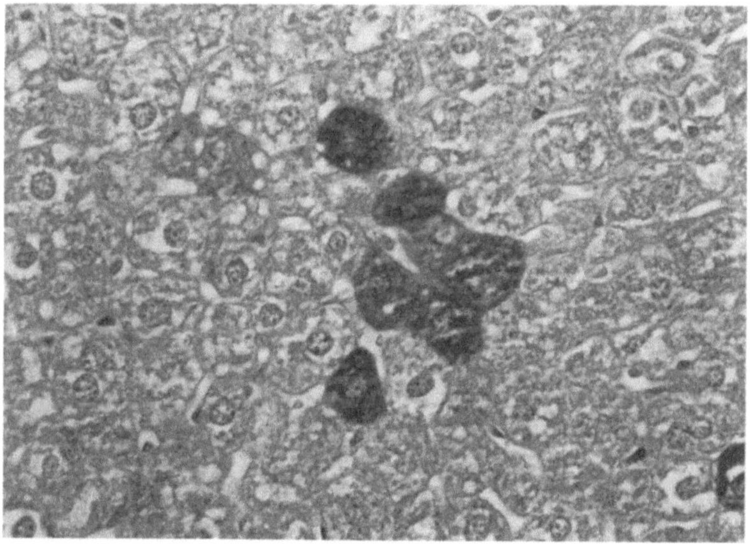

Fig 3 α_1-PAP containing hepatocytes within the liver parenchyma of a non-pregnant MF1 mouse (x 600)

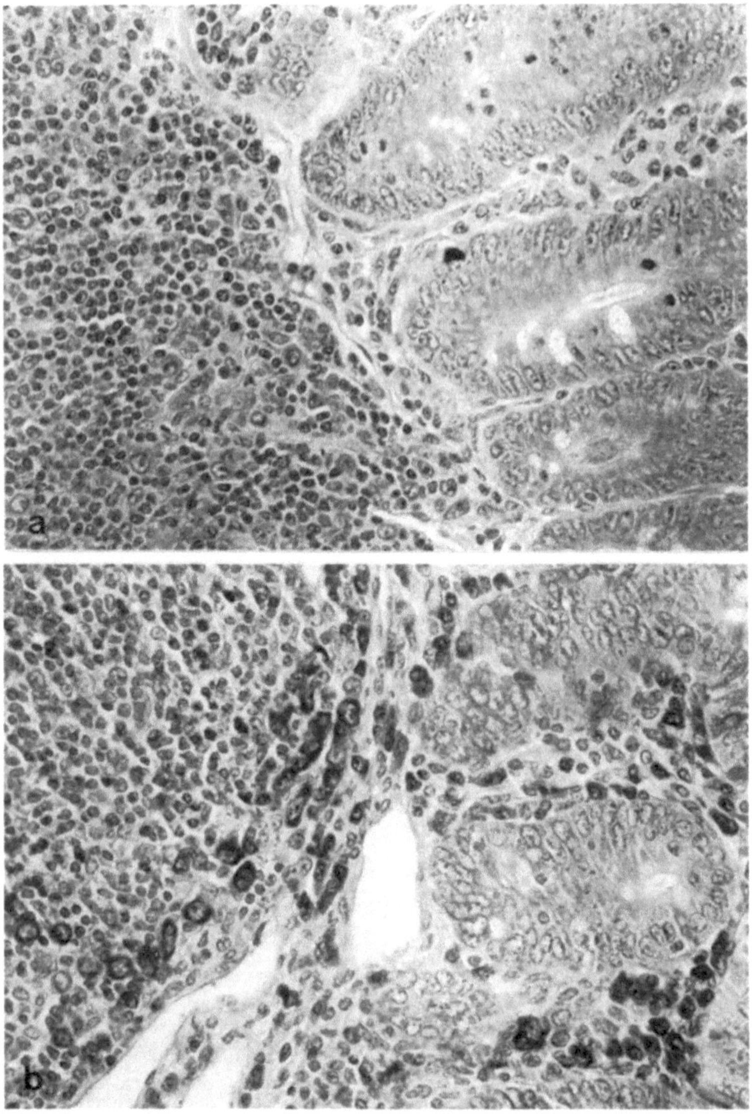

Fig 4 C57BL/10 mouse small bowel, stained by the indirect PAP technique for α_1-PAP. (a), virgin female; no positively staining cells are visible in either the mucosa (right) or the adjacent lymphoid tissue (left). (b), pregnant female (day 12); cytoplasmic staining for α_1-PAP is clearly visible in numerous mononuclear leucocytes within both the lamina propria and the adjacent lymphoid tissue. (a and b x 600)

TABLE 1 Incidence of α_1-PAP Positive Cells in Tissues of MF1 Mice

TISSUE	VIRGIN♀ (n = 4)	PREGNANT ♀ 12th day (n = 5)	PREGNANT ♀ 18th day (n = 4)
Small bowel mucosae	23.6 ± 5.3	†49.6 ± 7.6	25.0 ± 9.2
Large bowel mucosae	31.4 ± 5.5	*38.4 ± 7.0	15.9 ± 6.9
Peyer's patch	57.8 ± 13.1	#105.1 ± 34.2	59.8 ± 40.9
Liver	3.2 ± 0.6	3.6 ± 1.3	2.3 ± 0.8

Significance compared to virgin by Student's 't' test
†$p < 0.001$; # $0.01 < p < 0.02$; * $0.02 < p < 0.05$
Values are means ± 1S.D.

TABLE 2 Incidence of α_1-PAP Positive Cells in Tissues of C57BL/10 Mice

TISSUE	VIRGIN♀ (n = 3)	PREGNANT ♀ Day 8 (n = 1)	PREGNANT ♀ Day 12 (n = 5)	PREGNANT ♀ Day 17 (n = 2)
Small bowel mucosae	2.6 ± 1.9	†22.2 ± 3.9	†48.7 ± 6.6	†29.1 ± 6.1
Large bowel mucosae	2.6 ± 1.9	17.6 ± 7.5	†35.8 ± 12.2	†20.7 ± 1.3
Peyer's patch	very low	ND	82.7 ± 27.3	60.4 ± 11.8
Liver	*very low	2.0 ± 1.6	1.8 ± 1.0	2.0 ± 3.1
Placenta	ND	ND	++	+
Decidua	ND	++	+	ND

ND - not determined or non-applicable; +,++ - relative frequency and intensity of stained cells within one tissue; * - perivascular and periportal only

Significance compared to virgin by Student's 't' test † $p < 0.001$
Values are means ± 1S.D.

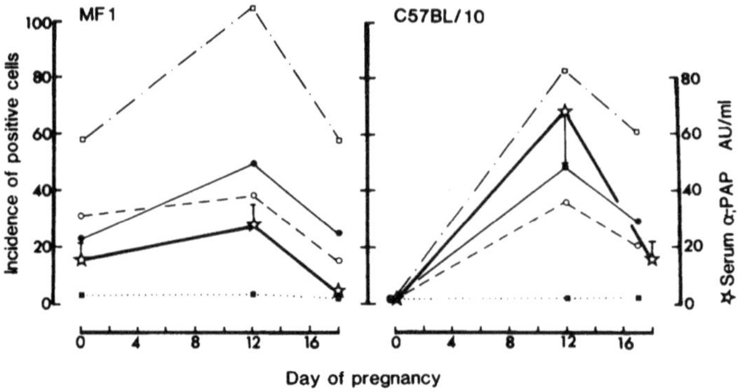

Figure 5 Mean incidence of α_1-PAP positive cells in various tissues of virgin and pregnant mice assessed by density counts of positive cells in 5 arbitrarily selected fields per tissue per animal, viewed under x 10 eyepiece, x 40 objective Tissues:
● small bowel mucosae
○ large bowel mucosae
□ Peyer's patch
■ liver
☆ serum α_1-PAP levels (mean ± 1 S.D.)

immunofluorescence procedure revealed that these cells were Ia positive. Moreover, on morphological criteria, they were distinct from lymphocytes and plasma cells and had the appearance of macrophages.

α_1-PAP positive cells within placenta

Examination of the placenta on day 12 of pregnancy revealed extensive staining for α_1-PAP on cells lining the maternal blood vessels within the labyrinthine placenta (Fig 6a & b). These cells appeared to represent cytotrophoblastic cells and similar

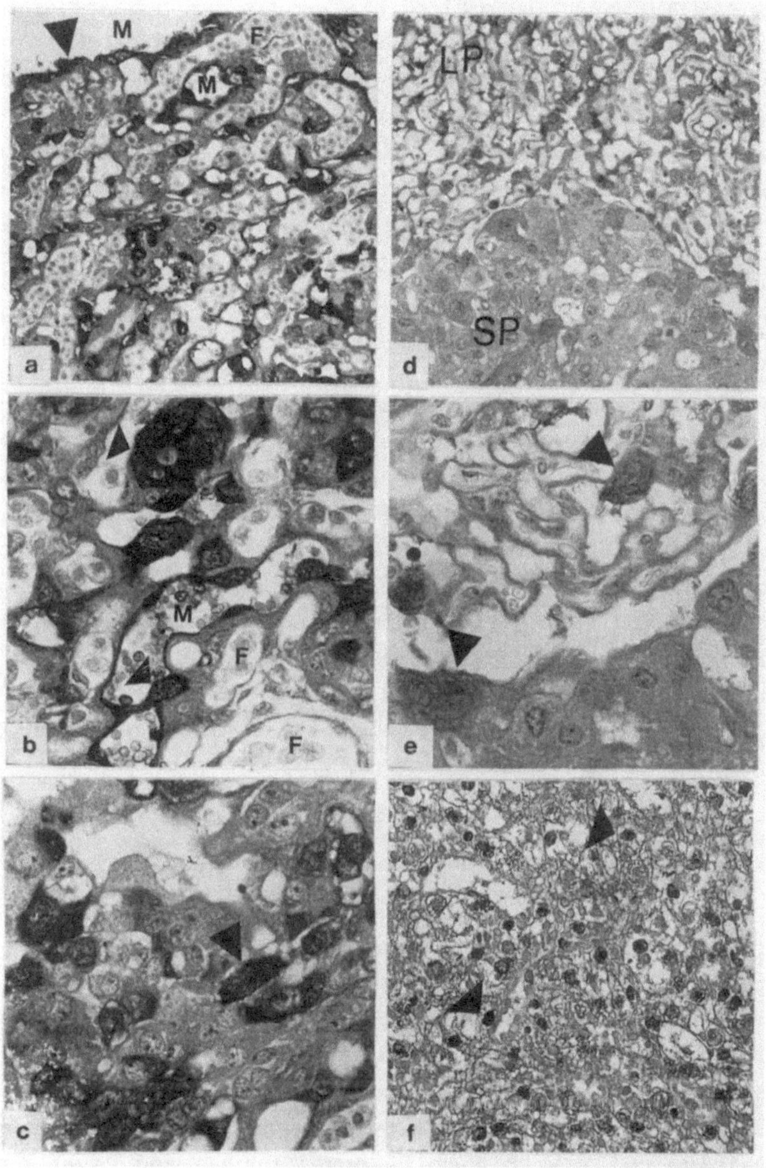

Fig 6 C57BL/10 feto-placental units and decidua during pregnancy.
(a) labyrinthine placenta, day 12 (x). M and F refer to maternal and fetal blood vessels. Arrowed α_1-PAP positive cells lining maternal blood vessel.
(b) same as 1 except x.
(c) spongiotrophoblast region of placenta, day 12 (x).
(d) labyrinthine (LP) and spongiotrophoblast (SP) regions of placenta, day 17 (x).
(e) same as 1 except x.
(f) maternal decidual tissue, day 8 (x). Arrowed numerous small round α_1-PAP positive cells.

positive cells were observed in the spongiotrophoblast region of the placenta (Fig 6c), cells with which cytotrophoblastic cells of the labyrinth are suggested to be contiguous. On day 17 of pregnancy, a dramatic fall in the level of staining for α_1-PAP was observed in both regions (Fig 6d & e). On day 8, numerous α_1-PAP positive cells were noted in the maternal region of the placenta, the decidua, and these had the morphological appearance of macrophages.

Cross-reactivity

Following the demonstration of immunological cross-reactivity between murine α_1-PAP and a protein in the sera of female rats (Waites and Bell, this volume) anti-mouse α_1-PAP was employed on tissues from non-pregnant female rats of the Sprague-Dawley strain. The localization of α_1-PAP positive cells was similar to that seen in equivalent MF1 mouse tissues, although with increased background staining. Furthermore, although murine α_1-PAP antiserum exhibited no cross reaction with human α_2-PAG in human tissue sections (gut, breast and renal carcinoma), anti-α_2-PAG serum did stain similar cells in sections of mouse colon.

Discussion

In this paper, we have confirmed our previous observations (3) that non-pregnant female mice of the MF1 and C57BL/10 strains exhibit respectively, 'high' and 'low' endogenous serum levels of α_1-PAP and that these levels correspond to the incidence of α_1-PAP staining leucocytes within gut-associated lymphoid tissue of these strains. We have further shown however, that as α_1-PAP serum levels increase during pregnancy, so too does the incidence of α_1-PAP positive cells within small bowel mucosa and Peyer's patches. In C57BL/10 mice, this increase in numbers of cells containing α_1-PAP is particularly dramatic, and at day 12 of gestation, the incidence of positive cells within the above sites matches that in MF1 females. In contrast, no significant change in the relatively small number of α_1-PAP positive hepatocytes was recorded during pregnancy in either strain.

In view of the suggested analogy between α_1-PAP and human α_2-PAG and the association between the latter protein and blood and tissue leucocytes in man, the identity of those mononuclear leucocytes containing cytoplasmic α_1-PAP is of particular interest. Human α_2-PAG has been demonstrated on the surface membrane of B lymphocytes and monocytes (13) and within IgA-secreting plasma cells in sites of IgA secretion (14). In our previous study (3), we found that, using a double staining procedure, α_1-PAP appeared to be associated with only a small proportion of IgA-producing plasma cells in non-pregnant MF1 mice. More recent immunohistochemical evidence obtained in this laboratory (Thomson and Sewell, unpublished) indicates that α_1-PAP-positive cells within murine lymphoid tissue are quite distinct from plasma cells. Indeed, further investigations are in progress to confirm whether, as indicated by the present study, α_1-PAP is synthesized by tissue macrophages.

In addition to tissue leucocytes, it is clear from these studies that other sites of α_1-PAP synthesis may be involved. It has been suggested that the murine pregnancy-associated protein, PAMP-1, which is physicochemically and immunochemically similar to α_1-PAP (see Waites and Bell, this volume) may be synthesized by the liver during pregnancy (15). However, on the basis of our findings, it seems unlikely that α_1-PAP is produced by the liver during gestation, although, as with other murine acute-phase reactants, synthesis of α_1-PAP by hepatocytes may occur following injury.

The demonstration of α_1-PAP positive staining in the rodent placenta, both membraneous and cytoplasmic, suggests that there is some contribution by the feto-placental unit to α_1-PAP synthesis during pregnancy. This has also been shown for PAMP-1 (15) although the apparent reduction in intensity of α_1-PAP staining towards parturition argues against the suggestion that the placenta binds the protein thereby causing serum levels to fall (16).

Our demonstration of α_1-PAP positive cells in decidua suggests an additional site of synthesis associated with the feto-placental unit. Recent studies have demonstrated α_1-PAP positive cells in GALT and deciduoma of pseudopregnant C57BL/10 females and furthermore, serum levels seem to parallel maturation of the deciduoma (Waites and Bell, unpublished). Further studies are in progress to elucidate the role of the decidua in α_1-PAP synthesis.

The differences between MF1 and C57BL/10 mice in their endogenous serum and tissue levels of α_1-PAP, together with the effects of pregnancy, raise important issues concerning the initiation/ regulation of α_1-PAP synthesis. Endogenous levels of the major murine acute-phase protein serum amyloid P-component are genetically controlled, both in normal animals and following injury (17-18). Moreover, there is evidence that synthesis of α_1-PAP, like human α_2-PAG is under steroid hormone control. However, the evidence tends to suggest that different regulatory mechanisms are operative in these different physiological situations (see Waites and Bell, this volume).

The finding of equivalent staining of α_1-PAP in tissues of the mouse and rat and of positive staining with anti-human α_2-PAG in murine colonic mucosa supports the suggested analogy between these proteins. Furthermore, a similar staining pattern has been demonstrated in rat tissues using anti-rat α_2-PAG (kindly provided by Dr E. J. Sarcione) (Udagawa, unpublished) and it is likely that a number of recently described rodent proteins all represent analogues of human α_2-PAG (see Waites and Bell, this volume). These particular murine pregnancy proteins may prove invaluable in determining the functional signficance of human α_2-PAG in fetal survival.

Acknowledgements

G.T.W. was in receipt of a Postgraduate Studentship from the Medical Research Council. We are grateful to Dr. H.F. Sewell for valuable discussion, to Mr. George King for excellent technical assistance.

References

1. Waites, G.T., S.C. Bell. 1984. Identification of a murine pregnancy-associated serum protein (α_1-PAP) as a female-specific acute-phase reactant. J. Reprod. Fert., 70, 581.

2. Thomson, A.W., C.H.W. Horne. 1980. Biological and clinical significance of pregnancy-associated α_2-glycoprotein (α_2-PAG): A review. Invest. Cell Pathol., 3, 295.

3. Udagawa, Y., S.S. Armstrong, G.T. Waites, S.C. Bell, C.H.W. Horne, A.W. Thomson. 1985. Immunohistochemical localization of murine α_1-pregnancy-associated protein (α_1-PAP) in non-pregnant females - a comparative study with human pregnancy-associated α_2-glycoprotein (α_2-PAG). Clin. exp. Immunol. In Press.

4. Stimson, W.H., J.C. Blackstock. 1975. Synthesis of a pregnancy-associated α-macroglobulin by human leucocytes. Experienta, 31, 371.

5. Horne, C.H.W., A.W. Thomson, C.M. Towler, F.K. Macmillan, L.M. Gibb. 1978. Relationship of pregnancy-associated α_2-glycoprotein (α_2-PAG) to peripheral blood leucocytes. Scand. J. Immunol. 8, 75.

6. Benjamin, D.C., H.E. Weimer. 1966. Synthesis of α_2-AP (acute phase) globulin of rat serum by the liver. Nature, 209, 1032.

7. Koj, A. 1974. Acute-phase reactants. Their synthesis, turnover and biological significance. In: Structure and Function of Plasma Proteins Vol. 1 (ed. by A.C. Allison) pp. 73-132. Plenum Press, London.

8. Baltz, M.L., R.F. Dyck, M.B. Pepys. 1980. Amyloid P-component in mice injected with casein: identification in amyloid deposits and in the cytoplasm of hepatocytes. Immunology, 41, 59.

9. Laurell, C.B. 1972. Electroimmunoassay. Scand. J. Clin. Lab. Invest., 29, Suppl. 125, 21.

10. Gallon, M.E., W.A. Reid, G.A. McHardie, R. Hardman, G.D. Smith, C.H.W. Horne, V.V. Kalashnikov, Y.S. Tatarinov. 1981. Localization of embryonic prealbumin in formalin-fixed human fetal and adult tissue. J. Clin. Pathol., 34, 764.

11. Waites, G.T. 1984. Studies on murine pregnancy-associated serum proteins. Ph.D. Thesis, University of Bristol.

12. Thomson, A.W., C.B.J. Hunter, N. Cruickshank, C.H.W. Horne. 1979. Study of pregnancy-associated α_2-glycoprotein in relation to populations of human blood leucocytes. Int. Arch. Allergy Appl. Immunol., 58, 251.

13. Horne, C.H.W., S.S. Armstrong, A.W. Thomson, W.D. Thompson.

1983. Detection of pregnancy-associated α_2-glycoprotein (α_2-PAG), an immunosuppressive agent, in IgA producing plasma cells and in body secretions. Clin. exp. Immunol., 51, 631.

14. Chemnitz, J., J. Hau, P. Svendson, J. Folkerson, J.G. Westergaard, B.C. Christensen. 1982. Immunohistochemical demonstration of human and murine pregnancy-associated serum proteins in maternal and placental tissue. Biblthca. anat., 22, 87.

15. Hau, J., B. Teisner, P.E. Svendsen. 1983. Placental affinity for pregnancy-associated murine protein-1 examined by crossed affino-immunoelectrophoresis. Electrophoresis, 4, 369.

16. Pepys, M.B., M. Baltz, K. Gomer, A.J. Davies, M. Doenhoff. 1979. Serum amyloid P-component: an acute phase reactant in the mouse. Nature, 278, 259.

17. Mortensen, R.F., K. Beisel, N.J. Zeleznik, P.T. Le. 1983. Acute-phase reactants of mice. II. Strain dependence of serum amyloid-P-component (SAP) levels and response to inflammation. J. Immunol., 130, 885.

18. Waites, G.T., S.C. Bell. 1984. Regulation of murine α_1-pregnancy-associated protein by gonadal steroids during the acute-phase response. J. Endocr., 101, 315.

ANIMAL PROTEINS IMMUNOLOGICALLY CROSS-REACTING WITH HUMAN PREGNANCY-ASSOCIATED α_2-GLYCOPROTEIN (PZP, α_2PAG, PAα_2G)

J. Hau

Department of Pathology, Laboratory Animal Unit, Royal Veterinary and Agricultural University, Copenhagen, Bülowsvej 13, DK-1870 Frederiksberg C, Denmark

Introduction

Since the discovery of the human "pregnancy zone protein" (PZP) in the late fifties (1,2) this protein has been given different names (Table 1), and the abbreviation PZP has been adopted in this paper.

Table 1. Synonyms for pregnancy zone protein

Synonym	Abbreviation	
pregnancy zone protein	PZP	(1,3)
Xh-protein	Xh	(4)
Xm-factor	Xm	(5)
pregnancy-associated globulin	Pa 1	(6)
Schwangerschafts-assoziierte α_2-Glykoprotein	SP-3	(7)
pregnancy-associated α_2-macroglobulin	PAM	(8)
α_2-pregnoglobulin	α_2P	(9)
pregnancy-associated α_2-globulin	α_2PAG	(10)
pregnancy-associated α_2-glycoprotein	PAAG	(11)

The detection and naming of potential animal analogues has not minimized the confusing nomenclature. Although some animal proteins were believed to be analogous to human PZP based on a mere physico chemical likeness and named accordingly (12,13) others were not claimed to be analogous, and they were naturally not given the name of the human protein (14-18). Finally there is a category of animal analogues which were detected and isolated utilizing a direct cross-reaction and the consequent affinity between antisera raised against human PZP and proteins present in serum of pregnant animals of various species, mainly primates (19-26).

The present paper deals with a comparison of the proteins analogous to human PZP found in the popular laboratory animal species mouse, rat, and rabbit.

Immunological Cross-Reaction between Human PZP and Murine Pregnancy Associated Murine Protein 1 (PAMP-1)

Immunologically cross-reacting rodent analogues to human PZP have been difficult to demonstrate (27) indicating major structural differences between PZP and the analogous proteins in rodents. However, using antisera raised in hens, it was possible to demonstrate immunological cross-reaction between pregnancy-associated murine protein-1 (PAMP-1) (14) and human PZP (27).

Figure 1 shows an example of this cross-reaction, which was demonstrated using counter current line immonoelectrophoresis.

Using hen anti-PZP in the antibody-containing gel and human pregnancy serum in the antigen-containing line gel a double line precipitate was formed indicating that the hen antiserum reacted with two different subpopulations of PZP molecules (Fig. 1).

When rabbit and goat antiserum against PZP were applied to the wells in the neutral gel they reacted with negative deflections of both lines, whereas purified PZP reacted with positive deflections of both lines. Hen antiserum against both PZP and PAMP-1 reacted with negative deflections of both lines indicating cross-reactions between PZP and PAMP-1. Serum fron non-immunized hens and rabbit and goat anti-PAMP-1 caused no deflections. PZP in serum of pregnant women caused positive deflections, whereas PAMP-1 in serum of mice did not affect the line precipitate.

PAMP-1 did not show any immunological cross-reaction with any of the following human pregnancy proteins: pregnancy-specific β_1-glycoprotein (SP-1), steroid-binding β-globulin (SP-2), placental lactogen (HPL), chorionic gonadotrophin (HCG), and pregnancy-associated plasma protein-A (PAPP-A). By contrast, human cortico-steroid-binding globulin (CBG) seemed to interact with PZP, when rabbit anti-CBG was tested in the assay, and thus a cross-reaction between CBG and PZP cannot be excluded.

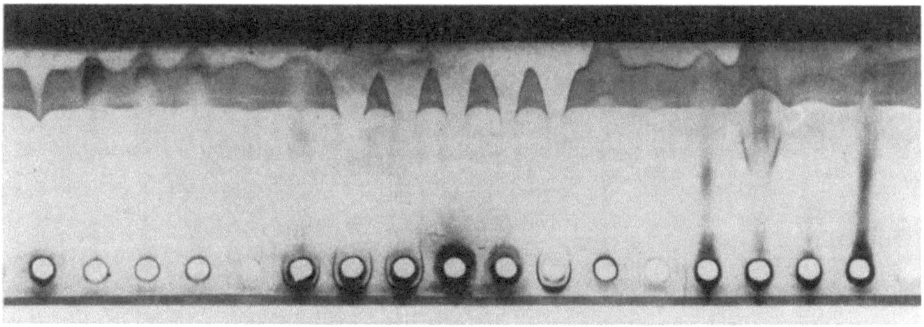

Fig. 1. Counter current line immunoelectrophoresis with hen anti-PZP in the upper gel and serum from pregnant women in the lower gel. The contents of the wells were: 1, hen anti-PAMP-1, 2, serum of male mice, 3, serum of non-pregnant female mice, 4, serum pregnant mice, 5, purified human CBG, 6, rabbit anti-CBG, 7, rabbit anti-PZP, 8, goat anti-PZP, 9, hen anti-PZP unabsorbed, 10, hen anti-PZP absorbed, 11, rabbit anti-PZP (DAKO Immunoglobulins, Copenhagen), 12, purified PZP, 13, rabbit anti-PAMP-1, 14, goat anti-PAMP-1, 15, serum of pregnant women, 16, serum of non-pregnant women, 17, serum of non-immunized hens, 18, buffer.

Cross-reaction studies of proteins in different species in general depend on the reaction between antibodies and antigens, where antibodies raised against a protein in one species react with a cross-reacting antigen from another species. This type of cross-reaction could not be found for PAMP-1 and PZP gel precipitation systems. The reaction in line immunoelectrophoresis and the absence of a reaction in direct gel precipitation experiments indicate that PAMP-1 and human PZP do not have sufficient antigen determinants in common to allow formation of precipitates when the antigen is reacting with antibodies raised against the antigen analogue. Consequently, the line immunoelectrophoretic method, which is more sensitive and not dependent on the presence of precipitating antibodies, was used. This method is based on a reaction between the antigen in the line gel and the antibodies in the wells raised against the antigen analogue prior to the formation of the line precipitate. The initial reaction prevents a number of the antigen molecules from reacting with its antibody in the antibody-containing gel resulting in a negative deflection of the line precipitate.

The cross-reaction found in this study indicates the advantages in using a species phylogenetically remote from man and rodents as antibody producer, and it should thus in theory be possible to produce a monoclonal antibody with specificity against an epitope common to the PZP of man and rodents. However, it will probably be impossible to produce a monoclonal antibody with specificity against a common epitope in a phylogenetically closely related species, since this species will only react with antibody production against PZP epitopes not present on the surface of its own PZP. This (or these) epitope(s) is (are) not likely to be shared by two different, but related mammal species.

Cross-Reaction between Murine Proteins Cross-Reacting with Human PZP and Pregnancy Proteins in the Rat and Rabbit

Following the first demonstration of immunological cross-reaction between PZP and PAMP-1 (26) it has been possible to demonstrate cross-reaction between PAMP-1 and pregnancy proteins of various rodent species isolated by other groups working in this field (28). Table 2 shows the result of the present cross-reaction studies conducted by line immunoelectrophoresis and earlier results obtained by direct precipitation techniques.

Table 2. PZP/PAMP-1 Cross-reacting proteins

Species	Protein	Method		Ref.
Monkey	Sp-3 (Bohn FRG)	Direct Precipitation		(20)
"	PZP (Lin et al. USA)	"	"	(22)
"	PZP (Teisner et al. DK)	"	"	(25)
Dog	PZP (von Schoultz et al. S)	"	"	(23)
"	SP-3 (Bohn, FRG)	"	"	(20)
Cow	PZP (von Schoultz et al. S)	"	"	(23)
Swine	PZP (Martinsson & Carlström S)	"	"	(24)
Rabbit	PALP-2 (Matsushita Jpn)	Line immunoelectrophoresis		this work
Rat	PAG (von Porstmann et al. GDR)	"	"	this work
"	α_2-PAG (Sarcione & Dellumo USA)	"	"	this work
Mouse	α_1-PAP (Waites & Bell UK)	"	"	this work

The proteins in the rabbit, rat, and mouse abbreviated PALP-2, PAG, α_2PAG, α_1PAP, and PAMP-1 appear to be immunochemically very similar and polyclonal antisera raised against one of the proteins are most likely usable for assays analysing the analogue in the other rodent species.

The biological and physico-chemical properties of the PAMP-1 and rat PAG have been found to be very similar (28).

By contrast, there are major differences between these rodent proteins and human PZP with respect to physico-chemical properties (28-31). It seems, however, that the hormonal regulation (32-34) and intracellular localization (35,36) of PAMP-1/α_1PAP, and PZP are quite similar.

To the extent that analogues to human PZP exist in rodents it seems reasonable, based on the findings described above, to suggest that PAMP-1/α_1PAP in the mouse, PAG/α_2PAG in the rat, and PALP-2 in the rabbit are the rodent analogues to human PZP.

References

1. Olivelli, F. and Ruggieri, P. 1958. Minerva Ginecologica 10, 953-958.
2. Smithies, O. 1959. Advances in Protein Chemistry 14, 65-113.
3. Afonso, J.F. and Farnham, N.G. 1962. Am.J.Obstet.and Gynec. 84, 199-205.
4. Bundschuh, G. 1966. Acta Biol. et Med.Germ. 17, 349-356.
5. Berg, K. and Bearn, A.G. 1966. J.Exp.Med. 123, 379-397.
6. MacLaren, J.A., Reid, D.E., Konugres, Angelyn, A. and Allen, F.H. 1966. Vox Sang. 11, 553-560.
7. Bohn, H. 1971. Arch.Gynäk. 210, 440-457.
8. Stimson, W.H. and Eubank-Scott, L. 1972. FEBS Letters 23, 298-302.
9. Berne, B.H. 1973. Fed.Proc. 32, 2616.
10. Than, G.N., Csaba, I.F., Szabõ, D.G., Karg, N.J. and Novák, P. 1975. Arch.Gynäk. 218, 125-130.
11. Berne, B.H. 1976. Prot.Biol. Fluids 24, 165-179.
12. Fischer, M., Bundschuh, G. and Aurich, I. 1978. Acta Biol.Med. Germ. 37, 1505-1507.
13. Sarcione, E.J. and Dellumo, D.S. 1981. Oncodev.Biol.Med. 2, 345-354.
14. Hau, J., Svendsen, P.E., Teisner, B. and Svehag, S.-E. 1978. J.Reprod. and Fert. 54, 239-243.
15. Hau, J., Svendsen, P., Teisner, B. and Thomsen Pedersen, G. 1980. J.Reprod.and Fert. 58, 389-393.
16. Porstmann, T. and Schmechta, H. 1980. Acta Biol. et Med.Germ. 39, 479-487.

17. Waites, G.T. and Bell, S.C. 1984. J.Reprod.Fert. 70, 581-589.

18. Matsushita, H. 1983. Jpn.J.Vet.Sci. 45, 85-89.

19. Hofmann, R., Straube, W., Klausch, B., Friemel, H. and Gunther, J. 1972. Arch. Gynäk. 212, 246-257.

20. Bohn, H., Ronneberger, H. 1973. Arch.Gynäk. 215, 277-284.

21. Lin, T.M., and Halbert S. 1978. Int.Archs. Allergy Appl. Immunol. 56, 207-223

22. Lin, T.M., Halbert, S.P. and Plasencia, R. 1976. Clin.Exp. Immunol. 26, 609-622.

23. Von Schoultz, B., Stigbrand, T., Martinsson, K. and Holmgren, N. 1976. Acta Endocrinol. 81, 379-384.

24. Martinsson, K. and Carlstrom, K. 1977. Zbl.Vet.Med.B. 24, 150-158.

25. Teisner, B., Folkersen, J., Svendsen, P., Hau, J. and Hindersson, P. 1979. J.Med.Primatol. 8, 298-304.

26. Matsushita, H. and Fijioka, F. 1980. Bull.Azabu Univ.Vet. Med. 1 (2), 237-240.

27. Hau, J., Westergaard, J.G., Svendsen, P., Bach, Annelise and Teisner, B. 1981. J.Reprod.Immunol. 3, 341-349.

28. Hau, J. and Porstmann, T. 1984. Lab.Anim. 18, 344-348.

29. Hau, J., Westergaard, J.G., Svendsen, P., Teisner, B., Bach, Annelise and Thomsen Pedersen, G. 1982. Placenta, Suppl. 4, 51-65.

30. Hau, J. 1982. In: Pregnancy Proteins: Biology, and Clinical Application (eds. J.G. Grudzinskas, B. Teisner and M. Seppälä). Academic Press, Sydney.

31. Hau, J., Svendsen, P., Chemnitz, J., Teisner, B. and Westergaard, J.G. 1983. Am.J.Reprod.Immunol. 3, 124-126.

32. Hau, J., Svendsen, P., Teisner, B. and Grudzinskas, J.G. 1982. J.Reprod.Fertil., 66, 273-275.

33. Westergaard, J.G. Bach, A., Teisner, B., Hau, J. and Grudzinskas, J.G. 1982. J.Reprod.Fertil. 66, 695-698.

34. Hau, J., Gidley-Baird, A.A., Teisner, B., Svendsen, P. and Westergaard, J.G. 1983. Acta Endocrinol. 104, 23-26.

35. Chemnitz, J., Hau, J., Svendsen, P., Folkersen, J., Westergaard, J.G. and Christensen, B.C. 1982. Bibl.Anat. 22, 87-92.

36. Udagawa, Y., Sylvia S. Armstrong and A.W. Thomson, These Proceedings.

PREGNANCY-SPECIFIC BETA$_1$-GLYCOPROTEIN IN RAT (RPG): ISOLATION, PHYSICO-CHEMICAL CHARACTERISTICS AND IMMUNOLOGIC STUDY.

Yu.S. Tatarinov, M.F. Kan, S.K. Krivonosov
USSR, Moscow 117861, 2nd Moscow Medical Institute, Department of Biochemistry

Introduction

For the first time human pregnancy-specific or trophoblast-specific beta$_1$-glycoprotein (TSG) was discovered in pregnant sera (1). Recently, this protein was found in human placenta (2) and pregnant blood plasma (3) and was named as SP-1 or PAPP-C. It is also designated as TBG and SBG. It was also revealed in blood serum of patients with trophoblastic tumours (4, 5), which surved the basis for the working out of immunochemical method of the diagnosis of this pathology.
By immunodiffusion, pregnancy-specific beta-globulins were also revealed in pregnant rat, mouse, rabbit and guinea-pig (6, 7). Discovery of functional analogs of human pregnancy-specific beta$_1$-glycoprotein in mammals will make it possible to create an experimental model for the study of its biological role during pregnancy, during the pathology of pregnancy, during the development of trophoblastic tumours.
In the present paper the results of the investigations on the isolation of rat pregnancy-specific beta$_1$-glycoprotein (RPG), on the study of its physico-chemical properties, and also the influence of this protein on the proliferative activity of human lymphocytes, stimulated by phytohemagglutinin (PhHA) and the reaction of the formation of E-rosettes are given.

Materials and Methods

The steps of the isolation and purification of rat pregnancy-specific beta$_1$-glycoprotein (RPG) are given in Table 1. Rat blood serum (2nd half of pregnancy) was obtained by decapitation under ether anaesthesia. Serum was mixed with the equal volume of 0,02 M solution of mercuric chloride and was left for 2 h by permanent mixing on magnetic mixer. Supernatant was fractionated with ammonium sulfate (50% of saturation). Precipitate was separated by centrifugation at 8000 rpm, dissolved in the minimum volume of distilled water and dialyzed against 0,05 M phosphate buffer pH 6.0 during twenty-four hours at +4°C.

Table 1
The Way of Isolation and Purification of RPG.

Steps of isolation and purification	Purity in %	Yield in %
1 Pregnancy blood serum of rats	-	100
2 Precipitation with 0,02 M $HgCl_2$	5	90
3 Treatment of supernatant with ammonium sulfate - 50% of saturation	20	55
4 Ionexchange chromatography on DEAE-32 Cellulose	45	30
5 Preparative isoelectrofocusing	90	24
6 Lyophilization	90	24

Then ionexchange chromatography on DEAE-32 Cellulose (firm "Whatman") was performed in 0,05 M phosphate buffer pH 6.0 at linear gradient of sodium chloride from 0 to 0,6 M, which was prepared on standard buffer. Gradient was set with the help of Ultragrade (firm "LKB", Sweden) at the rate of elution equal to 20 ml/h. For further work protein fraction, eluted from the column in the range of sodium chloride concentration from 0,2 to 0,4 M, was chosen. It was dialyzed against flowing water. Preparative isoelectrofocusing was carried out on Unifore ("LKB", Sweden) in a column with the working volume of 440 ml in density gradient of saccharose from 5% to 50% at pH gradient of ampholynes 3,5-10. Protein fraction with pH 5.8-6.6 was identified by us as purified RPG. Further the protein was purified from ampholynes and saccharose on DEAE-32 Cellulose, prepared on phosphate

buffer pH 6.0, ionic power 0,022. The washed preparation was eluted from Cellulose with 0,5 M solution of sodium chloride, prepared on standard buffer. It was lyophilized. The obtained protein has 90% purity with the yield about 24%.
Antiserum to RPG was obtained by immunization of rabbits with semi-purified preparation according to the scheme, described by us earlier (8). Molecular Weight of RPG was determined by the method of gel-filtration on the column 100x1,25 cm with Sephadex G-200, balanced with 0,01 M phosphate buffer pH 7.0, and also with the help of electrophoresis in polyacrylamide gel (PAAG) with SDS (sodium dodecyl sulfate) according to Laemmli (9). A set of proteins with known Molecular Weight ("Pharmacia Fine Chemicals", Sweden) were used as standards.
To study the relationship between RPG and acids and salts of some metals both organic (rivanol, trichloracetic, sulfosalicylic, picric acids, ethyl alcohol) and non-organic reagents (phosphoric tungstic, chloric acids, ammonium sulfate, copper sulfate) were used.
Analytic isoelectrofocusing was conducted in saccharose density gradient on the column of 110 ml and on standard plates ("Ampholyne PAG plate", "LKB", Sweden) on Multiphore ("LKB", Sweden) according to Osterman (10).
One-dimensional immunoeléctrophoresis was performed in agar gel ("Difco") in veronal-medinal buffer pH 8.5, ionic power 0,05 at tension 140 V and current 40 mA during 2 h. Relative electrophoretic mobility was determined according to the method (11). As standard protein serum albumin was used. Antiserum to albumin was obtained by immunization of rabbits with the corresponding protein preparation ("Reanal", Hungary), preliminary purified on Blue Sepharose. As electroneutral substance Dextran was used (M.W. 20 000 D).
Diffusion coefficient was determined as described (12) in 1% agar solution.
Amino acid analysis was carried out on Automatic Analyser Model G-2000, firm "Biotronic" (GFR).
Immunoelectrophoregrams of RPG were stained with Sudan Black in 70% ethanol and with Schiff's reagent.
Isolation of lymphocytes from human peripheral blood was performed according to Böym (13) in density gradient of Ficoll-Isopac.

Rosetteformation method was conducted according to modification, proposed by R.V. Petrov et al. (14). Reaction of blasttransformation of lymphocytes under the influence of mitogens (PhHA) was carried out according to the method by Ling (15), modified by N.L. Samoilina (16).
The obtained data were processed with the help of variation statistics, using Student's criteria.

Results

Molecular Weight of RPG, determined by the method of gel-filtration, is 130 000±4 000 D, and by electrophoresis with SDS - 85 000±2 000 D.
RPG during electrophoresis under non-denaturating conditions is found in the zone of $beta_1$-globulins, its relative electrophoretic mobility is 0,45±0,02, and diffusion coefficient - $2{,}7 \cdot 10^{-7}$ cm^2/sec. During isoelectrofocusing in saccharose density gradient RPG is focused as three stained bands 0,2-0,3 mm distant with pI: 5,85; 6,11 and 6,57. Similar results are received during analytic isoelectrofocusing on standard plates, where also three bands were revealed with pI: 5,83; 6,14 and 6,58. By disk electrophoresis of the samples, got during isoelectrofocusing, also three bands are revealed at a small distance from each other.

Table 2
Physico-Chemical Properties of RPG.

Physico-chemical parameters	Obtained results
Relative Molecular Weight:	
Gel filtration	130 000±4 000 D
Electrophoresis with SDS	85 000±2 000 D
Relative electrophoretic mobility	0,45±0,02
Molar coefficient of extinction	86 840±2 000
Isoelectric point	5,85; 6,11 and 6,57
Chemical nature	Glycoprotein

Table 3
Amino Acid Composition of RPG.

Amino acids	Molar percentage
Aspartic acid	9,53
Threonine	6,12
Serine	7,69
Glutaminic acid	13,58
Proline	6,95
Glycine	6,01
Alanine	7,65
Cystine	-
Valine	7,69
Methionine	traces
Isoleucine	2,71
Leucine	8,95
Tyrosine	4,09
Phenylalanine	4,45
Lysine	7,48
Tryptophan	1,14
Histidine	2,34
Arginine	3,63

The results of amino acid and carbohydrate analysis showed, that RPG molecule contains 90,16% of protein and 9,84% of carbohydrates, i.e. the antigen one can refer to the class of glycoproteins. The results of the determination of amino acid composition of RPG are given in Table 3. The constituent parts of carbohydrate component of the protein are: monosaccharides (mannose) - 7,47%, hexosamines (N-acetyl-D-glucosamine) - 2,37%. Neuraminic acid and its derivatives are not present. Therefore, it is clear that after incubation of the preparation with neuraminidase anodic electrophoretic mobility of the protein does not change.
As one can see from Table 4 during the treatment with organic and non-organic acids RPG is precipitated completely, while ethanol and rivanol (added in an equal proportion) do not precipitate it. Salting out of the indicated antigen begins at 20%

saturation of ammonium sulfate, and it precipitates completely at 50% saturation. High concentrations of copper sulfate (5%) precipitate this protein completely, while under the influence of zinc chloride (5%), barium chloride (10%), zinc sulfate (10%) it preserves immunochemical properties.

Table 4
The Relationship Between RPG and Organic and Non-Organic Chemical Reagents.

Chemical reagents[x]	Obtained results
The relationship between RPG and organic and non-organic acids:	
Sulfosalicylic acid 0,2 M	Precipitated irreversibly
Phosphoric tungstic acid 2%	- " -
Chloric acid 0,3 M	- " -
Trichloracetic acid 3%	- " -
Picric acid (saturated solution)	- " -
The relationship between RPG and salts of some metals:	
Ammonium sulfate - salting out range	20-50% of saturation
Copper sulfate 5%	Precipitated completely
Zinc chloride 5%	Precipitated not completely
Barium chloride 10%	- " -
Zinc sulfate 10%	- " -
Other reagents:	
Rivanol 0,5% 1:1	- " -
Ethanol 0°C 50%	- " -

[x]Final concentration of the reagent in the solution is given.

Taking into account that human TSG in physiological concentrations of 60 ug/ml has moderately expressed immunosuppressive action (18, 19), we decided to study the influence of RPG on the formation of E-rosettes and on the proliferative response of lymphocytes to PhHA. It was found out, that in the presence of RPG in physiological concentration (60 µg/ml) intensity of the reaction of E-rosettes formation lowered by 27% (Table 5), and what's more, the degree of the expression of the suppressive action depends on the way of RPG introduction into the reaction. So, after preliminary incubation of lymphocytes at 37°C during 60 min with RPG and with the following washing, the ability of lymphocytes to form rosettes with erythrocytes of the sheep (E-rosettes) lowered by 56% in comparison with the control ($P < 0,05$).

Table 5
Formation of E-Rosettes and Proliferative Activity of PhHA-Stimulated Lymphocytes under the Influence of RPG Preparation.

Conditions of the experiment	Amount of E-rosettes	Index of stimulation
Control	$51,85 \pm 2,48$	$49,93 \pm 3,84$
Introduction of RPG without incubation	$37,77 \pm 3,15^{x/}$	$33,84 \pm 4,21^{x/}$
Introduction of RPG after incubation at 37°C during 60 min	$22,71 \pm 1,16^{x/}$	$25,97 \pm 2,72^{x/}$

$^{x/}$Reliability coefficient $P < 0,05$.

Proliferative activity of lymphocytes in the presence of RPG (Table 5) lowered, moreover, the degree of the expression of inhibiting effect depended on the way of protein introduction into the reaction. After preliminary incubation of lymphocytes with RPG and the following washing the index of stimulation lowered nearly 2 times as compared to the control ($P < 0,05$).
Thus, the protein in physiological concentrations has moderately expressed inhibiting action: it lowers the ability of lymphocytes to form rosettes with sheep erythrocytes (E-rosettes) by 27-56% and suppresses proliferative response of lymphocytes to PhHA 1,5-2 times. Inhibiting effect of the protein increases after preliminary incubation of lymphocytes with RPG.

Discussion

RPG has $beta_1$-globulins mobility, its relative electrophoretic mobility is $0,45 \pm 0,02$. The fact, that relative Molecular Weight varies from 130 000 D (gel filtration) to 85 000 D (electrophoresis in PAAG with SDS), may be explained by a number of facts. Molecular Weight determination with the help of the method of gel filtration was conducted with semi-purified preparation, and during electrophoresis according to Laemmli RPG of 90% purity was investigated. Besides, such a difference in Molecular Weight is quite admissible, if we take into consideration the fact, that chromatographic method has a mistake in the range of 5 000-7 000 D. But taking into account, that the determination of the Molecular Weight by the method of gel filtration was carried out under the control of immunodiffusion analysis with standard test-system to the mentioned protein we are inclined to think, that such a difference is not connected with admixtures, but it indicates the presence of non-protein component in RPG molecule. The subsequent amino acid and carbohydrates analysis showed, that it contains carbohydrates (about 10%). Histochemically carbohydrate component is not determined, which is evidently connected with peculiarities of conformational structure of protein molecule of RPG or with low sensitivity of Schiff's reaction. During isoelectrofocusing purified RPG concentrates in the form of three bands in mild acid medium; it indicates, that the antigen refers to neutral proteins and testifies to its heterogeneity. Protein is salted out with ammonium sulfate from blood serum of pregnant rats in the range of 20-50% of saturation, practically does not precipitate under the influence of the salts of some metals, under the influence of ethanol, rivanol, while non-organic and organic acids precipitate it completely. It is necessary to take it into account when one isolates RPG.
The same is with TSG (20, 21, 22, 23). Its Molecular Weight, described by different methods varies from 120 000 to 90 000 D, during isoelectrofocusing it is situated in the zone of pI 4.0 and 6.12, it is precipitated by ammonium sulfate of 20-50% saturation.
The comparative analysis of amino acid composition of TSG (24) and RPG showed, that the content of a number of amino acids

(aspartic acid, proline, leucine, tryptophan and etc.) makes the proteins similar, as regards other amino acids slight difference is observed. The most significant difference is connected with carbohydrate component, both in amount (TSG contains 29,3% of carbohydrates, RPG - 9,8%) and in quality (RPG molecule lacks fucose and neuraminic acid).

The study of immunological properties of RPG showed, that the protein suppresses the activity of T-lymphocytes, blocking, evidently, their receptor system.

The obtained results will find application during isolation of RPG in pure form, which one can use to get high-affinity, high-avidity antisera, to study its properties to determine biological function of TSG during pregnancy.

Acknowledgement

We thank L.Cch. Mazor for her highly appreciated assistance in translating and typing the manuscript.

References

1. Tatarinov, Yu.S., V.N. Masyukevich. 1970. Immunochemical identification of new $beta_1$-globulin in the blood of pregnant women. Byull. eksp. Biol. Med. 6, 66-68.

2. Bohn, H. 1971. Nachweis und Characterisierung von Schwangerschaftsprotein in der menschlichen Placenta, sowie ihre quantitative immunologische Bestimmung im Serum schwangerer Frauen. Arch. Gynäk. 210, 440-457.

3. Lin, T.M., S.P. Halbert, D. Kiefer, W.N. Spellacy. 1974. Three pregnancy-associated human plasma proteins: purification, monospecific antisera and immunological identification. Int. Arch. Allergy and Appl. Immunol. 47, 35-53.

4. Tatarinov, Yu.S., N.V. Mesnyankina, D.M. Nikoulina, L.A. Novikova, B.O. Toloknov, D.M. Falaleeva. 1974. Immunochemical identification of $beta_1$-globulin of the "pregnancy zone" in serum of patients with trophoblastic tumors. Int. J. Cancer. 14, 548-554.

5. Tatarinov, Yu.S. 1978. Trophoblast-specific $beta_1$-glycoprotein as a marker for pregnancy and malignancies. Gynaec. Obstet. Invest. 9, 161-166.

6. Tatarinov, Yu.S., S.K. Krivonosov, D.D. Petrunin, O.P. Shevchenko. 1976. Comparative immunochemical analysis of specific b-globulins of "pregnancy zone" in humans and mammals. Byull. eksp. Biol. Med. 10, 1123-1125.

7. Shevchenko, O.P., D.D. Petrunin, Yu.S. Tatarinov. 1978. Comparative study (immunochemical and physico-chemical) in specific pregnancy beta-globulins. Byull. eksp. Biol. Med. 8, 213-215.

8. Krivonosov, S.K., D.D. Petrunin, L.A. Ivkova. 1978. An immunochemical study of the specific "pregnancy protein" in the ontogenesis of mammals. Ontogenez. 5, 481-486.

9. Laemmli, U.K. 1970. Cleavage of structural proteins during the assembly of the head of bacteriophage T-4. Nature. 227, 680-682.

10. Osterman, L.A. 1983. Study of Biological Macromolecules with the help of Electrofocusing, Immunoelectrophoresis and Radioisotope Methods. Nauka, Moscow. pp. 3-74.

11. Uriel, J. 1963. In: Immunoelectrophoretic Analysis. (P. Grabar, P. Burtin). Moscow. pp. 70-93.

12. Barabanov, V.M. 1973. In: Manual in Immunology. Meditsina, Moscow. pp. 240-251.

13. Böym, A. 1968. Separation of leucocytes from blood and bone marrow. Scand. J. Clin. Lab. Invest. 21, 1-109.

14. Petrov, R.V., M.A. Stenina, K.A. Lebedev. 1976. Peculiarities attending assessment of the number of T-lymphocytes and of other rosette-forming cells in the blood of the healthy and sick persons. Byull. eksp. Biol. Med. 2, 197-199.

15. Ling, N.R. 1971. Stimulation of Lymphocytes. Meditsina, Moscow. pp. 288.

16. Samoilina, N.L. 1970. Methode morphologique d'evaluation de la transformation blastique des lymphocytes en culture avec phyto-hemagglutinine. Lab. Delo. 8, 445-463.

17. Krivonosov, S.K., M.F. Kan, Yu.S. Tatarinov, L.A. Ivkova. 1985. Physical-chemical description of rat pregnancy-specific protein (PSP). Ontogenez. 3, 265-269.

18. Cerni, C., G. Tatra, H. Bohn. 1977. Immunosuppression by human placental lactogen (HPL) and pregnancy-specific b_1-glycoprotein (SP_1). Arch. Gynäk. 233, 1-8.

19. Tatarinov, Yu.S., I.N. Golovisticov, N.K. Gorlina, D.D. Petrunin, T.M. Tsagaraeva, O.P. Shevchenko. 1980. Effect of trophoblastic b_1-glycoprotein on cultured lymphocyte proliferation, induced by mitogens and allogenic cells. Immunologiya. 5, 14-16.

20. Bohn, H. 1972. Isolierung und Charakterisierung des SP_1. Blut. 24, 292-302.

21. Bohn, H. 1974. Studies on the pregnancy-specific $beta_1$-glycoprotein (SP_1). Arch. Gynäk. 216, 347-358.

22. Nikulina, D.M., A.V. Sokolov, Yu.S.Tatarinov. 1977. Physicochemical characteristics of specific $beta_1$-globulin from blood serum of pregnant women. Vopr. Med. Khim. 3, 324-326.

23. Sokolov, A.V., G.A. Kozlyaeva, N.V. Mesnyankina, Yu.S. Tatarinov. 1978. Isolation and purification of specific b_{1-G} globulin. Vopr. Med. Khim. 2, 240-243.

24. Bohn, H. 1975. III International Symposium on Immunology of Reproduction. Varna, Bulgaria. pp. 111-125.

MURINE MODELS TO HUMAN PREGNANCY SPECIFIC β_1-GLYCOPROTEIN AND α-FETOPROTEIN AND THEIR APPLICATION IN TERATOGENIC STUDIES

J. Hau

Department of Pathology, Laboratory Animal Unit, Royal Veterinary and Agricultural University, Copenhagen, Bülowsvej 13, DK-1870 Frederiksberg C, Denmark

Introduction

During the past five years murine models for the study of pregnancy specific β_1-glycoprotein (SP-1, PSβG) and α-fetoprotein (AFP) have been developed (1,2,3).

Immunologically cross-reacting proteins to human AFP have been described in all species examined (4), but an immunologically cross-reacting mouse analogue to human SP-1 was first described in 1980 (1). The murine analogue to human SP-1 was termed pregnancy-associated murine protein-2 (PAMP-2) - a protein which was first detected in 1978 (5).

The present paper deals with the usefulness of murine AFP and PAMP-2 in teratogenic studies. In mice the maternal serum levels of PAMP-2 are significantly correlated with placental mass during pregnancy (6), and the maternal serum levels of AFP are significantly correlated with fetal mass during pregnancy (2). These correlations can be utilized when assessing possible effects of various treatments on placental and fetal growth.

Pregnancy Proteins in Animals

Immunological Cross-Reactions

Immunological cross-reaction between m-AFP and h-AFP

Cross-reaction was observed using line immunoelectrophoresis. Cathodic to the antibody-containing gel a neutral gel with wells was placed. Cathodic to this gel was the antigen-containing gel which received a pool of serum from pregnant female mice.

Monospecific rabbit anti-human AFP (kindly donated by DAKOPATTS, Denmark) reacted by causing negative deflections of the m-AFP line when placed in wells below the line (Fig. 1).

Neither pooled serum samples from pregnant women nor samples of amniotic fluid caused any deflection of the line in line immunoelectrophoresis when antiserum to m-AFP was used, nor did any stainable precipitate appear when murine pregnancy serum was tested in a direct immunoelectrophoresis system using monospecific rabbit anti-human AFP.

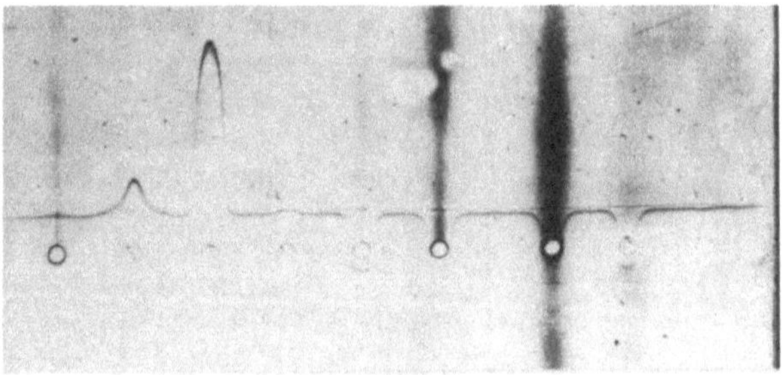

Fig. 1. Line immunoelectrophoresis with goat anti-m-AFP in the antibody-containing gel and murine fetal homogenate in the antigen-containing line gel.
The contents of the wells were: (1) serum of non-pregnant mice, undiluted, (2) serum of pregnant mice, diluted 1:10, (3) murine fetal homogenate, diluted 1:20, (4) human amniotic fluid, undiluted, (5) goat anti-m-AFP, undiluted, (6) rabiit anti-rat AFP, undiluted, (7) goat anti-human AFP, undiluted, (8) rabbit anti-human AFP (DAKO), undiluted, and (9) buffer.

Immunological cross-reaction between PAMP-2 and SP-1

Using human pregnancy serum as the antigen and rabbit or goat anti-PAMP-2 as antibody no precipitates were formed. Similarly no precipitates were formed when murine pregnancy serum was used as the antigen and rabbit and goat anti-SP-1 as the antibody in immunoelectrophoresis.

Immunological cross-reaction between SP-1 and PAMP-2 was demonstrated by line immunoelectrophoresis. The precipitation line was formed by a complex between human SP-1 and rabbit anti-SP-1. Hen antiserum against both SP-1 and PAMP-2 reacted, when placed in wells in the neutral gel, by negative deflection of the precipitation line, whereas serum from non-immunized rabbits and hens caused no deflection. SP-1 antigen present in a pool of serum from pregnant women caused a positive deflection, whereas PAMP-2 antigen present in serum of pregnant mice caused no deflection (Fig. 2).

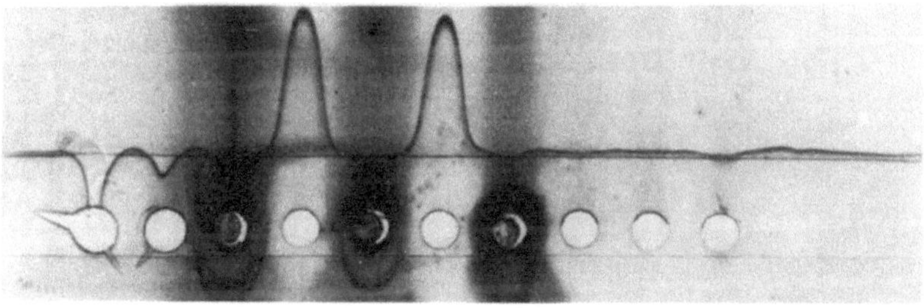

Fig. 2. Line immunoelectrophoresis with rabbit anti-SP-1 in the antibody-containing gel (D), serum from pregnant women in the line gel (B) and two neutral gels (A, C). The contents of the wells were: (1) rabbit anti-SP-1 (DAKO), diluted 1:50, (2) rabbit anti-SP-1 (DAKO), diluted 1:100, (3) hen anti-PAMP-2, undiluted, (4) serum of pregnant women, diluted 1:8, (5) hen anti-SP-1, undiluted, (6) serum of pregnant women, diluted 1:8, (7) serum of control hen, undiluted, (8) serum of pregnant mice, undiluted, (9) serum of pregnant mice, diluted 1:10, (10) serum of non-pregnant mice, undiluted, (11) buffer.

Physico-chemical and biological properties of SP-1 and PAMP-2 and of h-AFP and m-AFP

Table I shows that the main β-mobile SP-1 component and PAMP-2 are quite similar in virtually all of their physico-chemical properties compared.

Although the molecular weight of PAMP-2 is found to be lower than that of SP-1, when estimated by gel filtration, the two proteins migrate to the same extent in SDS-PAAG gel indicating identical molecular weights (Fig. 3). The isoelectric points of the two proteins are identical, and the electrophoretic migration velocities of the two proteins are quite similar.

Both proteins exist in two molecular forms, when analyzed in charge-shift immunoelectrophoresis.

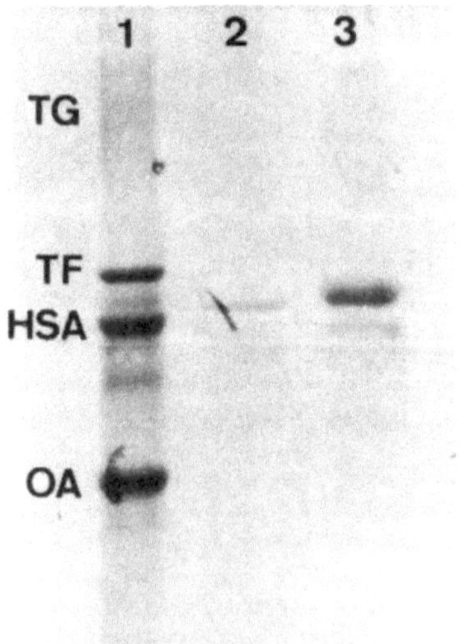

Fig. 3. Polyacrylamide slab gel stained with Coomasie Brilliant Blue (8 %). Lane 1: Marker proteins, TG = thyroglobulin, TF = transferrin, HSA = human serum albumin, OA = ovalbumin, Lane 2: Purified PAMP-2, Lane 3: Purified SP-1.

Both proteins seem to be pregnancy-specific, however, ectopic production of SP-1 has been observed (7). Neither SP-1 nor PAMP-2 are detectable in serum of non-pregnant healthy individuals using electroimmunoassays.

Both proteins have been localized intracellularly in the synctiotrophoblast of the human placenta and in the analogous structure, spongiotrophoblast in murine placenta. The half-life of PAMP-2 is shorter than that of SP-1. This may be due to the method of estimation of this property. The half-life of SP-1 has been measured by the changes in maternal serum concentration after delivery, whereas PAMP-2 concentration changes were measured following injection of the protein into non-pregnant mice.

Both proteins appear in serum following implantation and increase during pregnancy reaching maximum levels late in pregnancy. PAMP-2 decline in concentration prior to parturition, whereas SP-1 appears to remain at a high level until parturition.

SP-1 was found intracellularly in the cytoplasm of the villous syncytiotrophoblast, and PAMP-2 was found intracellularly in trophoblastic cells in the spongiotrophoblast (8,9).

Both proteins contain sialic acid, but PAMP-2 is more retarded in electrophoretic migration velocity following neuraminidase treatment than SP-1. This might be due to a difference in the carbohydrate structure of the two proteins. Whereas SP-1 is completely precipitated by Con A, PAMP-2 shows heterogeneity in this respect.

Table 1. Physico-chemical and Biological Properties of SP-1 (main component with β_1-mobility) and PAMP-2

	SP-1β	Ref.	PAMP-2
Occurrence - in males	-	10	-
In non-pregnant females	-	10	-
Molecular weight	90,000	10	78,000
Isoelectric point	3.8 ± 0.2	11	4.0
Electrophoretic mobility	β_1	10	α_2
Surface properties (charge-shift)	amphiphilic	6,12	heterogeneous
Sialic acid contents	+	6	+
Reaction with Con A[a)]	++	13	-/+
Intracellular localization	syncytio-trophoblast	8,9	spongio-trophoblast
Half-life	30 h	14	13 h

a) Reaction with Con A in crossed affino immunoelectrophoresis. - = no affinity, + = retardation in migration velocity, and ++ = complete precipitation.

Human and murine AFP are pregnancy-specific proteins not detectable in the circulation of healthy non-pregnant individuals by electro-immunoassays. Table 2 shows a comparison between the physico-chemical properties of human and murine APF.

The molecular weight of m-AFP is 2,500 Daltons higher than that of h-AFP, but in exact agreement with earlier measurements of the molecular weight of murine AFP (15). Although the isoelectric point of h-AFP is higher thant that of m-AFP, h-AFP migrates slightly faster than m-AFP in the electrophoresis.

Both proteins contain sialic acid, and they are both hydrophilic when present in serum or amniotic fluid. In fetal tissue, however, an additional form of h-AFP and m-AFP was detectable. This molecular form was amphipilic.

m-AFP is more heterogeneous with respect to reactivity with Con A than h-AFP. Four molecular forms can be distinguished, whereas h-AFP is present in two forms only. The Con A-binding form of h-AFP binds Con A with a lower affinity ($K = 5 \times 10^{-6}$) than molecular forms 1 and 2 of m-AFP ($K = 4.5 \times 10^{-6}$ and $K = 3.3 \times 10^{-6}$ resp.).

Human and murine AFP appear similar in their maternal serum levels during pregnancy. Both proteins reach maximum levels late in pregnancy, and they both decrease in concentration prior to parturition.

The half-life of h-AFP is longer than that of m-AFP. This is probably due to the lower metabolic rate of humans compared to mice. However, the half-life of h-AFP was estimated as changes in levels following delivery, whereas the half-life of m-AFP was estimated by injection of protein into non-pregnant animals. Fractions of m-AFP have affinity for oestradiol, and m-AFP synthesis by adult liver cells may be induced by administration of oestrogen (16), but although a similar affinity for human AFP has been postulated this could not be confirmed.

Table 2. Physico-chemical and Biological Properties of Human and Murine AFP.

	Human AFP	Ref.	Murine AFP
Occurrence - in males	-	4	-
In non-pregnant females	-	4	-
Molecular weight	70,000	17	72,500
Isoelectric point	4.9 - 5.2	18	4.4
Electrophoretic mobility	α_1	19	α_1
Sialic acid contents	+		+
Surface properties of protein in fetal tissue	heterogeneous	20	heterogeneous
Surface properties of protein in serum and amniotic fluid	hydrophilic	20	hydrophilic
Reaction with Con A	heterogeneous 1 form not binding (0) 1 form binding (1)	21	heterogeneous 1 form not binding (0) 3 forms binding(1,2,3)
Dissociation constants of Con A-AFP complex			
Mol.form-1	5.0×10^{-6}	21	4.5×10^{-6}
Mol.form-2	absent	21	3.3×10^{-6}
Half- life	5 days	22	15 h
Affinity for oestrogens	+	22	+

Maternal Serum Levels and Correlations with Placental and Fetal Mass

Litter size and serum levels of PAMP-2 and m-AFP in pregnant mice

Extirpation of one of the uterine horns results in a reduction of the number of fetuses and consequently in total fetal and placental weight (Table 3).

The maternal serum levels of pregnancy-associated murine protein-1 and -3 in pregnant semihysterectomized mice were not significantly different from those of normal, pregnant mice.

By contrast, the serum levels of PAMP-2 and m-AFP were lower in pregnant semihysterectomized mice than in normal, pregnant mice (Table 4).

Table 3. Weight of fetuses and placentae in normal and semi-hysterectomized mice on Day-17 of pregnancy. Values represent means ± s.e.m. of 10 animals.

	Total fetal weight (g)	Total placental weight (g)	N
Normal, pregnant mice	5.78 ± 0.47	1.440 ± 0.12	10
Semihysterectomized, pregnant mice	2.72 ± 0.45	0.609 ± 0.07	10

Table 4. Maternal serum levels of PAMP-2 and m-AFP in 17-day-pregnant normal and semihysterectomized mice. Values represent means ± s.e.m. of 10 mice.

	PAMP-2	PAMP-4	N
Normal, pregnant mice	194 ± 12	55 ± 6	10
Semihysterectomized, pregnant mice	108 ± 18	27 ± 4	10
p	$p < 0.01$	$p < 0.01$	

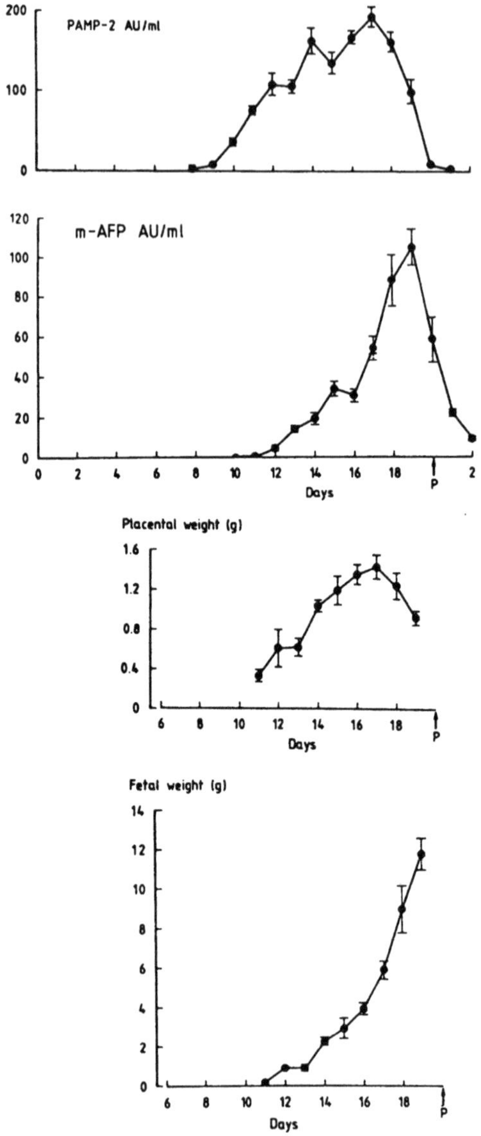

Fig. 4. Daily serum levels of PAMP-2 and m-AFP and total placental and fetal weight during pregnancy. Each value represents the mean $\pm$ s.e.m. for 10 animals. P = parturition.

These results indicate that the serum levels of PAMP-2 and m-AFP are correlated to the weight of the feto-placental unit.

Maternal serum levels of PAMP-2 and m-AFP and correlation to placental and fetal weight

The placentae and fetuses of the mice measured for maternal serum levels during pregnancy were weighed. The results are shown in Fig. 4. The maternal PAMP-2 serum level was significantly correlated ($p < 0.001$) to placental weight on each day of pregnancy, but not to fetal weight. No significant correlation between serum levels and placental weights could be found in 54 primigravid mice on Day-18 of pregnancy (Fig. 5) (6) indicating that single measurements of PAMP-2 levels cannot be used to monitor the exact mass of placental tissue on a given day of pregnancy due to the biological variation.

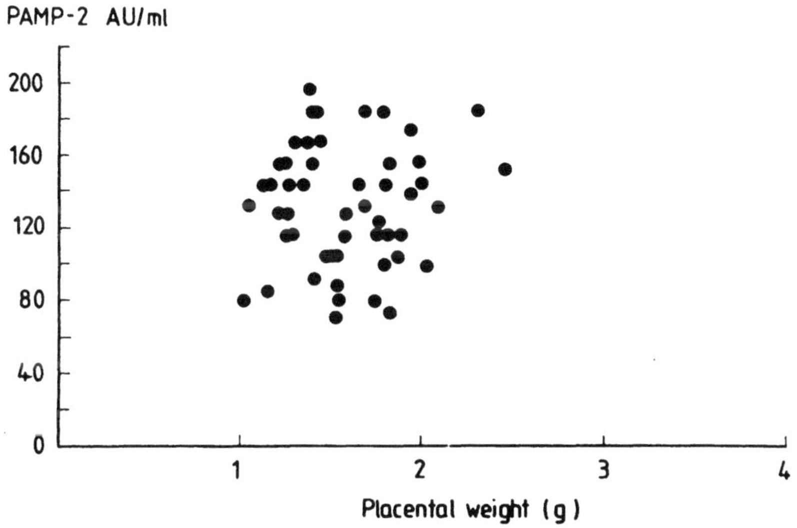

Fig. 5. The PAMP-2 level on Day-18 of pregnancy and placental weight in 54 primigravid NMRI mice.

The m-AFP level in maternal serum was significantly correlated to fetal weight during pregnancy ($p < 0.001$), but not correlated to placental weight. A significant correlation was found between the m-AFP levels and fetal weights of 54 mice on Day-18 of pregnancy ($p < 0.001$) (Fig. 6).

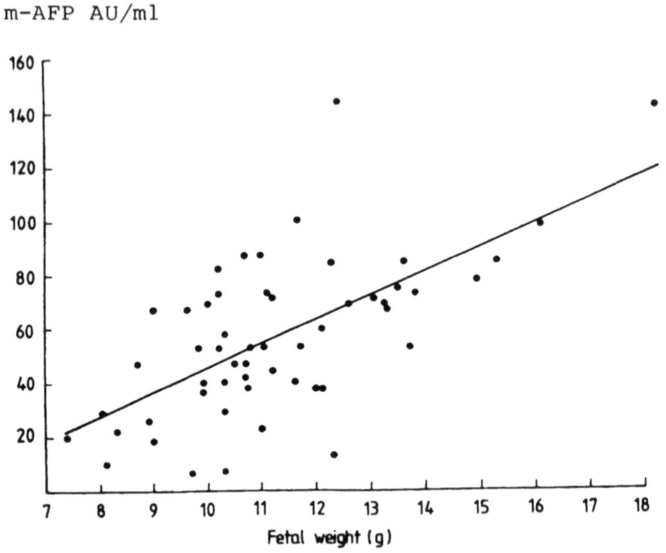

Fig. 6. The m-AFP level on Day-18 of pregnancy and fetal weight in 54 NMRI mice.

The positive correlation between maternal PAMP-4 concentration and fetal weight points to the mouse as an obvious model for the study of fetal growth retardation and the study of factors affecting fetal growth.

Examples of the Usefulness of PAMP-2 and m-AFP in Studies of Factors Affecting Placental and Fetal Growth

Active and passive immunization as well as termination of natural tolerance to fetal and pregnancy specific proteins in various animal species have been demonstrated to result in a reduction of fertility, failing ability to conceive, prevention of implantation or abortion (23, 24, 25, 26, 27, 28).

However, systemic administration of purified monospecific rabbit anti PAMP-2 antibodies did not affect the outcome of pregnancy in mice (28).

By contrast, intrauterine administration of antibodies against PAMP-2 resulted in prevention of implantation (28). These results indicate that PAMP-2 is present on the surface of the implanting embryo.

The correlations between murine maternal serum levels of PAMP-2 and AFP and placental and fetal mass respectively have been used in studies of factors affecting the growth of the fetoplacental unit.

The correlations make it possible to screen the effect of materials potential hazardous to the fetoplacental unit, and this relatively simple experimental design may make more expensive and laborious analyses obsolete.

References

1. Hau, J., Westergaard, J.G., Svendsen, P., Bach, Annelise & Teisner, B.: 1980. J.Reprod.Fertil. 60, 115-119.

2. Hau, J., Svendsen, P., Teisner, B., Thomsen Pedersen, G. & Kristiansen, B.R. 1981. Biol.Reprod. 24, 683-689.

3. Hau, J. 1982. In: Pregnancy Proteins: Biology Chemistry and Clinical Application (J.G. Grudzinskas, B. Teisner & M. Seppala, eds.), 435-446, Academic Press, Sydney.

4. Gitlin, D. 1975. Ann. N.Y. Acad.Sci. 259, 7-16.

5. Hau, J., Svendsen, P., Teisner, B. & Svehag, S.-E. 1978. J.Reprod.Fert. 54, 239-243.

6. Hau, J., Westergaard, J.G., Svendsen, P., Teisner, B., Bach,A. & Thomsen Pedersen, G. 1982. In: Immunology of Human Placental Proteins, Placenta 'Suppl. 4), 51-65.

7. Rosen, S.W., Kaminska, J., Calvert, I. and Aaronson, S.A. 1979. Am.J.Obstet.Gynecol. 134, 734-738.

8. Hau, J., Westergaard, J.G., Svendsen, P., Teisner, B. & Chemnitz, J. 1985. In: And Animals, Archibald, J., Ditchfield, J. & Rowsell, H.C. (Eds.), Gustav Fisher Verlag, New York, 365-384.

9. Chemnitz, J., Hau, J., Svendsen, P., Folkersen, J., Westergaard, J.G. & Christensen, B.C. 1982. Bibl.Anat. 22, 87-92.

10. Bohn, H. 1971. Arch.Gynäk. 210, 440-457.

11. Lin, T.M., Halbert, S.P, Kiefer, D., Spellacy, W.N. and Gall, S. 1974. Am.J.Obstet.Gynecol. 118, 223-236.

12. Hau, J., Grudzinskas, J.G., Teisner, B. & Svendsen, P. 1983. Placenta 4, 291-294.

13. Westergaard, J.G., Teisner, B., Folkersen, J., Hindersson, P., Schultz-Larsen, P. and Svehag, S.-E. 1979. Scand.J.Clin.Lab. Invest. 39, 351-359.

14. Bohn, H. 1974. Arch. Gynäk. 216, 347-358.

15. Kerckaert, J.P., Bayard, B. and Biserte, G. 1979. Biochim. Biophys. Acta 576, 99-108.

16. Hau, J., Chemnitz, J., Teisner, B., Tornehave, D. & Svendsen, P. 1984. Acta Endocrinol. 106, 141-144.

17. Ruoslahti, E. and Seppälä, M. 1971. Int.J.Cancer 7, 218-225.

18. Lester, E.P., Miller, J.B., Baron, J.M. and Yachnin, S. 1978. Immunol. 34, 189-198.

19. Bergstrand, C.G. and Czar, B. 1956. Scand.J.Clin.Lab.Invest 8, 174.

20. Hau, J., Teisner, B. & Bøg-Hansen, T.C. 1982. In: Lectins, Biology, Biochemistry and Clinical Biochemistry, Vol. 2 (T.C. Bøg-Hansen, ed.), W. de Gruyter, Berlin, 467-474.

21. Bøg-Hansen, T.C. & Hau, J. 1983. Chromatography Library, vol. 18B, 219-252.

22. Uriel, J., de Nechaud, B. and Dupiers, M. 1972. Biochem.Biophys.Res.Commun. 3, 1175-1180.

23. Stevens, V.C. 1978. Bulletin of the World Health Organization 56 (2), 179-192.

24. Bohn, H. 1975. 3rd International Symposium on Immunology of Reproduction, Varna, Bulgaria, 111-125.

25. Bohn, H. and Weinmann, E. 1974. Arch.Gynäk. 217, 209-218.

26. Bohn, H. and Weinmann, E. 1976. Arch.Gynäk. 221, 305-312.

27. Gusev, A.I. and Yazova, A.K. 1970. Biochem. U.S.S.R. 35, 1, 142-149.

28. Hau, J., Gidley-Baird, A., Westergaard, J.G. & Teisner, B. 1985. Biomed.Biochem.Acta 44, 1255-1259.

EFFECT OF SEX STEROIDS AND TROPHOBLAST CULTURE SUPERNATANTS ON THE CYTOTOXIC T-CELL ACTIVITY IN MICE

P. Van Vlasselaer, M. Vandeputte
University of Leuven, Rega Institute, B-3000 Leuven, Belgium

Introduction

The semi-allogeneic conceptus developing in the uterus of a female mammal is a natural and successful allograft since it is not rejected by the otherwise immunocompetent mother (1). One of the several mechanisms proposed to explain this phenomenon is based on the existence of potent immunosuppressive factors at the mother-fetal interface. Trophoblast cells, which form the boundary zone of fetal territory, display a definite immunosuppressive activity (2). This property is mainly due to the progesterone secreted by these cells (3). Trophoblast cells and their culture supernatants as well as progesterone were found to inhibit allogeneic lymphocyte interactions and the generation of cytotoxic lymphocytes (CTL) *in vitro*. They did not, however, suppress the CTL activity of previously *in vitro* sensitized lymphocytes for their target cells (2-5). Schleimer *et al*. (6) described a similar finding with CTL derived from mice immunized *in vivo* and which were not suppressed by an acute exposure to glucocorticoids. Preincubation of these CTL for several hours with different glucocorticoids caused, however, a marked inhibition of their activity. Since preimmunization of the mother against paternal antigens and hence the presence of cytotoxic lymphocytes directed against antigens also present on the fetal allograft does not interfere with the fetal survival (7), we examined the effect of preincubation in the presence of trophoblast supernatants of progesterone and other steroids on the CTL activity.

Materials and Methods

Animals

Mice of the inbred strains C57Bl ($H-2^b$) and Balb/c ($H-2^d$) as well as mice from the outbred NMRI strain and rats of the inbred R/A Pfd($RI-I^u$) strain were used (Proefdierencentrum, K.U.Leuven, Belgium).

Pregnancy Proteins in Animals

Trophoblast preparation and culture

Ectoplacental cones (EPC) from NMRI mice were dissected at day 7.5 of pregnancy and washed twice in NCTC-135 medium (Gibco Europe) supplemented with 1 % Na-pyruvate, 0.1 % glucose, 10 % fetal calf serum (FCS) (Gibco Europe), 3 % Na-bicarbonate and antibiotics. The day of copulation plug was taken as day 0 of pregnancy. Trophoblast culture supernatants were obtained from EPC cultured in MEM, 10 % FCS, 3 % Na-bicarbonate, antibiotics and 5×10^{-5}M 2-mercaptoethanol in 24 multiwell plates (NUNC). All trophoblast supernatants were centrifuged and stored at -40°C until use.

Preparation of the lymphocytes and the mixed lymphocyte culture (MLR)

Balb/c splenocytes were obtained from adult mice following gentle teasing of the spleens with forceps in MEM, 10 % FCS and antibiotics. The lymphocytes were separated from the total cell population by centrifugation on a Ficoll-Metrizoate gradient for 20 min at 1000 g and 20°C. The splenocytes were treated with mitomycin C (MitC) at a concentration of 40 µg for 4-6 x 10^6 cells/ml at 37°C for 20 min in MEM, 10 % FCS and antibiotics. Before cultivation the cells were washed five times in MEM and antibiotics.

C57Bl lymphocytes were obtained from adult mice following gentle teasing of the lymph nodes with forceps in MEM, 10 % FCS and antibiotics. The cells were washed once in MEM with antibiotics before use.

The MLR was cultured in 96 well-round-bottomed microtiter plates (Greiner) in MEM, 10 % FCS, 3 % Na-bicarbonate, antibiotics and 5×10^{-5}M 2-mercaptoethanol. In each well 5×10^5 C57Bl responders were cultured with 10^5 Balb/c stimulators in a total volume of 200 µl/cup for 96 hr at 37°C in a 5 % CO_2-95 % air mixture. After this culture period the CTL were harvested by pipetting, counted and suspended in a final concentration of 5×10^6 viable cells/ml.

Preparation of the cytotoxic lymphocytes (CTL)

CTL (5×10^6 cells/tube) were put in plastic tubes (Falcon) and supplemented either with normal MLR medium, with serial diluted EPC supernatants (1:2; 1:4; 1:8 and 1:16 diluted) or with different steroids (cortisol, Hc, estradiol, E_2, estriol, E_1 or progesterone, PG) at final concentrations ranging from 10 to 1 µg/ml. All CTL, the steroid or EPC supernatants treated as well as untreated control CTL were preincubated for various times ranging from 1 to 16 hr at 37°C in a 5 % CO_2-95 % air mixture. Prior to the target challenge, the preincubated CTL were centrifuged and resuspended in a final concentration of 5×10^6 cells/ml using the medium they were preincubated in. In order to verify the effect of IL-1 or IL-2 on the CTL-treated with different steroids or with EPC supernatants, the interleukins were added to the medium together with the steroids or EPC supernatants and the CTL were further incubated for 16 hr prior to the target challenge. All steroids were dissolved in ethanol and subsequently diluted with the MLR culture medium to an ethanol-medium concentration of 0.2 %. The ethanol concentration as well as the steroids in the used concentrations (10 to 1 µg/ml) did not affect the viability of the in vitro generated lymphocytes. Some of the trophoblast culture supernatants were incubated with serial diluted antiprogesterone serum (APS) (Sanbio) for 1 hr at room temperature under gentle shaking and subsequently added to the CTL and incubated for 16 hr prior to the target challenge. The antiserum was diluted in the MLR culture medium until final concentrations ranging from 10^{-2} to 10^{-5} were obtained.

The cell-mediated lympholysis assay (CML)

As described, the CTL were preincubated with steroids or trophoblast supernatants prior to the effector phase of the CML (^{51}Cr-release).

A cell culture of P815 mastocytoma cells (DBA/2 origin) was used as a source of $H\text{-}2^d$ target cells. The cells were maintained at 37°C in a 5 % CO_2-95 % air mixture in RPMI-1640 containing 20 % FCS. Upon reaching confluence the cells were passaged and fresh medium was added three times a week. Tumor cells (10^6) were washed and resuspended in 0.1 ml MEM, 10 % FCS containing 200 µCi (^{51}Cr, Amersham) sodium chromate. The cells were incubated at 37°C for 60 min in a 5 % CO_2-95 % air mixture. After this incubation the target cells were washed three times and resuspended in the MLR medium.

Preincubated effector cells (5 x 10^5/cup) and target cells 10^4/cup) were mixed in a final volume of 200 µl in 96 well V-bottomed microtiter plates (Greiner). After incubation for 4 hr at 37°C in a 5 % CO_2-95 % air mixture an aliquot of the supernatant (100 µl) was removed and radioactivity was measured in a gamma counter. Spontaneous release of ^{51}Cr was obtained by monitoring radioactivity released during the incubation of 10^4 target cells alone. Maximum release of the ^{51}Cr was obtained by measuring the radioactivity released during the incubation of 10^4 target cells in 150 µl saponine (10 %). The percentage lysis was calculated according to the formula :

$$\text{percentage specific lysis} = \frac{\text{cpm experimental - cpm spontaneous release}}{\text{cpm maximum - cpm spontaneous release}} \times 100$$

Interleukin-1 and interleukin-2 preparations

As described, the rat and mouse IL-2 preparations were both obtained by concanavalin A (ConA) stimulation of rat R/A and mouse C57Bl splenocytes respectively (8). Briefly, the splenocytes were obtained as described for the MLR and suspended in a final concentration of 2.5 x 10^6 cells/ml in the MLR medium supplemented with ConA (2.5 μg/ml) and 2-mercaptoethanol (5 x 10^{-5}M). A total volume of 40 ml cell suspension was cultured in plastic tubes (Falcon) for 48 hr at 37°C in a 5 % CO_2-95 % air mixture. After incubation the supernatants were collected and treated twice with Sephadex G-25 (Pharmacia) (0.2 g/100 ml, 15 min at room temperature) in order to remove the remaining ConA. All ConA supernatants were put through a 2 μm filter (Gelman) and assayed for their IL-2 content on a CTL-L2 tumor-specific cytotoxic T-cell line (9). As well rat as mouse ConA supernatants had a definite stimulatory effect on CTL-L2 cell growth up to a final dilution of 1:60. An IL-1-containing supernatant was obtained from lipopolysaccharide (LPS, 20 μg/ml) (Sigma) stimulated P388D1 cells as described (10). Cultured supernatants were collected 24 hr after stimulation and 5-fold concentrated with an amicon concentrator (model 8050) equiped with a YM10 Diaflo membrane. The supernatants were filtered (Gelman, 0.2 μm) and assayed for their IL-1 content on suboptimal phytohemagglutinin stimulated C3H/He thymocytes as described (11). All batches used had a stimulatory effect on the thymocytes up to a final dilution of 1:500.

Results

When in vitro C57Bl anti-Balb/c generated CTL were incubated at 37°C for 16 hr with progesterone at concentrations ranging from 0.5 to 10 μg/ml, their cytotoxic activity was markedly inhibited in a dose dependent way (Fig. 1). In contrast, cortisol used at similar doses displayed already a maximal effect at the lowest dose (0.5 μg/ml) used in these experiments. Estradiol and estriol did not inhibit the CTL activity in the same assay. However, when the CTL were exposed to identical concentrations of progesterone or cortisol just prior to the target challenge, they displayed a cytotoxic activity equal to non-treated CTL (results not shown).

A comparable result as for the progesterone was observed with trophoblast supernatants of EPC giant cell cultures. Whereas a preincubation period of 16 hr with EPC supernatants induced a dose-dependent inhibition of the CTL activity (Fig. 2A), the latter supernatants remained inactive when added just prior to the target challenge (results not shown). The percentage suppression of the CTL activity varied with the dilution of EPC supernatants used. In contrast, the supernatant of mouse kidney fibroblasts (1:2 final dilution) remained ineffective on CTL activity (results not shown).

Pretreatment of the trophoblast supernatants for 1 hr with serial diluted APS decreased the inhibitory effect of the supernatants on the CTL activity

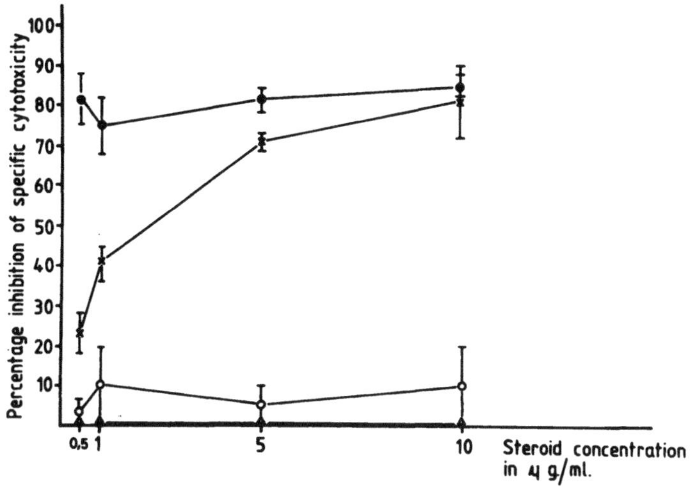

Figure 1. Inhibition of the CTL-mediated cytotoxicity induced by different steroids : ●, cortisol; X, progesterone; O, estriol; Δ, estradiol.

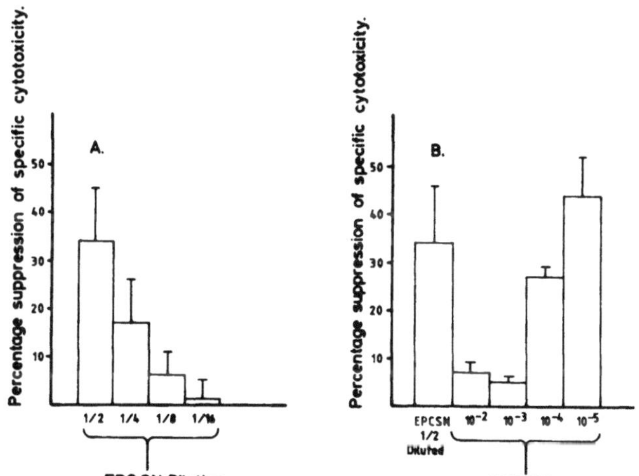

Figure 2. A. Effect of serial diluted trophoblast supernatants of ectoplacental cone (EPC) origin on the CTL activity.
B. Effect of different antiprogesterone serum (APS) dilutions on the EPC supernatants (1:2 diluted) induced inhibition of the CTL activity.

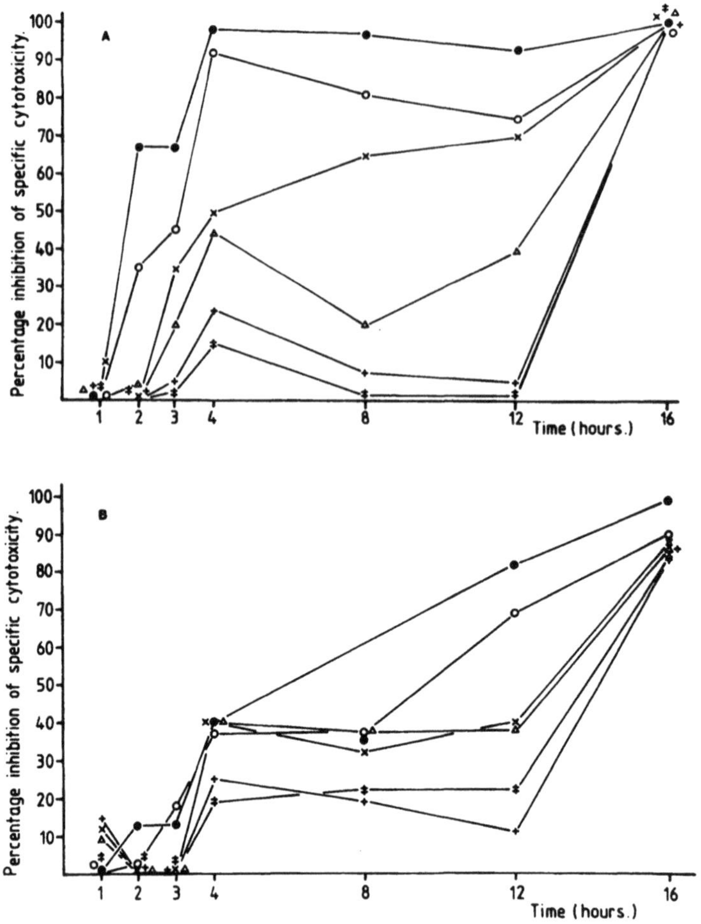

Figure 3. Kinetics of progesterone (A) and cortisol (B) induced inhibition of the CTL-mediated cytotoxicity. Symbols representing the different steroid concentrations : ●, 100 µg/ml; O, 50 µg/ml; X, 40 µg/ml; Δ, 30 µg/ml; +, 20 µg/ml; ‡, 10 µg/ml.

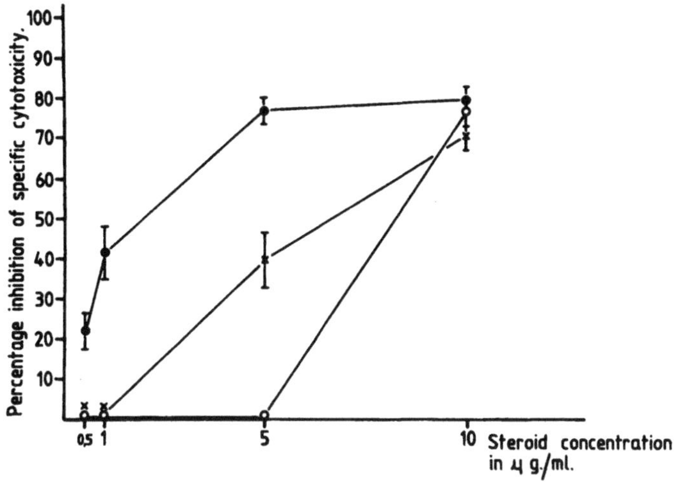

Figure 4. Effect of rat (O) and mouse (X) interleukin-2 (IL-2) (1:2 diluted) on the inhibition of the CTL activity induced by progesterone alone (●).

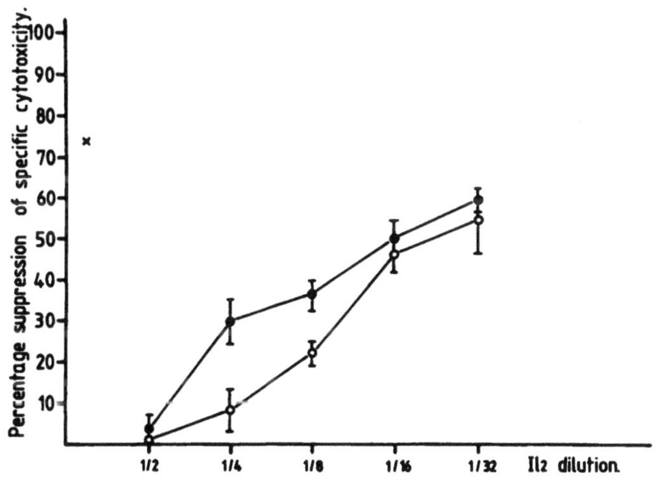

Figure 5. Effect of serial diluted interleukin-2 (IL-2) of rat (O) and mouse (●) origin on the progesterone (X, 5 μg/ml) induced inhibition of the CTL activity.

using the 16 hr preincubation assay. A marked decrease of the EPC supernatants inhibitory effect on the CTL activity was observed when the latter EPC supernatants were pretreated with 10^{-2} or 10^{-3} diluted APS. Decreasing concentrations of the antiserum led to the gradual restoration of the inhibitory effect of EPC supernatants on CTL activity (Fig. 2B).

The kinetic study on the influence of progesterone on the CTL revealed that the inhibitory effect on the CTL activity can be seen within 2 hr of preincubation provided high progesterone concentrations are used. After a preincubation time of 4 hr, progesterone inhibits completely all CTL activity at concentrations of 100 or 50 µg/ml. At lower concentrations (40 to 10 µg/ml) the dose relationship was less clear and maximal response was only obtained after a 16 hr incubation period. Cortisol revealed a comparable behaviour but necessitated longer incubation periods to be effective. Indeed, even after 4 hr incubation with high doses of cortisol (100 and 50 µg/ml) only 50 and 40 % inhibition was obtained (Fig. 3). Steroid concentrations ranging from 10 to 100 µg/ml were not toxic for the CTL as tested by the trypan blue exclusion test after the appropriate incubations times.

Based on the results of Schleimer et al. (6) on the IL-2 induced reversion of the cortisol effect on CTL activity, the capacity to reverse the inhibitory effect of progesterone and EPC supernatants by rat and mouse IL-2 preparations was investigated. Using a 1:2 diluted crude rat or mouse IL-2, the inhibitory effect of progesterone at concentrations of 0.5 to 1 µg/ml was completely abolished (Fig. 4). The rat IL-2 abolished the effect of progesterone up to a concentration of 5 µg/ml (Fig. 4). The effect of both mouse and rat IL-2 was dose-related (Fig. 5). Comparable results as for progesterone were obtained with the same mouse and rat IL-2 preparations added to EPC supernatants. The inhibitory effect of the EPC supernatant was completely abolished by mouse and rat IL-2 (data not shown).

Whereas rat and mouse IL-2 preparations reversed the inhibitory effect of low doses of progesterone and of trophoblast supernatants, IL-1 containing supernatants from LPS-stimulated P388D1 cells did not influence this inhibitory effect (results not shown).

Discussion

The present study demonstrates that cortisol, progesterone and trophoblast supernatants inhibit the cytotoxic activity of in vitro generated CTL. Other studies have demonstrated no or only minimal inhibitory effects of similar concentrations of either cortisol and progesterone or EPC supernatants on the CTL effector function (3,4). However, in these studies the effect on the CTL activity was only tested after acute exposure and not after preincubation of the CTL with steroids or with EPC supernatants. Without preincubation only extremely high doses have been shown to inhibit CTL activity as measured *in vitro* (6).

As described by Pavia *et al.* (4) sex steroids inhibit the inductive and proliferative phaze in the allogeneic lymphocyte interaction. This leads to a decreased generation of CTL. Trophoblast giant cells of EPC or placental origin or their supernatants inhibit in comparable ways the CTL generation during cocultivation with allogeneic mixed lymphocyte cultures (3).

Schleimer *et al.* (6) recently described the inhibitory effect of different corticosteroids on CTL activity after preincubation for 24 hr prior to the effector-target challenge. In parallel with their work we investigated the effect of progesterone and trophoblast supernatants on CTL activity after different periods of preincubation. Preincubation with cortisol or progesterone for 16 hr induced a marked inhibition of the CTL activity when concentrations ranging from 10 to 1 μg/ml were used. Cortisol was found to be much more potent at lower concentrations than progesterone but required a much longer preincubation with the CTL to exert its suppressive activity. Identical concentrations of estradiol or estriol had no effect.

One effect of the glucocorticoids on lymphocyte function has been attributed to the blockade of the production of IL-2 (12). Since at the concentrations used by us (10 to 1 μg/ml) the effect of cortisol could not be reversed by either mouse or rat IL-2 (data not shown), it is likely that other mechanisms than the inhibition of IL-2 production are responsible for the inhibition of the CTL activity at these rather high concentrations of cortisol (6).

In contrast to cortisol, progesterone does not affect the IL-2 production (13). Yet, our results indicate the reversible character of the progesterone-induced CTL activation by IL-2 preparations. In comparison to cortisol, the effect of higher concentrations of progesterone is still reversible provided the concentration does not exceed 5 μg/ml. These results together with the inability of progesterone to inhibit IL-2 production (13) suggest that the latter steroid interferes with the IL-2 receptor-binding rather than with the IL-2 production by the CTL. However, the mechanism by which progesterone alters lymphocyte activity is still not clear. The difference between the CTL-steroid interaction by cortisol and progesterone may be related to the presence or absence of receptors for these steroids. Lymphocytes possess a receptor for corticoids through which e.g. nucleoside transport is mediated (14), whereas progesterone probably affects lymphocyte activity by direct membrane interactions since a receptor for this steroid could not be demonstrated (15). Indeed, studies on synthetic membranes demonstrate the ability of progesterone to alter synthetic phospholipid layers (16). Moreover, Alfsen (17) described the high affinity of the latter steroid for cell membranes and its interaction with membrane proteins.

Trophoblast supernatants derived from ectoplacental cone cultures clearly inhibit the CTL activity in a dose-dependent way. Like in previous work (3), addition of APS to these EPC supernatants abolished nearly completely their inhibitory effect. This indicates that the EPC supernatants inhibition of CTL inactivation during preincubation is mainly due to progesterone.

From these and previous experiments, one can conclude that progesterone and EPC supernatants not only interfere with the induction and proliferative phaze but also with the effector phaze of allogeneic lymphocyte reactions. Inhibition of the effector phaze necessitates, however, a continuous and lasting contact with progesterone. It can thus be envisaged that the entire sequence of events for potential cell-mediated rejection of the fetus is suppressed by the continuous exposure of immunocompetent maternal cells within the placenta to high free levels of progesterone. Upon returning to the general circulation, normal reactivity of these cells is assured since the concentration of free progesterone is considerably lowered through binding of the steroids to proteins present in the blood.

Abbreviations used : EPC, ectoplacental cone; MLR, mixed lymphocyte reaction; CML, cell mediated lympholysis; MitC, mitomycin C; APS, rabbit anti-human progesterone serum; MEM, minimum essential medium; FCS, heat-inactivated fetal calf serum; Na-bicarbonate, sodium bicarbonate; cpm, count per minute; Ci, Curie; CTL, cytotoxic lymphocyte.

Acknowledgments : This work was supported by the National Fund for Scientific Research (Belgium). We are indebted to C. Seghers for technical assistance.

References

1. Beer, A.E., Billingham, R.E. 1976. The Immunobiology of Mammalian Pregnancy. Prentice-Hall, Englewood Cliffs, N.J., USA.
2. Pavia, C.S., Stites, D.P. 1981. Cell. Immunol. 58, 202.
3. Van Vlasselaer, P., Vandeputte, M. 1984. Cell. Immunol. 83, 422.

4. Pavia, C.S., Siiteri, P.K., Perlman, J.D., Stites, D.P. 1979. J. Reprod. Immunol. 1, 33.

5. Clemens, J., Siiteri, P., Stites, D. 1979. J. Immunol. 122, 1978.

6. Schleimer, R., Jaques, A., Shin, H., Lichtenstein, L., Plaut, M. 1984. J. Immunol. 132, 266.

7. Chaouat, G., Monnot, P. 1984. Am. J. Reprod. Immunol. 6, 5.

8. Gillis, S., Ferm, M., Ou, W., Smith, K. 1978. J. Immunol. 120, 2027.

9. Gillis, S., Smith, K.A. 1977. Nature 268, 154.

10. Lachman, L.B., Hacker, M.P., Blyden, G.T., Handschumacher, R. 1977. Cell. Immunol. 34, 416.

11. Mizel, S. 1980. Mol. Immunol. 17, 571.

12. Gillis, S., Crabtree, G., Smith, K. 1979. J. Immunol. 123, 1632.

13. Gillis, S., Crabtree, G., Smith, K. 1979. J. Immunol. 123, 1624.

14. Marone, G., Lichtenstein, L.M., Plaut, M. 1980. J. Pharmacol. Exp. Therap. 215, 469.

15. Neifeld, J., Lippman, M., Tormey, D. 1977. J. Biol. Chem. 252, 2972.

16. Heap, R., Symons, A., Watkins, J. 1970. Biochim. Biophys. Acta 218, 482.

17. Alfsen, A. 1983. Progr. Biophys. Molec. Biol. 42, 79.

IN VITRO FERTILIZATION (IVF) OF CATTLE OOCYTES - AN ANALYSIS OF THE IMPORTANCE OF THE LH-SURGE, CUMULUS EXPANSION AND IN VITRO INSEMINATION TIME

T. Greve, H. Callesen, P. Hyttel [(a)] and K.P. Xu
Institute for animal reproduction and [(a)] Department of anatomy,
Royal Veterinary and Agricultural University, 13 Bülowsvej, DK-1870 Frederiksberg C - Denmark

Introduction

In vitro fertilization (IVF) of cattle oocytes recovered either from slaughterhousematerial or from the living bovine (laparoscopy, laparotomy or ovariectomy) has not been very successfull. Only in vivo matured oocytes have given rise to normal pregnancies following non-surgical transfer of the fertilized eggs (13) and live born calves following surgical transfer (2). Conversely, in vitro matured (20-24 h) oocytes aspirated from the slaughtered animal (1) or from non-ovulated follicles of the superovulated heifer (6) has never produced pregnancies in spite of pronuclei formation and cleavage of the 4- or 8-cell stages.

It is yet to be elucidated why cattle oocytes are so resistant to IVF when e.g. human oocytes are conceived in such high percentages in vitro (5). Both the female, the male, and environmental factors (culture methods e.g.) must be taken into consideration. It is generally believed that a normal, synchronous follicular and cumulus-oocyte-complex maturation is a prerequisite for normal fertilization and embryonic development both in vivo and in vitro (10), (14). In the gonadotrophin stimulated cow the normal endocrine balance (LH-discharge e.g.) and maturational process (follicular microenvironment) will often be disturbed resulting in development of oocytes with poor developmental capability (11), (6), (4), (7). The influence of especially the preovulatory LH-discharge on the IVF-rates has not, however, been fully evaluated and it needs further studies in the cow. Certain properties of the cumulus-oocyte-complexes (activation, ooplasma morphology and prescence of polarbodies) at the time of in vitro insemination is important at least for the human IVF-results (15) and may also influence the bovine IVF-results.

A certain culture time prior to insemination in vitro is required for human oocytes in order to avoid polyspermia (16). It is not known if a similar condition exists in cattle IVF-programmes.

Thus, there are several essential points that need further detailed analyses,

Pregnancy Proteins in Animals

and it was the aim with the present experiment to examine if (1) the LH-peak per se and the interval from LH to the oocyte aspiration, (2) the cumulus-oocyte-complex, and (3) the interval from oocyte aspiration to in vitro insemination would influence the in vitro fertilization rates of cattle oocytes matured in vivo.

Material and methods

Heifers and cows (n = 26) of the indigenous breeds were stimulated with either 1500-2500 I.U. PMSG (pregnant mares' serum gonadotrophin) or 32 mg FSH-p (follicular stimulating hormone) administered in decreasing doses in their mid-luteal phase. Two to 2½ days after the initial gonadotrophin treatment (Gn) they were injected with 0,75 mg cloprostenol (PG) to induce luteolysis and estrus. Jugular blood samples were obtained daily for progesterone analysis (12) and every four h for LH-determination (8).

Oocyte aspiration (OA) was performed 2-3-4 days after the PG-treatment via mid-line laparotomy under general anesthesia, or after transvaginal ovariectomy. Following isolation from the follicular fluid (FF) the oocytes were placed in Hepe's buffered Ham's F-10 medium. At this stage the cumulus-oocyte-complexes (COC) were evaluated under an inverted microscope (phasecontrast) and the cumulus dissociation (activation), ooplasma morphology and occurrence of polarbodies (PB) was described. Following this procedure the oocytes were placed in individual wells in multidishes containing 1,0 ml defined medium (DM, (2)) or Ham's F-10 supplemented with 10% fetal calf serum (FCS) and 50 I.U. penicillin and 50 µg streptomycin. Incubation occurred at 38°C in 5% CO_2/air and 100% humidity. FF was analyzed for progesterone (P4) and oestradiol-17β (E2) as described earlier (7).

Frozen or fresh semen was treated as described in earlier studies (6) and in vitro insemination (IVI) was performed 2-7 h after OA with 10^6 live sperm/ml. The co-culture time varied from 18-24 h and following this period the oocytes were cultured either in RAS (6) or in Ham's F-10 supplemented with 10% homologous donor-serum or 10% FCS, and 50 I.U. penicillin - 50 µg streptomycin/ml for a period of 12-96 h in which period the oocytes were observed for fertilization defined in this study as:

(a) division into 2-4- and 8-cell stages with symmetrical blastomeres and nuclei

or (b) syncharyosis (the 1-cell egg)

or (c) pronucleus formation (PN) of both the male and the female.

(a) and (b) was observed following fixation in 1:3 acetic acid: methanol and staining with orcein (9)

Results

A total of 23 out of 94 oocytes was considered fertilized, yielding an all over fertilization rate of 24,5% (table I). The fertilized oocytes originated from 13 of the 26 animals and within this group the fertilization rate was 43 ± 22% (mean ± standard deviation). Polyspermia or fragmentation of the female PN-occurred in another 3 oocytes recovered from animals where normal fertilization of other oocytes had also taken place. The number of different developmental stages are presented table I.

Table I. IVF Results Showing the Number of Different Developmental Stages.

2 Pronuclei	1-celle stage or syncharyosis	2-cell stage	4-cell stage	8-cell stage	Total
7	2	8	3	3	23 (24,5%)

The oocytes aspiration (OA) was performed 59.8 ± 6.9 h (mean ± standard deviation) following the PG-injection. The mean (± SD) interval from the LH-peak-level to OA was 17.4 ± 7.6 (11 animals).
Ten animals had no LH-surge and in the remaining 5 heifers the LH-determinations have not yet been done.

The IVF-rate was extremely poor in animals without a descernible LH-discharge. Only one animal (No. 44) produced a 2-cell egg, that developed from an oocyte with a visible first polarbody.

It was not possible, however, to find any direct relation between the interval from the LH-peak level to the oocyte aspiration and the subsequent IVF-rate as it would have been anticipated.

The mean (± SD) interval from estrus to OA was 10.1 ± 3.4 h. Only 16 animals exhibited clear signs of heat, whereas 10 had no or very vague estrus. This feature invariably coincided with abscence of the LH-surge. Consequently, animals with poor signs of heat constituted an inferior oocyte donor- and IVF-group.

In vitro insemination (IVI) occurred 6.1 ± 3.2 h (mean ± standard deviation) following OA IVI-OA interval and within this narrow time span no relation existed between IVI-OA and the IVF-rates.

Oocytes with partially activated and fully activated cumulus cells had an even chance for fertilization if they originated from an animal with a LH-discharge. Oocytes with tight cumulus layers never activated in vitro and were never fertilized. The ooplasma score had no direct influence on the IVF-rates. The incidence of fertilization was equal in oocytes with (23.8%) and without (25.3%) a discerviable PB 1 regarding both the LH-positive and LH-negative groups of animals.

Concentrations of E2 and P4 was assayed in 26 FF, but no relation between their actual levels, their ratio, and IVF could be established.

Fungus infections occurred in two experiments and resulted in rapid sperm and COC-degeneration.

Discussion

The IVF-rates were lower than reported from other laboratories using a similar methodology (2). The reason for this discrepancy is not easily explicable, but the semen quality (motility and longitivity) may well be a causative element. In the above cited studies the authors found a significant difference in IVF% between bulls. Fresh semen from certain animals consistently produced high IVF-rates, whereas semen and especially frozen semen from others was unsuitable for IVF use. The intermale variation was not specifically tested in this experiment, but it was noticed that semen from certain bulls had a shorter longitivity and low penetration rates.

The COC were recovered approx. 10 h after standing estrus. According to previous experiments this interval may be too short to allow adequate in vivo maturation (6).

Fertilization was never achieved in COC from animals that did not show heat. Thus, estrus per se and the interval from estrus to OA may be used as an easily accessible biological parameter for assessing the quality of an oocyte donor.

The LH-assays constituted a valuable parameter. Animals without a descernible LH-discharge i.e. animals with periovulatory and follicular endocrinological disturbances (3), (4) constituted an inferior COC-donor group. Abscence of the LH-peak does not, however, preclude oocyte nuclear maturation. According to previous experiments (Goff et al. 1985, unpublished data and (4)) the meiosis can resume in superovulated heifers even without the prescence of the preovulatory LH-discharge. In the present study only one COC belonging to the LH-negative group of animals cleaved in vitro. This emphasizes the importance of the LH-peak for the normal and complicated maturational process that includes both nuclear, cytoplasmatic, and biochemical changes (11).

Previous (6), (9) and current experiments (4), (7) indicated that the oocyte maturational (nuclear) stages are directly proportional to the in vivo maturation time i.e. the interval from LH-peak level to the oocyte aspiration (OA). Therefore, it was anticipated that the IVF rate would increase with increasing LH-OA time. The between animal variation and the all over low IVF-rates did, however, preclude to establish any relationship between those two parameters. It was also remarkable that the between follicles IVF-rate within an animal exceeded the between animal IVF-rates. This may reflect a considerably variation in maturation between follicles and between COC revealed as it has been in recent ultrastructural and endocrine experiments (7), (4).

This study indicated that COC with tight cumulus upon recovery could not be fertilized in vitro. It also showed that in vitro activation could not be achieved even after a prolonged cultivation period in DM. This is contrary to observations from human IVF where in vitro maturation of slightly immature COC increased the IVF-rates (15).
The preparation of IVF-eggs that developed in vitro was low. In order to overcome this problem the eggs must be transferred to the oviducts of the recipient already at the 2- and 4-cell stage (2), or they may be temporarily cultured in an intermediate recipient such as rabbit and sheep oviducts for 3-5 days prior to non surgical transfer to the final recipient (13). Both techniques are elaborous and development of monolayer cultures would be a great advantage for the technique.

Most eggs were arrested in their development prior to or at the 4-cell stage, which is believed to be the time when the fetal genome is turned on (First 1985, personal communication).

Parthenogenetic division (activation) may be more widespread than anticipated at least when oocytes are matures in vitro and cultured in rabbit oviducts (17), and it can not be excluded that some eggs included in this study were activated and not actually ferlitized in vitro.

In conclusion the present experiment was undertaken to study the influence of the LH-discharge per se and the COC-morphology on the IVF-rates.

The present experiment allowed several conclusions. The IVF-rates of oocytes recovered from animals with endocrine disturbances and especially without a discernible LH-discharge were never fertilized in vitro. It was not possible, however, to establish a direct relationship between the fertilization rates and the interval from the LH-peak level and the oocyte aspiration time.

The COC-morphology could to a certain extent be used to predict the outcome of the IVF. Only COC with partially or totally expanded cumulus cell layers fertilized in vitro whereas any other cumulus type (tight, 1-cell layer, nude) was incompatible with IVF. It was also striking that in vitro maturation of COC was impossible, but it may reflect that the COC was degenerated prior to aspiration.

In vitro insemination time in relation to aspiration had no direct effect on the IVF-rates.

Routine use of this biotechnology will require more specific analysis of factors affecting the normal in vivo and in vitro maturation of COC. Also it will need to define fertilization more precisely. Parthenogenesis may occur frequently after IVF and proof for normal fertilization will be: development of the fertilized egg into a liveborn calf.

Acknowledgement

The work was supported by a grant from the Danish Veterinary- and Agricultural Research Council and Leo Research Foundation. The valuable assistance of Inger Heinze, MT and Karen Ejstrup, MT is appreciated.

References

1. Ball, G.D., M.L. Leibfried, R.W. Lenz, R.L. Ax, B.D. Bavister, N.L. First. 1983. Biol. Reprod. 28, 717-725.
2. Brackett, B.G., D. Bousquet, M.L. Boise, W.L. Donawick, J.F. Evans, M.A. Dressel. 1982. Biol. Reprod. 27, 147-158.
3. Callesen, H., T. Greve, P. Hyttel. 1984. J. Vitro Fertil. and Embryo Transf. 2, 101.
4. Callesen, H., T. Greve, P. Hyttel. 1986. Theriogenology (in press).
5. Edwards, R.G. 1985. Proc. R. Soc. Lond. B 223, 417-448.
6. Greve, T., D. Bousquet, W.A. King, J.K. Betteridge. 1984. Theriogenology 22, 151-165.
7. Hyttel, P., H. Callesen, T. Greve. 1985. J. Reprod. Fert. (accepted for publication).
8. Jensen, Mikêl A., T. Greve, A. Madej, L.-E. Edqvist. 1982. Theriogenology 18, 33-41.
9. King, W.A., D. Bousquet, T. Greve, A.K. Goff. 1984. Submitted to Gamete Research.
10. Moor, R.M., J.C. Osborne, I.M. Crosby. 1981. In: Roland R.,Van Hall E.V., S.G. Hiller, K.P. McNatty, J. Schoemaker (eds): Follicular Maturation and Ovulation. Amsterdam: Excerpta Medica. Elsevier/North Holland. 249-264.
11. Moor, R.M., Th. A.M. Kruip, D. Green. 1984. Theriogenology 21, 103-116.
12. Solti, L., T. Greve, H.H. Koefoed-Johnsen. 1978. Acta vet.scand. 19, 298-309.
13. Sirard, M.A., R.D. Lambert, P. Guay, D.P. Ménard, M. Bedoya. 1985. Theriogenology 23, 230.
14. Staigmiller, R.B., R.M. Moor. 1985. Gamete Reserach 9, 221-229.

15. Testart, J., R. Frydman, D. de Mouzon, B. Lasalle, J.C. Belaisch. 1983. Biol. Reprod. 28, 415-424.

16. Trounson, A.O., L.R. Mohr, C. Wood, J.F. Leeton. 1982. J. Reprod. Fert. 64, 285-294.

17. Xu, K.P., T. Greve, S. Smith, P. Liehman, H. Callesen, P. Hyttel. 1986. Theriogenology (in press).

STUDIES ON EARLY PREGNANCY FACTOR (EPF) IN THE PIG

E. Koch* and F. Ellendorff

Institut für Tierzucht und Tierverhalten (FAL), Mariensee 3057 Neustadt 1, F.R. Germany.

*Present address: AFRC Institute of Animal Physiology, Babraham, Cambridge CB2 4AT, U.K.

Introduction

The fertilized ovum seems to release a signal, termed ovum factor (OF), which subsequently results in the appearance of an early pregnancy factor (EPF) in the serum (1). Thus, diagnosis of pregnancy and detection of embryonic mortality within days after fertilization or after embryonic death appears to be possible (2). EPF is detected by the rosette inhibition test (3), in which it causes an increase in the titre of the antilymphocyte serum (ALS) required to inhibit the formation of spontaneous rosettes between lymphocytes and heterologous red blood cells.

It was the purpose of the present investigation to study characteristics of EPF production in the pig. To this end, rosette inhibition titres (RIT) were determined during pregnancy and after removal or transfer of embryos.

Methods

Serum samples were collected from 70 non-pregnant (female, male and gonadectomized) and 205 pregnant (day 1 to 118 after insemination) German Landrace pigs. Pregnancy was confirmed at the time of parturition and all sows farrowed between day 113 and 119 after the first insemination.

From 8 pigs embryos were removed surgically 4 to 6 days after mating. Serum samples were taken from these animals before and 24 h after embryo recovery.

Pregnancy Proteins in Animals

From 2 chronically catheterized sows serum was collected at different time intervals after transfer of 14 or 15 embryos, respectively. Both animals were slaughtered 22 days after the transfer and the uteri examined for the presence of conceptuses.

EPF activity was determined in the serum by the rosette inhibition test as described previously (4). The rosette inhibition titre (RIT) is defined as the highest dilution of ALS which causes an inhibition of rosette formation by at least 25%, as expressed by log 2.

Results

After preincubation of lymphocytes in serum of non-pregnant pigs, the upper limit of the 95%-confidence level of the RITs was calculated to be 11.2. Therefore, all titres > 12 were defined as indicating the presence of EPF activity.

Problems of the assay system were particularly evident due to the absence of elevated RITs after preincubation of lymphocytes in serum of pregnant pigs. For this reason all EPF-negative samples were reanalysed up to two items. Following this modification a correct pregnancy diagnosis occurred in 88.7% for all cases (n = 205). A total of 8.6% false positive statements was contrasted by 12.4% false negative results during the whole gestation period.

During the first half of pregnancy, RITs displayed periodic fluctuations, which resembled the physiological cycle interval of the pig. Between weeks 5 to 9 of pregnancy an enhanced number of EPF-negative animals was recorded. Subsequently, a continuous increase in mean RITs occurred, which then declined gradually from day 80 of pregnancy on.

Before removal of embryos, all 8 tested sows showed an EPF-positive RIT (19.3 $\pm$ 1.07), while 24 h after the recovery of embryos the mean RIT decreased to 11.8 $\pm$ 0.53. EPF activity was then detected in serum of only one pig with a markedly reduced titre.

After the transfer of embryos into 2 sows an EPF-positive titre was evident within 2 to 4 h. Later embryonal loss as confirmed at autopsy was preceded by a disappearance of elevated RITs in both recipients.

Discussion

The present investigation confirms earlier studies in the pig (5), which showed that EPF activity is detectable throughout the entire gestation in this species. However, the orignally suggested biphasic pattern of EPF-production has now been found to be a polyphasic phenomenon. Because of the periodic fluctuations of RITs during pregnancy, detection of EPF activity showed different rates of coincidence with the pregnancy status during the course of pregnancy. Nevertheless, a correct pregnancy diagnosis occurred in nearly 90% of all cases.

Although a transient unspecific detection of EPF-positive titres has been observed after mating oestrous sows with a vasectomized boar (6), these experiments strongly suggest that the sustained production of EPF is closely related to the presence of viable conceptuses. Thus, EPF appears within a few hours after transfer of embryos and usually is not detectable 24 h after removal of fertilized ova.

The rosette inhibition test, however, proved to be a cumbersome and time-consuming assay system. It was found to be basically suitable for limited investigations under well defined experimental conditions, but a practical application for pregnancy diagnosis is not possible at present.

Acknowledgements

We thank Mrs I. Stelter for expert technical assistance and Mrs J.A. Tickner for typing the manuscript. This work was supported by the Deutsche Forschungsgemeinschaft.

References

1. Cavanagh, A.C., H. Morton, B.E. Rolfe and A.A. Gidley-Baird. 1982. Am. J. Reprod. Immunol. 2, 97.

2. Nancarrow, C.D., B.M. Evison, R.J. Scaramuzzi and K.E. Turnbull. 1979. J. Reprod. Fert. 57, 385.

3. Morton, H., H.R. Tinneberg, B. Rolfe, M. Wolf and L. Mettler. 1982. J. Reprod. Immunol. 4, 251.

4. Koch, E. and F. Ellendorff. 1985. J. Reprod. Fert. 74, 29.

5. Morton, H., D.J. Morton and F. Ellendorff. 1985. J. Reprod. Fert. 68, 437.

6. Koch, E. and F. Ellendorff. 1985. J. Reprod. Fert. 74, 39.

PLACENTAL LACTOGEN AND OTHER PREGNANCY-SPECIFIC PROTEINS IN FARM ANIMALS

Isabel A. Forsyth, Glenys A. Bloomfield

Endocrinology and Animal Physiology Department, Animal and Grassland Research Institute, Church Lane, Shinfield, Reading, RG2 9AQ, UK

Introduction

In the cow pregnancy can be detected by measurement of progesterone in milk or serum. However, this is only of use at 24 days after insemination, and, since progesterone is not pregnancy specific, the timing of sample collection must be exact or misleading results will be obtained. Due to the occurrence of embryo loss in as many as 10% of animals before day 60 of pregnancy the progesterone test is a more reliable test of non-pregnancy than of pregnancy. Oestrone sulphate measurements can also be used, and have the advantage of being pregnancy specific. However, this steroid is only reliably detected after day 100 of pregnancy, which is too late to be of use in the practical management of reproduction in a herd. There is therefore a real need to identify pregnancy-specific factors in the cow and other ruminants which will be of use both as aids to farm management and as research tools in the study of reproduction.

Placental lactogens

One class of pregnancy-specific proteins which are now well known to occur in ruminants are the placental lactogens (also known as chorionic somatomammotrophins) secreted by trophoblast tissue of the fetus from early in pregnancy (1). Their precise physiological role(s) are still unknown, but effects on mammary gland development, maintenance of the corpus luteum and on maternal metabolism are indicated by experiments in sheep and goats (2).

Pregnancy Proteins in Animals

As information on amino acid sequences has accumulated, it has become apparent that many proteins can be grouped into families of molecules with related structures. Placental lactogens are recognised as members of the prolactin/growth hormone gene family, having probably arised by duplication and divergence from a common ancestral gene. Each hormone now has different cells of origin and is differently controlled. Molecular evolution within the gene family has clearly been a complex process and the complexity is particularly apparent in the case of the placental lactogens, which, unlike prolactin and growth hormone, appear to be present in only certain mammals (3) (4).

Detection of placental lactogen has so far relied largely on bioassays (5) or on radioreceptor assays (6) which measure its prolactin-like or growth hormone-like activities. As assessed by these assays, the hormone has been observed only in primates including man, in rodents and in ruminants. Among farmed mammals, for example, it has been detected in goats, sheep, cows and deer, but not in pigs or horses (3). Human placental lactogen has a very high degree of homology with human growth hormone, differing from it at only about 15% of all amino acid residues (4). By contrast, a structure more closely related to prolactin is indicated from preliminary structural observations on ovine placental lactogen (7) and from the immunological cross reactions of one form of rat placental lactogen (8).

Even among ruminants, there are differences between species in the physicochemical properties of the hormone and its circulating plasma concentrations through pregnancy (Table 1). An observation of interest is the lack of immunological cross-reaction between the placental lactogens of different ruminants. Using a specific radioimmunoassay for ovine placental lactogen (9), we have been unable to detect any appreciable cross reaction between ovine, caprine, bovine or red deer placental lactogens. A similar result has been reported by Chan et al. (10). This is in marked contrast with the virtual immunological identity of the prolactins and also the growth hormones of sheep, goats and cows.

Table 1. Ruminant Placental Lactogens

	Sheep	Goat	Cow
Molecular weight	23 000	23 000	40 000
Isoelectric point	8.6 - 8.9	8.8	5.4 - 5.9
Maternal plasma concentration* (ng/ml; mean ± SEM)	150 ± 18	218 ± 43	< 10

* in late pregnancy with 1 fetus.
Compiled from various sources. See references in Ref. 3.

Because of the relatively late appearance of placental lactogen in the maternal peripheral circulation (reliably after day 60 in sheep and goats and day 100 in cows), placental lactogen is, like oestrone sulphate, only of confirmatory value in pregnancy diagnosis. Fluctuating concentrations and variability between individuals make it an unreliable guide to fetal numbers (2) (9) (10).

Other pregnancy-specific proteins

In an attempt to discover other pregnancy-specific proteins in the cow, the cross-reactions between bovine serum or placental extracts and a number of antisera to human pregnancy proteins have been studied. The antisera used were all kindly provided by Professor T. Chard and were raised in rabbits to the following human proteins: human chorionic gonadotrophin (hCG), human placental lactogen (hPL), schwangerschaftsprotein 1 (SP1) and placental protein 5 (PP5). Using standard immunodiffusion techniques no cross-reaction could be detected between any of the antisera and Sephadex G200 fractions of serum from pregnant cows, from either early or late pregnancy, or with placental extracts, also purified on Sephadex G200. No cross reaction was obtained with antiserum to PZP (Dakopatts) under similar conditions.

The majority of pregnancy associated proteins from other species appear to be glycoproteins. Glycoprotein enriched fractions of serum from pregnant cows and bovine placentae have therefore been prepared by chromatography on Concanavalin A-Sepharose (Pharmacia), as previously described (11). Three fractions of differing affinity for the lectin were produced. The first, designated I, was not bound, and passed straight through the column. The second, designated II, was eluted from the gel by 0.2M a-D-methylmannoside in 0.01M Tris buffer, 0.3M sodium chloride, pH 7.5, while the third, designated III, was more firmly bound and was only eluted by 0.3M a-D-methylmannoside and 0.3M ammonium sulphate in 0.5M acetate buffer, 0.3M sodium chloride, pH 5.0.

After dialysis the fractions were freeze-dried and their cross-reactivity with the antisera to human pregnancy proteins was assessed by immunodiffusion. Glycoprotein fractions III prepared from pools of bovine serum from cows in the second or third trimester of pregnancy and from two placental extracts all showed strong cross-reaction with all four rabbit antisera (Plate 1). This reaction

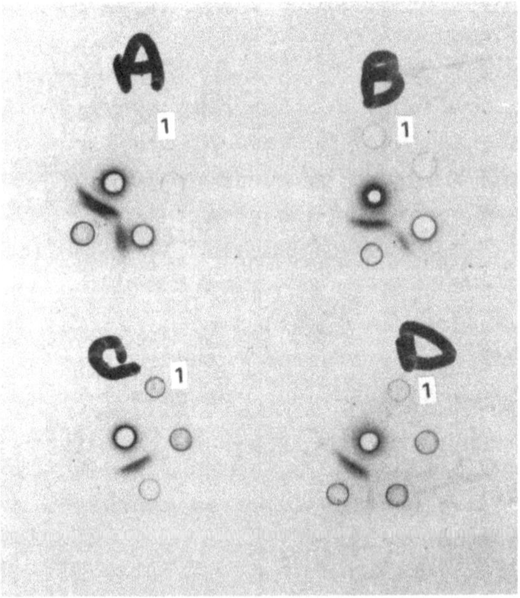

Plate 1.
Cross reactions between Concanavalin A fractions (see text) and a rabbit antiserum to human PP5. All wells contain 4 µl; Concanavalin A fractions are 10mg/ml solutions in 0.9%NaCl, antiserum is undiluted. A: serum from cows in late pregnancy; fractions (clockwise from 1) Ia, Ib, II, III; B: serum from cows in mid pregnancy Ia, Ib, II, III; C: placental extract I, II, III; D: placental extract I, IIa, IIb, III. Note the cross reaction between fractions II and III in A, B and D.

was not related to any pregnancy specific phenomena, however, since an identical reaction was obtained with the same glycoprotein fraction from serum of non-pregnant animals. There appears to be some non-specific reaction between the preparations from the two species.

The same response was found with a number of other rabbit sera from animals which had been immunised in our laboratory to Sephadex G200 fractions of bovine pregnancy sera. Absorption of these latter rabbit antisera with the non-glycoprotein fraction from bovine pregnancy serum after Concanavalin A-Sepharose chromatography (fraction I) removed all cross-reaction between them and fractions I and II from Concanavalin A chromatography of bovine pregnancy sera, but the strong cross-reaction with fraction III remained.

These rabbit antisera were produced using Freund's incomplete adjuvant supplemented by BCG vaccine (Glaxo). The reaction between the rabbit and bovine preparations did not appear to be due to reactions to the tubercule bacilli, however, as no reaction was obtained between the rabbit sera and BCG vaccine when this was tested directly in immunodiffusion experiments. Such non-specific reactions could lead to misleading results if proper controls are not included in immunological experiments such as those described.

An additional phenomena which was noted in these immunodiffusion studies was that there appeared to be a weak cross-reaction between the fraction II and III preparations from bovine pregnancy serum, when these were placed in adjacent wells on the immunodiffusion plate. This precipitin reaction is presumably due to the presence of some of the small subclass of IgG molecules which are glycosylated in either fractions II or III.

Conclusion

These results indicate that, even in closely related species like the sheep and goat, failure to detect immunological cross-reactions is not necessarily good evidence for the absence of homologous proteins. It is clear that a wide range of biological, immunological and increasingly of molecular biology techniques must be used to determine the existence and the homologies of pregnancy-specific proteins.

References

1. Flint, A.P.F., A. Henville, W.B. Christie. 1979. J.Reprod.Fert. 56, 305.
2. Hayden, T.J., C.R. Thomas, S.V. Smith, I.A. Forsyth. 1980. J.Endocr. 86, 279.
3. Forsyth, I.A.: J.Dairy Sci. (in press)
4. Wallis, M. 1984. In: Prolactin Secretion: A Multidisciplinary Approach (F. Mena and C. Valverde-R., eds.). Academic Press Orlando, p.1.
5. Forsyth, I.A. 1974. In: Lactogenic Hormones, Fetal Nutrition and Lactation (J.B. Josimovich, M. Reynolds and E. Cobo, eds.). Wiley New York, p.49.
6. Kelly, P.A., T. Tsushima, R.P.C. Shiu, H.G. Friesen. 1976. Endocrinology 99, 765.
7. Hurley, T.W., F.E. Grissom, S. Handwerger, R.E. Fellows. 1977. Biochemistry 16, 5605.
8. Robertson, M.C., H.G. Friesen. 1981. Endocrinology 108, 1288.
9. Thordarson, G. 1984. Ph.D. thesis. University of Reading.
10. Chan, J.S.D., H.A. Robertson, H.G. Friesen, 1978. Endocrinology 102, 1606.
11. Bloomfield, G.A., M.R. Faith, H.G. Pierce. 1978. Biochim. Biophys. Acta 533, 371.

PREGNANCY DETECTION IN FARM ANIMALS BY RADIOIMMUNOASSAY OF A PREGNANCY-SPECIFIC PROTEIN IN SERUM

R.G. Sasser, C.A. Ruder, K.A. Ivani

Department of Animal Science, College of Agriculture, University of Idaho
Moscow, Idaho 83843

Introduction

Pregnancy is routinely detected in cattle by inserting the hand into the rectum and palpating through the rectal wall for presence of a conceptus within the uterus. Proteins produced by the placenta have been measured in maternal serum for detection of pregnancy for many years in some species. Pregnant mare serum gonadotropin (PMSG) was discovered in 1930 (4) and measurement provides a serological test for pregnancy (5). Human chorionic gonadotropin (hCG) was discovered in urine of pregnant women (1) and can be assayed in urine or blood as early as 8 to 10 days after conception (12). In farm animals other than equidae, similar pregnancy markers have not been found. However, researchers (11, 15) have found proteins in the uterus of pregnant cattle which may be involved in maintenance of pregnancy.

Methods

Pregnancy-specific protein B (PSPB) was isolated and partially purified by us at the University of Idaho (2). The protein was found after immunization of a rabbit with homogenates of placenta, collection of rabbit antisera and adsorbtion of antisera with somatic tissues to remove antibodies to proteins not specific to the placenta. Remaining antibodies were against alpha fetoprotein and PSPB of the placenta. Immunoelectrophoresis and immunodiffusion techniques using the adsorbed antisera were used to isolate PSPB. Placental extracts were subjected to ionexchange and gel filtration chromotography and isoelectric focusing. A preparation of PSPB (R-37) was used to immunize rabbits to obtain an antiserum for development of a double antibody radioimmunoassay after the methods of Davis et al. (6) for prolactin. The second antibody against rabbit gamma globulin was developed in a sheep. This assay was used to develop a serological method for detection of pregnancy in ruminant animals. In addition PSPB

Pregnancy Proteins in Animals

antiserum was used to localize PSPB in placental tissue by immunocytochemical techniques.

Results and Discussion

Radioimmunoassay development

Rabbit antiserum to PSPB (RGS 38-1) was used at a final dilution of 1:300,000. This dilution bound 40 percent of radiolabeled (8) PSPB (specific activity of 90 to 110 μCi/μg) in absence of unlabeled PSPB.

Intra- and interassay coeficients of variations were 4.5 and 9.6 percent. Measurement of PSPB added to serum with no PSPB resulted in a regression equation for PSPB measured (Y) compared to that added (X) of $Y=.006 + 1.01X$. The correlation coeficient was .99. The assay was sensitive to less than 100 ng PSPB (preparation R-37) per assay tube when 200 μl serum was assayed. It was established that the assay was specific for bovine PSPB with minimal cross-reactivity (< 0.5%) with bovine (b) LH and ovine FSH and partial crossreactivity with ovine placental lactogen (<.05%). There was no crossreactivity with hCG, PMSG, bPRL, bTSH or bovine growth hormone. Serum pools from rams, wethers and steers and nonpregnant and pregnant ewes were assayed in the PSPB RIA. There was no PSPB in ram, wether, steer, or nonpregnant ewe sera. Pregnant ewe serum inhibited binding in the RIA in a nonparallel manner, suggesting that pregnant ewe serum contained a crossreacting antigen. Serum pools from pregnant cows inhibited binding in a manner parallel to the standard curve whereas serum pools from nonpregnant cows did not inhibit binding. This was evidence that not only could PSPB be measured in serum of pregnant cows, but that PSPB was pregnancy-specific and therefore the RIA might be useful for pregnancy detection.

Detection of pregnancy

Pregnancy detection by measurement of PSPB was compared to detection by rectal palpation and slaughter. A commercial herd of 102 primi- and multiparous beef cows was sampled by jugular venipuncture 27 days after the end of a 75-day breeding period. A local veterinarian rectally examined all cows at the time that blood samples were collected. Ninety-seven cows subsequently calved and five did not. Comparision of detection of pregnancy by RIA with calving and slaughter data from these cows resulted in complete agreement whereas detection by rectal palpation missed 9% of the pregnant animals (9). These data demonstrated that measurement of PSPB by RIA could be used successfully to detect pregnancy in cows managed under a commercial situation.

Because PSPB appeared to be a pregnancy marker and because it was not known whether it was present in serum throughout gestation, a study was conducted with a two-fold objective: a) to determine circulating levels of PSPB throughout gestation and b) to determine the earliest time that PSPB could be detected in the circulation after conception.

To accomplish the first phase of the study, five dairy cows were bled from the tail vein twice weekly from breeding through 21 days postpartum. Serum PSPB levels increased as gestation progressed. PSPB levels ($\bar{x} \pm$ SE) were 1.20 $\pm$ 0.34 ng/ml by 28 days post-breeding. Concentrations increased gradually to 9.3 $\pm$ 0.6 ng/ml by 91 days, 35.4 $\pm$ 5.6 ng/ml by 195 days and 149.8 $\pm$ 74.8 ng/ml by 269 days. PSPB levels were highest two days prior to calving (541.3 $\pm$ 144.5 ng/ml) and declined sharply during the postpartum period. By 21 days postpartum, PSPB levels were 78 $\pm$ 21.1 ng/ml (10). Further study is necessary to determine the postpartum time when PSPB is cleared from the serum. It would also be of interest to determine if PSPB is involved in postpartum anestrus in cattle. This time profile showed that PSPB was present from early pregnancy and continuously throughout gestation. The peak in PSPB concentration just prior to calving may be associated with the physiological changes related to the onset of parturition. The role that PSPB may have in parturition remains to be studied.

The second phase of this objective was to define the earliest time at which PSPB could first be detected in serum of cows following conception. Data from the profile of PSPB levels throughout gestation suggested that day 24 post-breeding might be the earliest time that it could be measured. Twenty-one beef and dairy cows were bled daily from day 15 through day 30 post-breeding and sera were assayed for PSPB. Results of this assay were in agreement with the data from the time profile and showed that PSPB could be measured reliably at 24 days post-breeding (10).

Another study was conducted during the early phases of pregnancy in beef cattle to determine reliability of the RIA in pregnancy detection. Blood samples and reproductive tracts from 429 cows were collected at slaughter. All parous females were at least 7 months postpartum and all had been exposed to multiple sires for 61 days or less. Examination of the reproductive tracts, which included a microscopic search of flushing media if a conceptus was not obvious, indicated that 242 females were pregnant and 187 were not. Of 51 cows pregnant between 2 to 23 days, the RIA detected 15.7%. Ninety-nine percent of those pregnant greater than 23 days (191 cows) were detected. The nonpregnant animals were detected at a rate of 94.7 percent. These data indicate that the PSPB RIA is highly effective in detecting pregnant or nonpregnant cows early in gestation.

As stated previously, serum from pregnant ewes cross-reacted in a nonparallel manner in the bovine PSPB RIA; however, serum from nonpregnant ewes did not inhibit binding. A study was conducted to determine if the bovine RIA was useful for detection of pregnancy in sheep. Currently, electronic devices are used to detect pregnancy in sheep. To compare the reliability of pregnancy detection in sheep by RIA with that of electronic devices, two studies were conducted.

In the first study, 33 ewe lambs of Suffolk and Panama breeds were bred over a 56-day breeding period in late fall. At 40, 70 and 97 days past the breeding season, blood samples were collected for RIA and pregnancy detection was performed with a Pregmatic 3 ultrasound device. Thirty of these ewes lambed and three did not. The Pregmatic 3 predicted 47%, 90% and 93% of the pregnancies on the respective days whereas RIA predicted 107%, 103% and 100% (16). Overall the RIA was able to detect pregnancies more accurately earlier in gestation than was the Pregmatic 3.

In the second sheep study, 190 multiparous ewes of the Polypay, Rambouillet or Targhee breeds were bred out of season over a 60-day period in summer. At 38 days past the end of the breeding season, blood samples were collected for RIA and a real-time ultrasound scan was performed. Of these 190 ewes, 169 lambed and 21 did not. The RIA detected 171 ewes as pregnant and the real-time scan detected 173 as pregnant (16). These data would suggest that the PSPB RIA is an accurate and acceptable alternative to ultrasound devices for the detection of pregnancy in sheep.

Due to the success encountered with the detection of pregnancy in sheep by means of the PSPB RIA, other ruminant species have been studied for this same purpose. Assay of serum samples from white-tailed and mule deer (17) demonstrated that the RIA was accurate in detecting pregnancy in these species. Further testing with domestic goats, mountain goats and elk has shown that the RIA was equally accurate in the detection of pregnancy in these animals. Sera from pregnant sows, mares, dogs, rats and humans sera have been tested, but no cross-reactive antigen was found. It may be that PSPB is found only in species with syndesmochorial (cotyledonary) placentae.

Site of synthesis of PSPB

An immunohistochemical study using light microscopy was conducted to determine which cells of the placentome contained PSPB. An unlabelled antibody-enzyme method with soluble glucose oxidase-anti-glucose oxidase complexes was used as the immunohistochemical stain (13). Large binucleate cells of the trophoblastic ectoderm stained positively while other cells of the placentomal or interplacentomal region did not stain. Cells in maternal tissue did not stain. This study suggested that the binucleate cells were responsible for synthesis and secretion of PSPB (7).

Isolated binucleate cells of the trophoblastic ectoderm were collected (14) from two placentae of 200 and 280 days of age. Cell fractions containing 80 to 100 percent binucleate cells contained from 300 to 1700 ng per 1×10^6 cells. In addition incubation of 8×10^5 cells from a 200-day old placenta in 1 ml media 199 for 18 hours resulted in presence of 234.3 ± 7.1 ng/ml ($\bar{x} \pm$ SE) PSPB in media (13).

These studies show that PSPB is of fetal origin. Techniques used provide a means for study of regulation of secretion of PSPB.

Acknowledgements

This study was supported by the Idaho Agricultural Experiment Station, the United Dairymen of Idaho and the National Association of Animal Breeders. The authors wish to acknowledge Pamela Panlasigui for technical assistance, Harold Zenner and Sons for use of their cattle, R.C. Bull, T.J. Reimers, W.P. Eckblad, R.R. Mauer, J D Mankin, J.J. Ney, R. Turner, J.J. Dahmen, J. Stellflug, D.B. Houston, A.K. Wood and R.E. Short for professional assistance and NIAMDD for cited hormone preparations.

References

1. Aschheim, S. and B. Zondek. 1927. Hypophysenvorderlappenhormon und ovarialhormone im harn von schwangeren. Kin. Wschr. 6, 1322.

2. Butler, J.E., W.C. Hamilton, R.G. Sasser, C.A. Ruder, G.M. Hass and R.J. Williams. 1982. Detection and partial characterization of two bovine pregnancy-specific proteins. Biol. Reprod. 26, 925-933.

3. Clark, C.A., E.C. Downs and F.J. Primus. 1982. An unlabeled antibody method using glucose oxidase-antioxidase complexes (GAG): A sensitive alternative to immunoperoxidase for the detection of tissue antigens. J. Hist. Cytoch. 30, 27-34.

4. Cole, H.H. and G.H. Hart. 1930. The potency of blood serum of mares in progressive stages of pregnancy in effecting the sexual maturity of the immature rat. Amer. J. Physiol. 93, 57-68.

5. Cole, H.H. and G.H. Hart. 1942. Diagnosis of pregnancy in the mare by hormonal means. J. Amer. Vet. Med. Assoc. 101, 124-128.

6. Davis, S.L., L.E. Reichert, Jr. and G.D. Niswender. 1971. Serum levels of prolactin in sheep as measured by radioimmunoassay. Biol. Reprod. 4, 145-153.

7. Eckblad, W.P., R.G. Sasser, C.A. Ruder, P. Panlasigui and T. Kuczynski. 1985. Localization of pregnancy-specific protein B (PSPB) in bovine placental cells using a glucose oxidase-anti-glucose oxidase immunohistochemical stain. Proc. West. Sect. Amer. Soc. Anim. Sci. 36:396-397.

8. Greenwood, F.C., W.M. Hunter and J.S. Glover. 1963. The preparation of ^{131}I-labeled human growth hormone of high specific activity. Biochem. J. 89, 114-123.

9. Ivani, K.A., C.A. Ruder, R.G. Sasser. 1984. Diagnosis of pregnancy by radioimmunoassay of a pregnancy-specific protein in serum of the cows. Biol. Reprod. 30, 51, Suppl. 1 (abstract).

10. Ivani, K.A., C.A. Ruder, and R.G. Sasser. 1984. Serum concentration of a bovine pregnancy-specific protein during gestation. J. Anim. Sci. 59, 367, Suppl. 1 (abstract).

11. Laster, D.B. 1977. A pregnancy-specific protein in the bovine uterus. Biol. Reprod. 16, 682-690.

12. Marshall, J.R., C.B. Hammond, G.T. Ross, A. Jacobson, P. Rayford and W.D. O'Dell. 1968. Plasma and urinary chorionic gonadotropin during early human pregnancy. Obstet. Gynecol. 32, 760-764.

13. Reimers, T.J., W.P. Eckblad, C.A. Ruder and R.G. Sasser. 1985. Production of pregnancy-specific protein B by bovine binucleate trophoblastic cells. Biol. Reprod., submitted.

14. Reimers, T.J., M.D. Ullmann and W. Hansel. 1985. Progesterone and prostanoid production by bovine binucleate trophoblastic cells. Biol. Reprod., in press.

15. Roberts, G.P. and J.M. Parker. 1976. Fractionation and comparision of proteins from bovine uterine fluid and bovine allantotic fluid. Biochem. Biophys. Acta. 446, 69-76.

16. Ruder, C.A., R.G. Sasser, K.A. Ivani, P.M. Panlasigui, J.J. Dahmen and J.N. Stellflug. 1984. Diagnosis of pregnancy in sheep by measurement of a blood antigen that cross reacts in a radioimmunoassay for a bovine pregnancy-specific protein. J. Anim. Sci. 59, 367, Suppl. 1 (abstract).

17. Wood, A.K., R.E. Short, A.E. Darling, G.L. Dusek, R.G. Sasser and C.A. Ruder. 1985. Serum assay for detecting pregnancy in mule and white-tailed deer. J. Wildlife Management (submitted).

SUBJECT INDEX

AUTHOR INDEX

Bei Fragen zur Produktsicherheit wenden Sie sich bitte an:
If you have any questions regarding product safety,
please contact:

Walter de Gruyter GmbH
Genthiner Straße 13
10785 Berlin
productsafety@degruyterbrill.com